Proteomic Studies for the Identification and Characterization of Marine Bioactive Molecules

Proteomic Studies for the Identification and Characterization of Marine Bioactive Molecules

Guest Editor

Nedeljka Rosic

MDPI

Basel • Beijing • Wuhan • Barcelona • Belgrade • Novi Sad • Cluj • Manchester

Guest Editor
Nedeljka Rosic
Faculty of Health
Southern Cross University
Gold Coast
Australia

Editorial Office
MDPI AG
Grossspeteranlage 5
4052 Basel, Switzerland

This is a reprint of the Special Issue, published open access by the journal *Marine Drugs* (ISSN 1660-3397), freely accessible at: https://www.mdpi.com/journal/marinedrugs/special_issues/747IJQ0OF9.

For citation purposes, cite each article independently as indicated on the article page online and as indicated below:

Lastname, A.A.; Lastname, B.B. Article Title. *Journal Name* **Year**, *Volume Number*, Page Range.

ISBN 978-3-7258-6386-0 (Hbk)
ISBN 978-3-7258-6387-7 (PDF)
https://doi.org/10.3390/books978-3-7258-6387-7

Cover image courtesy of Nedeljka Rosic

Contents

About the Editor

Nedeljka Rosic

Nedeljka Rosic is a multidisciplinary researcher with a PhD in Biomedical Sciences (University of Queensland) working at Southern Cross University (Australia). She has established expertise in xenobiotic enzymes, directed evolution, and drug discovery, and is the recipient of multiple competitive fellowships and research grants. Her research program encompasses marine bioproducts, environmental stress, biomarkers, host–microbe interactions, genomics, proteomics, and protein engineering. She maintains active collaborations with academic researchers and industry partners at the national and international levels and has won a NASA award for innovation.

Dr. Rosic supervises PhD, MSc, and Honours students in the disciplines of biomedical science, molecular biology, and omics-based research. She is strongly committed to research excellence and scientific leadership and contributes to STEM outreach through invited lectures and educational engagement with primary and secondary school communities, as well as public events, promoting scientific literacy and community impact.

marine drugs

Editorial

Proteomic Studies for the Identification and Characterization of Marine Bioactive Molecules

Nedeljka Rosic [1,2]

[1] Faculty of Health, Southern Cross University, Gold Coast, QLD 4225, Australia; nedeljka.rosic@scu.edu.au
[2] Marine Ecology Research Centre, Southern Cross University, Lismore, NSW 2480, Australia

The marine environment is a rich source of natural products that, as promising bioactive compounds, demonstrate environmentally friendly potential for application across various industries [1,2]. These bioactive molecules may be used in environmental protection, supporting the transition to green sectors, and promoting the sustainable use of natural resources [3–6]. Proteomic studies play a crucial role in this process by analyzing the activity, stability, degradation, modification, and interactions of proteins, as well as their influence on metabolic pathways [7–9]. This approach also assesses variations in protein expression levels and the impact of differential gene expression on proteomic profiles [10,11]. Advances in proteomics, particularly through the integration of cutting-edge technologies such as mass spectrometry-based protein interactome studies, artificial intelligence (AI)-based approaches, and machine learning techniques are opening new frontiers in modern-day research explorations that were not possible in the past [12–14]. This rapid progress is especially vital in the medical field, where there is an urgent demand for novel and more effective therapeutic agents, disease biomarkers and addressing the needs of the aging population [15,16]. Moreover, proteomic insights are deepening our understanding of the bioactive potential of marine-derived bioproducts, paving the way for innovative applications in health and biotechnology.

Over the last few decades, proteomics research has tremendously expanded, including advancements via high-throughput techniques resulting in faster and more reliable proteome profiles, a better understanding of biological mechanisms, and the discovery of new natural products [6,17,18]. The use of proteomics to facilitate the application of algal natural products in commercial purposes requires the effective use of specific proteomics methodologies combining assessments of bioactivity analyses, such as anti-bacterial, anti-proliferation, anti-viral, anti-aging, anti-fungal, and anti-inflammatory activities [19–22]. This Special Issue, entitled "Proteomic Studies for the Identification and Characterization of Marine Bioactive Molecules" (https://www.mdpi.com/journal/marinedrugs/special_issues/747IJQ0OF9, accessed on 24 October 2025) focuses on the application of proteomics in the utilization of marine species as a source of valuable pharmacologically active compounds. As a result of growing interest in proteomics research, this Special Issue has gathered ten original articles and reviews.

This editorial aims to provide an overview of recent advances in proteomics and its application for the discovery and characterization of marine bioactive molecules, as published in this Special Issue "Proteomic Studies for the Identification and Characterization of Marine Bioactive Molecules". The contributions from this issue are listed below within the List of Contributions.

Contribution 1: https://doi.org/10.3390/md23030102

Sun et al. analyzed protein components using sodium dodecyl sulfate–polyacrylamide gel electrophoresis (SDS–PAGE) on crude venoms from five vermivorous cone snails and compared their composition and bioactivity. In addition, reverse-phase HPLC (RP-HPLC) fingerprinting and mitochondrial cytochrome c oxidase I (COI) gene clustering revealed that venom diversity was genetically correlated among species. All venoms were lethal to insects and zebrafish, with *Conus quercinus* showing the strongest insecticidal activity (LD_{50} = 0.6 μg/mg) and *C. tessellatus* exhibiting the highest zebrafish lethality (LD_{50} = 0.2 μg/mg). Four species' venoms displayed cytotoxicity against ovarian cancer cells, while only *C. caracteristicus* showed significant analgesic effects. These findings identify several cone snail species with promising bioactivities, underscoring their potential for developing novel marine-derived peptide drugs and highlighting the importance of exploring the South China Sea's biodiversity for pharmaceutical discovery.

Contribution 2: https://doi.org/10.3390/md23020082

Cunha et al. used high-performance size-exclusion chromatography to explore bioactive peptides from fish body mucus of *Halobatrachus didactylus*, a species inhabiting pollutant-exposed intertidal zones. In silico computational assessment of peptide aggregation propensity was used to further assess the aggregation propensities of selected peptides. In vitro studies revealed peptide aggregation at specific ionic strengths (344 and 700 mM) and neutral pH. Though none of the analyzed peptides showed antimicrobial activity, they inhibited *Pseudomonas aeruginosa* biofilm formation. HdVLPN and HdLPN peptides showed strong antioxidant activity, decreasing at acidic pH, while HdPPP peptide displayed antihypertensive and antidiabetic potential through ACE and α-glucosidase inhibition. The study highlights the importance of validating peptide bioactivities in their native mucus environment and conserving this species as a potential source of novel biotechnological compounds.

Contribution 3: https://doi.org/10.3390/md22100470

Hua et al. explored the South China Sea's sea anemone resources, including *Macrodactyla doreensis*, containing toxins rich in proteins and peptides with potential biotechnological value. This study conducted comprehensive omics analyses of different tissues (i.e., the tentacles and column), including transcriptomic and proteomics in silico modelling to construct putative protein and peptide databases and compare their profiles utilizing multiple bioinformatics tools (e.g., the Pfam database, Multiple sequence alignment MUSCLE, AlphaFold2 Modeling). Among 4239 identified transcripts, putative proteins accounted for 81.53%, mainly immunoglobulins and proteases concentrated in the column, linked to biological functions. Putative peptides (18.47%) were primarily found in the tentacles, featuring ShK-domain and Kunitz-type peptides associated with predatory behavior. Forty peptides across eight typical families were identified, and their 3D structures and potential targets were predicted using AlphaFold2 and molecular docking. The findings highlight the molecular diversity and functional complexity of *M. doreensis* toxins, demonstrating an effective approach for predicting peptide structure and activity, supporting future discovery of bioactive compounds from marine organisms.

Contribution 4: https://doi.org/10.3390/md22100468

Tassara et al. employed proteomic and biochemical methods to analyze the crude nematocyst extract of *Velella velella*, a common hydrozoan harmless to humans. The analysis identified 783 proteins, including structural components, enzymes, and potential toxins, revealing a venom composition similar to more toxic cnidarians. Biochemical assays confirmed active hydrolytic enzymes such as proteases, phospholipases, hyaluronidases,

DNases, and chitinases within the extract. Proteomics allowed the detection of trace compounds with minimal animal collection, important for ecosystem preservation. These findings provide a valuable resource for future research exploring the pharmacological applications of *V. velella* venom components through recombinant production and functional studies, advancing the development of novel marine-derived biotechnological tools.

Contribution 5: https://doi.org/10.3390/md22030113

Kureara et al. investigated the protective effects of oyster (*Crassostrea nippona*) extracts against dexamethasone-induced muscle atrophy in C2C12 cells. Extracts were enzymatically hydrolyzed using alcalase (AOH), flavourzyme (FOH), and protamex (POH). Among them, AOH showed the strongest effects, enhancing cell proliferation and promoting dose-dependent myotube formation. The conventional polyacrylamide gel-based method, SDS-PAGE, was used to evaluate the molecular weight and amino acid composition of AOH. Western blot and RT-qPCR analyses revealed that AOH downregulated atrophy-related proteins (MuRF-1, Atrogin, Smad2/3, and Foxo-3a) while upregulating myogenesis-related markers (myogenin, MyoD, myosin heavy chain, and mTOR). These findings indicate that AOH modulates the ubiquitin–proteasome and mTOR signaling pathways to mitigate muscle degradation and promote muscle growth. Overall, *C. nippona* hydrolysate, particularly AOH, shows promise as a bioactive ingredient for developing functional foods aimed at preventing or alleviating muscle atrophy and related conditions such as sarcopenia.

Contribution 6: https://doi.org/10.3390/md22020071

Li et al. targeted sea anemone venoms, which contain diverse toxins that may aid in both prey capture and predator defense. This study applied integrated transcriptomic and proteomic "venomics" to characterize the toxins of *Heteractis magnifica*. Transcriptomic analysis identified 728 putative toxin sequences (442 from tentacles and 381 from the column), which were classified into 68 gene superfamilies. Proteomic analysis confirmed 101 proteins and peptides (91 detected in tentacles and 39 in the column). Several toxin types, including ShK-like peptides and defensins, were co-expressed in both tissues. Homology-based modeling further predicted the 3D structures and potential functions of seven representative toxins. Collectively, this comprehensive venomics approach revealed the molecular diversity and tissue distribution of *H. magnifica* toxins, enhancing understanding of cnidarian venom systems and providing a valuable foundation for future marine drug discovery and development.

Contribution 7: https://doi.org/10.3390/md22010012

Ramirez-Carreto et al. explored sea anemones as valuable sources of bioactive polypeptides by performing a proteomic analysis of crude venom extracts from *Anthopleura dowii* and *Lebrunia neglecta*, identifying 201 polypeptides, 39% of which were common to both species. The identified proteins included hydrolases, oxidoreductases, transferases, heat shock proteins, adhesion proteins, and protease inhibitors. Functional annotation and interaction analyses revealed that these proteins are mainly involved in endoplasmic reticulum-related metabolic processes, such as carbon metabolism and protein processing. Additionally, several oxidative-stress-related enzymes, including superoxide dismutase, peroxiredoxins, thioredoxin, and glutathione oxidase, were detected. Overall, the findings provide new insights into the complex polypeptide composition of sea anemone venoms and highlight their potential as a source of novel bioactive molecules for developing therapeutic agents and biotechnological applications.

Contribution 8: https://doi.org/10.3390/md21020111

Jiang et al. assessed *Tachypleus tridentatus* (horseshoe crab), a marine organism used in traditional Chinese medicine, with its plasma known for various health benefits. This study investigated the effects of *T. tridentatus* plasma on bone growth in rats. Plasma components were separated by ultrafiltration and identified using mass spectrometry; samples with different molecular weight ranges were then tested. The 10–30 kDa plasma fraction showed the strongest bone-growth-promoting activity, outperforming whole plasma. Transcriptomic, proteomic, and other bioinformatics analyses revealed that peptides in this fraction, particularly those rich in specific amino acid residues, were associated with enhanced growth. Differentially expressed proteins were primarily enriched in the PI3K-AKT signaling pathway, a key regulator of cell proliferation and bone development. These findings suggest that *T. tridentatus* plasma, especially the 10–30 kDa fraction, holds promise as a bioactive growth-promoting agent with potential clinical applications in bone repair and regenerative medicine.

Contribution 9: https://doi.org/10.3390/md23040165

This review investigated the increasing demand for sustainable natural products in food, supplements, and cosmetics to slow aging and promote longevity. Both macro- and microalgae are rich sources of proteins, amino acids, fatty acids, vitamins, and minerals with potential health benefits. The biotechnological industry increasingly uses high-throughput omics technologies, especially proteomics, to identify anti-aging compounds from these natural resources. Proteomics enables detailed analysis of protein content, structure, function, and interactions, aiding understanding of molecular aging processes. Mass spectrometry-based and chemical proteomics techniques help identify and quantify proteins and natural products involved in aging regulation. This review highlights marine algae's anti-aging compounds and novel proteomics methods, supporting their application in innovative and sustainable biotechnological solutions for healthy aging.

Contribution 10: https://doi.org/10.3390/md21120633

Rosic and Thornber explored current evidence on the growing patterns of marine macroalgae, or seaweeds, facing varying abiotic (light, temperature, nutrients, salinity) and biotic (grazing, pathogens) stresses, and their adaptive strategies, such as synthesis of secondary metabolites like Mycosporine-Like Amino Acids (MAAs). MAAs, as small, water-soluble compounds with UV-absorbing properties, have been noted for their antioxidative, anti-inflammatory, and photoprotective properties, which are valuable for biotechnological applications. However, their broader use is limited by low bioavailability, challenges in heterologous production, and scarcity in natural sources. Bloom-forming macroalgae from Chlorophyta, Phaeophyceae, and Rhodophyta clades occasionally produce massive biomass during blooms, offering a rich source of pharmacologically active compounds, including MAAs. This review highlights these bloom-forming species and explores how proteomics can advance their utilization, addressing current environmental and biotechnological hurdles to harness their full potential for sustainable applications.

In summary, this Editorial's concluding remarks include gratitude to the Editorial Assistant, Managing Editors, and Editorial Board. Furthermore, the authors who have contributed their findings to this Special Issue are highly appreciated for their submissions and follow-up modifications that have increased knowledge and scientific understanding in the areas of proteomics and marine products. Lastly, we appreciate the help of the reviewers who thoroughly examined all the submitted manuscripts, allowing further progress in the field of proteomics and marine active compounds.

Conflicts of Interest: The author declares no conflicts of interest.

List of Contributions:

1. Sun, S.; Liao, Y.; Fu, J.; Liang, Y.; Chen, Y.; Mao, K.; Gao, B. Fingerprint Analysis and Comparison of Activity Differences of Crude Venom from Five Species of Vermivorous Cone Snail in the South China Sea. *Mar. Drugs* **2025**, *23*, 102. https://doi.org/10.3390/md23030102.
2. Fernandez Cunha, M.; Coscueta, E.R.; Brassesco, M.E.; Almada, F.; Gonçalves, D.; Pintado, M.M. Bioprospecting Bioactive Peptides in *Halobatrachus didactylus* Body Mucus: From In Silico Insights to Essential In Vitro Validation. *Mar. Drugs* **2025**, *23*, 82. https://doi.org/10.3390/md23020082.
3. Hua, Z.; Liao, Y.; Fu, J.; Li, X.; Xu, Q.; Lin, L.; Huang, M.; Gao, B. Revealing the Diversity of Sequences, Structures, and Targets of Peptides from South China Sea *Macrodactyla doreensis* Based on Transcriptomics. *Mar. Drugs* **2024**, *22*, 470. https://doi.org/10.3390/md22100470.
4. Tassara, E.; Mikšík, I.; Pompach, P.; Mariottini, G.L.; Xiao, L.; Giovine, M.; Pozzolini, M. Proteomic Analysis and Biochemical Characterization of the Nematocyst Extract of the Hydrozoan *Velella velella*. *Mar. Drugs* **2024**, *22*, 468. https://doi.org/10.3390/md22100468.
5. Kurera, M.J.M.S.; Nagahawatta, D.P.; Liyanage, N.M.; Jayawardhana, H.H.A.C.K.; Dissanayake, D.S.; Lee, H.-G.; Kim, Y.-S.; Kang, S.I.; Jeon, Y.-J. Exploring the Potential of *Crassostrea nippona* Hydrolysates as Dietary Supplements for Mitigating Dexamethasone-Induced Muscle Atrophy in C2C12 Cells. *Mar. Drugs* **2024**, *22*, 113. https://doi.org/10.3390/md22030113.
6. Li, M.; Mao, K.; Huang, M.; Liao, Y.; Fu, J.; Pan, K.; Shi, Q.; Gao, B. Venomics Reveals the Venom Complexity of Sea Anemone *Heteractis magnifica*. *Mar. Drugs* **2024**, *22*, 71. https://doi.org/10.3390/md22020071.
7. Ramírez-Carreto, S.; Miranda-Zaragoza, B.; Simões, N.; González-Muñoz, R.; Rodríguez-Almazán, C. Marine Bioprospecting: Enzymes and Stress Proteins from the Sea Anemones *Anthopleura dowii* and *Lebrunia neglecta*. *Mar. Drugs* **2024**, *22*, 12. https://doi.org/10.3390/md22010012.
8. Jiang, S.; Qu, X.; Liu, S.; Wei, J.; Yi, X.; Liu, Y.; Gao, C. Proteomic Identification of Plasma Components in *Tachypleus tridentatus* and Their Effects on the Longitudinal Bone Growth Rate in Rats. *Mar. Drugs* **2023**, *21*, 111. https://doi.org/10.3390/md21020111.
9. Rosic, N. Unveiling the Anti-Aging Potential of Marine Natural Bioproducts. *Mar. Drugs* **2025**, *23*, 165. https://doi.org/10.3390/md23040165.
10. Rosic, N.; Thornber, C. Biotechnological Potential of Macroalgae during Seasonal Blooms for Sustainable Production of UV-Absorbing Compounds. *Mar. Drugs* **2023**, *21*, 633. https://doi.org/10.3390/md21120633.

References

1. Carroll, A.R.; Copp, B.R.; Grkovic, T.; Keyzers, R.A.; Prinsep, M.R. Marine natural products. *Nat. Prod. Rep.* **2025**, *42*, 257. [CrossRef]
2. Rosic, N.N. Recent advances in the discovery of novel marine natural products and mycosporine-like amino acid UV-absorbing compounds. *Appl. Microbiol. Biotechnol.* **2021**, *105*, 7053.
3. Lauritano, C.; Ferrante, M.I.; Rogato, A. Marine natural products from microalgae: An -omics overview. *Mar. Drugs* **2019**, *17*, 269. [CrossRef] [PubMed]
4. Kumar, M.S.; Adki, K.M. Marine natural products for multi-targeted cancer treatment: A future insight. *Biomed. Pharmacother.* **2018**, *105*, 233. [CrossRef]
5. Desriac, F.; Jégou, C.; Balnois, E.; Brillet, B.; Le Chevalier, P.; Fleury, Y. Antimicrobial peptides from marine proteobacteria. *Mar. Drugs* **2013**, *11*, 3632. [CrossRef] [PubMed]
6. Ambrosino, L.; Tangherlini, M.; Colantuono, C.; Sangiovanni, A.M.; Miralto, M.; Sansone, C.; Chiusano, M.L. Bioinformatics for Marine Products: An Overview of Resources, Bottlenecks, and Perspectives. *Mar. Drugs* **2019**, *17*, 576. [CrossRef] [PubMed]
7. Zhang, Y.; Fonslow, B.R.; Shan, B.; Baek, M.-C.; Yates, J.R., III. Protein Analysis by Shotgun/Bottom-up Proteomics. *Chem. Rev.* **2013**, *113*, 2343. [CrossRef] [PubMed]
8. Graves, P.R.; Haystead, T.A. Molecular biologist's guide to proteomics. *Microbiol. Mol. Biol. Rev.* **2002**, *66*, 39. [CrossRef] [PubMed]
9. Al-Amrani, S.; Al-Jabri, Z.; Al-Zaabi, A.; Alshekaili, J.; Al-Khabori, M. Proteomics: Concepts and applications in human medicine. *World J. Biol. Chem.* **2021**, *12*, 57. [CrossRef] [PubMed]
10. Ryu, J.; Park, S.-J.; Kim, I.-H.; Choi, Y.H.; Nam, T.-J. Protective effect of porphyra-334 on UVA-induced photoaging in human skin fibroblasts. *Int. J. Mol. Med.* **2014**, *34*, 796. [CrossRef] [PubMed]

11. Rong, D.; Su, Y.; Jia, D.; Zeng, Z.; Yang, Y.; Wei, D.; Lu, H.; Cao, Y. Experimentally validated oxidative stress -associated prognostic signatures describe the immune landscape and predict the drug response and prognosis of SKCM. *Front. Immunol.* **2024**, *15*, 1387316. [CrossRef] [PubMed]
12. Pashaei, E.; Liu, S.; Li, K.; Zang, Y.; Yang, L.; Lautenschlaeger, T.; Huang, J.; Lu, X.; Wan, J. DiCE: Differential centrality-ensemble analysis based on gene expression profiles and protein–protein interaction network. *Nucleic Acids Res.* **2025**, *53*, gkaf609. [CrossRef] [PubMed]
13. Tan, C.; Zhang, Z.; Liu, H.; Liu, X.; Ai, Y.; Wu, X.; Jian, E.; Song, Y.; Yang, J. Progress and trends on machine learning in proteomics during 1997-2024: A bibliometric analysis. *Front. Med.* **2025**, *12*, 1594442.
14. Wu, S.; Zhang, S.; Liu, C.-M.; Fernie, A.R.; Yan, S. Recent Advances in Mass Spectrometry-Based Protein Interactome Studies. *Mol. Cell. Proteom.* **2025**, *24*, 100887. [CrossRef]
15. Argentieri, M.A.; Xiao, S.; Bennett, D.; Winchester, L.; Nevado-Holgado, A.J.; Ghose, U.; Albukhari, A.; Yao, P.; Mazidi, M.; Lv, J.; et al. Proteomic aging clock predicts mortality and risk of common age-related diseases in diverse populations. *Nat. Med.* **2024**, *30*, 2450. [CrossRef] [PubMed]
16. Gafken, P.R.; Paczesny, S. Blood proteomics for quantitative biomarkers of cellular therapies. *Biomark. Res.* **2025**, *13*, 120. [CrossRef] [PubMed]
17. Wink, M. Bioprospecting: The Search for Bioactive Lead Structures from Nature. In *Medicinal Plant Biotechnology: From Basic Research to Industrial Applications*; Wiley: Hoboken, NJ, USA, 2008; p. 97.
18. Jagadeesh, D.S.; Kannegundla, U.; Reddy, R.K. Application of proteomic tools in food quality and safety. *Adv. Anim. Vet. Sci.* **2017**, *5*, 213.
19. Rosic, N.N.; Huang, W.; Johnston, W.A.; DeVoss, J.J.; Gillam, E.M. Extending the diversity of cytochrome P450 enzymes by DNA family shuffling. *Gene* **2007**, *395*, 40. [CrossRef] [PubMed]
20. Rosic, N.N. Mycosporine-Like Amino Acids: Making the Foundation for Organic Personalised Sunscreens. *Mar. Drugs* **2019**, *17*, 638. [CrossRef] [PubMed]
21. Rocha-Martin, J.; Harrington, C.; Dobson, A.D.; O'Gara, F. Emerging strategies and integrated systems microbiology technologies for biodiscovery of marine bioactive compounds. *Mar. Drugs* **2014**, *12*, 3516. [CrossRef] [PubMed]
22. Bedoux, G.; Hardouin, K.; Burlot, A.S.; Bourgougnon, N. Chapter Twelve—Bioactive Components from Seaweeds: Cosmetic Applications and Future Development. In *Advances in Botanical Research*; Bourgougnon, N., Ed.; Academic Press: Amsterdam, The Netherlands, 2014; p. 345.

marine drugs

MDPI

Article

Fingerprint Analysis and Comparison of Activity Differences of Crude Venom from Five Species of Vermivorous Cone Snail in the South China Sea

Shibo Sun, Yanling Liao, Jinxing Fu, Yanxia Liang, Yurong Chen, Kailin Mao * and Bingmiao Gao *

Engineering Research Center of Tropical Medicine Innovation and Transformation of Ministry of Education, Hainan Key Laboratory for Research and Development of Tropical Herbs, International Joint Research Center of Human-Machine Intelligent Collaborative for Tumor Precision Diagnosis and Treatment of Hainan Province, School of Pharmacy, Hainan Medical University, Haikou 571199, China; sunshibo@hainmc.edu.cn (S.S.); liaoyanling@hainmc.edu.cn (Y.L.); hy0207145@muhn.edu.cn (J.F.); 15708919030@163.com (Y.L.); hy0215005@muhn.edu.cn (Y.C.)

* Correspondence: maokailin@muhn.edu.cn (K.M.); gaobingmiao@hainmc.edu.cn (B.G.)

Abstract: The South China Sea is rich in cone snail resources, known for producing conotoxins with diverse biological activities such as analgesic, anticancer, and insecticidal effects. In this study, five vermivorous cone snail samples were collected from the South China Sea and their crude venom was extracted to investigate the variations in venom components and activities, aiming to identify highly active samples for further research. Cluster analysis using reverse-phase high-performance liquid chromatography (RP-HPLC) fingerprints and mitochondrial cytochrome c oxidase I (COI) gene sequences revealed that the diversity of venom components across different conotoxin species is genetically correlated. Activity assays demonstrated that all five cone snail venoms exhibited lethal effects on insects and zebrafish. Notably, the crude venom of *Conus quercinus* showed the highest insecticidal activity with an LD_{50} of 0.6 μg/mg, while *C. tessellatus* venom exhibited the most potent zebrafish lethality with an LD_{50} of 0.2 μg/mg. Furthermore, the crude venom from four cone snail species demonstrated toxicity against ovarian cancer cells, and only *C. caracteristicu* venom displayed significant analgesic activity. This study systematically identifies cone snail samples with promising insecticidal, anticancer, and analgesic properties, paving the way for the development and utilization of cone snail resources from the South China Sea and offering a novel approach for advancing marine peptide drug research.

Keywords: cone snail; conotoxins; insecticidal activity; analgesic activities; anti-proliferative

1. Introduction

Cone snails, marine gastropods belonging to the family Conidae, are distributed across the world's tropical coastal waters, with a significant presence in the South China Sea. These marine gastropods exhibit a remarkable diversity in their distribution, with some species found across the entire tropical Indo-Pacific, while others are restricted to specific bays or seamounts [1–3]. It is estimated that there are around 900 species within the Conus genus, making it the largest genus among marine invertebrates [4]. The greatest species richness for cone snails is observed in the Philippines and countries towards Papua New Guinea, the Solomon Islands, New Caledonia, and Fiji [5]. However, the South China Sea also harbors a significant number of these species, highlighting its importance as a biodiversity hotspot for cone snails. Cone snails are renowned for their complex venom delivery systems and

 https://doi.org/10.3390/md23030102

the array of potent toxins they produce [6]. These fascinating creatures can be categorized into three distinct trophic types based on their predation habits: vermivorous (worm-eating), piscivorous (fish-eating), and molluscivorous (mollusk-eating) [7,8]. Among the vast diversity of cone snails, the South China Sea boasts an abundant resource of genus Conus, with the vermivorous cone snails being particularly prevalent. The *Conus betulinus*, or "bucket snails", are the most numerous in this region. Recent advancements in genomics, transcriptomics, and peptidomics have shed light on the intricate composition of these venoms, revealing over 133 distinct cone snail toxins [7,9–11]. However, predicting their pharmacological activities remains a challenge due to the vast diversity and complexity of these bioactive peptides.

Cone snail toxins, also known as conotoxins or conopeptides, are a diverse group of bioactive peptides produced by predatory marine cone snails [12–14]. These toxins are characterized by their potency, specificity, and unique pharmacological properties [15–17]. Conotoxins are rich in disulfide bridges, which contribute to their stability and diversity. They are known for their extraordinary potency and high specificity, with the ability to target specific ion channels, receptors, and transporters with remarkable precision [15,18,19]. Conotoxins have been classified into pharmacological families based on their targets, such as α, δ, μ, ω, κ, γ, etc., [18,20–25]. For instance, ω-conotoxins are antagonists of voltage-gated calcium channels, and some are effective against neuropathic pain [21]. The ω-conotoxin MVIIA, derived from the fish-hunting species *C. magus*, is a synthetic version known as Ziconotide and is used to treat severe pain [26]. Different cone snail species exhibit a remarkable diversity in their venom composition. Each of the approximately 900 cone snail species expresses a unique library of hundreds of peptide toxins. This diversity is a result of evolutionary adaptation, with each species developing its own set of conopeptides to target the specific physiological circuits of their prey [4]. It is estimated that there may be over 50,000 different conopeptides across all cone snail species [27], highlighting the vast potential for novel drug discovery and therapeutic applications.

In this study, we aim to address the central question of how the venom composition and biological activities of five species of vermivorous cone snail from the South China Sea differ, with a particular focus on their insecticidal, anticancer, and analgesic properties. This research is crucial as it explores the potential of cone snail venoms as a source of novel marine drugs, which could lead to significant advancements in pharmaceutical applications. To achieve this, we utilized reverse phase high-performance liquid chromatography (RP-HPLC) for venom fingerprinting and mitochondrial cytochrome c oxidase I (COI) gene clustering to analyze the genetic diversity of the cone snails, followed by conducting a series of bioactivity assays. This comprehensive approach not only enhances our understanding of vermivorous cone snail venom diversity but also lays the groundwork for future drug development initiatives.

2. Results

2.1. Protein Components in the Crude Venoms

The South China Sea is rich in cone snail biodiversity. In the Sanya region, we collected specimens and identified five predominant vermivorous cone snail species: *C. generalis*, *C. caracteristicus*, *C. betulinus*, *C. quercinus*, and *C. tesselatus* (Figure 1a) The schematic diagram of dissecting the venom tube shows that the radula morphology of the five species of vermivorous cone snails is different, and there are significant differences in the size and length of the venom tube and muscle vesicles among different cone snails (Figure 1b). Specifically, the venom tube of the *C. generalis* is the most compact, measuring a mere 8.3 cm in length, while the *C. betulinus* possesses the most elongated venom tube among the species studied, stretching to an impressive 15.6 cm. The protein content of the crude venom was separated using Sodium Dodecyl Sulfate–PolyAcrylamide Gel Electrophoresis (SDS–PAGE); as shown in Figure 1c, the protein molecular bands are primarily concentrated below 17 kDa. Notably, the venom of *C. tesselatus* exhibits a broader distribution of molecular bands, with some bands extending up to 100 kDa. The complexity of crude venom composition is further underscored by the results of RP-HPLC, as shown in Figure 1d. The primary peaks, indicative of venom composition, are concentrated between 15 and 40 min, highlighting the complex characteristics of these venom compounds.

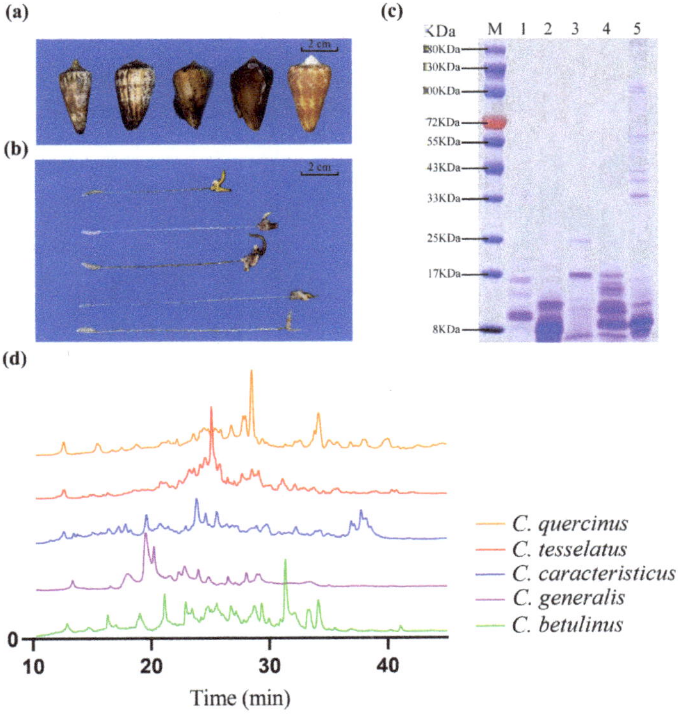

Figure 1. Comparison of external characteristics and crude venom composition of five species of vermivorous cone snails. (**a**) External characteristics of common vermivorous cone snails found in the South China Sea. From left to right: *C. generalis*, *C. caracteristicus*, *C. betulinus*, *C. quercinus*, and *C. tesselatus*. (**b**) Schematic representation of venom tube dissection for the five species of cone snail. From top to bottom: venom glands and ducts of *C. generalis*, *C. caracteristicus*, *C. betulinus*, *C. quercinus*, and *C. tesselatus*. (**c**) SDS-PAGE analysis of crude venom from the five species of cone snail. M indicates the standard marker. Lanes 1–5 correspond to the crude venom of *C. tesselatus*, *C. caracteristicus*, *C. quercinus*, *C. betulinus*, and *C. generalis*, respectively, with a loading volume of 10 μL. (**d**) RP-HPLC analysis of crude venom from the five species of cone snail.

2.2. Similarity Evaluation and Phylogenetic Tree Analysis

Ten batches of crude venom solutions across five species of vermivorous cone snails were taken for analysis. Using the sample chromatogram as the reference fingerprint, multi-point calibration was performed to generate overlay and control fingerprints (see Figures S1–S5). Analysis revealed 16 common peaks in the crude venom of *C. generalis* and *C. caracteristicus*, 18 in *C. betulinus*, 14 in *C. tesselatus*, and 15 in *C. quercinus*. The similarity evaluation results can be found in Tables S1–S5, where all the similarity values of samples from different species are larger than 0.917. This indicates that the quality of five species of vermivorous cone snail samples from the South China Sea is relatively stable, with similar chemical composition but varying content.

Given the genetic disparities in the chemical composition of cone snail venom, a phylogenetic tree was constructed using mitochondrial COI gene sequences. This analysis aimed to correlate venom compositional variations with genetic divergence. The morphologically defined species were segregated into discrete clusters within the phylogenetic tree, as inferred from the cone snail COI gene fragment. As shown in Figure 2a, the phylogenetic analysis revealed that among the five species of vermivorous cone snail, *C. caracteristicus* and *C. quercinus* exhibited the closest genetic relationship, sharing a common branch in the Maximum Likelihood (ML) tree derived from COI sequences. *C. generalis* emerged as the earliest diverging lineage among the other three species, with *C. tesselatus* and *C. betulinus* displaying a strong genetic similarity. Additionally, the crude venom RP-HPLC cluster analysis depicted in Figure 2b indicated that at a squared Euclidean distance threshold of 25, the species formed distinct clusters. Notably, *C. betulinus* and *C. caracteristicus* were the first to cluster, followed by the grouping of *C. tesselatus* and *C. quercinus*, which eventually merged with the cluster that included *C. generalis*.

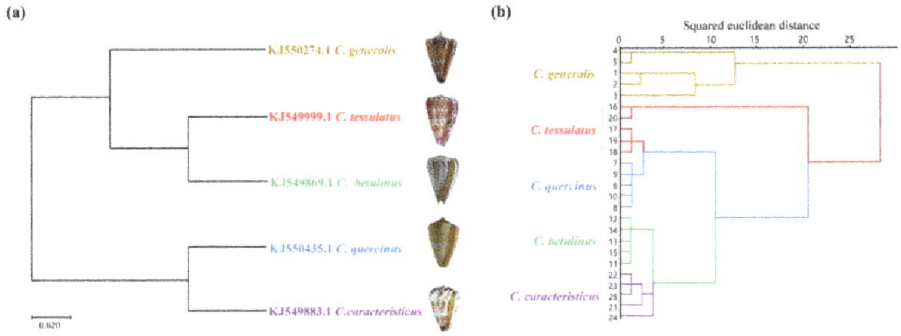

Figure 2. Cluster analysis of mitochondrial COI gene sequence and HPLC fingerprint of five species of vermivorous cone snails. (**a**) Based on COI gene sequences of five species of vermivorous cone snails, a phylogenetic tree was constructed using the ML method. (**b**) Cluster diagram based on RP-HPLC data of five species of crude venom from vermivorous cone snails.

The cluster analysis of the COI gene and RP-HPLC fingerprinting largely align, except for a distinction where *C. betulinus* and *C. caracteristicus* are grouped differently. This suggests a correlation between venom patterns and genetic diversity. Furthermore, these findings offer evidence for a reliable experimental approach to developing chemical fingerprint spectral clustering trees.

2.3. Insect Toxicity

To evaluate the lethal effects of venom from five species of vermivrous cone snails (*C. generalis*, *C. caracteristicus*, *C. betulinus*, *C. quercinus*, and *C. tesselatus*) on *Spodoptera litura* larvae in vivo, crude venom extracts were administered into the lower abdomen of the larvae. As illustrated in Figure 3, crude venom significantly increased the mortality rate of *Spodoptera litura* larvae in a concentration-dependent manner. Specifically, *C. quercinus* and *C. tesselatus* exhibited significant insecticidal activity even at low venom concentrations (at a concentration of 1.00 μg/mg, *C. quercinus* and *C. tesselatus* venom resulted in a mortality rate of 64% ± 10% and 48% ± 10%, respectively), with a marked increase in mortality rates as the venom dosage increased. In contrast, the venom of *C. generalis* and *C. betulinus* showed lower insect toxicity (at a concentration of 1.00 μg/mg, *C. generalis* and *C. betulinus* venom resulted in a mortality rate of 18% ± 10% and 18% ± 10%, respectively), requiring higher concentrations to achieve statistically significant lethal effects compared to the other three species. The median lethal dose (LD_{50}) values are presented in Figure 3f. Notably, *C. quercinus* exhibits the highest toxicity, with an LD_{50} value of 0.6 μg/mg, while *C. generalis* shows the lowest toxicity, with an LD_{50} value of 4.2 μg/mg.

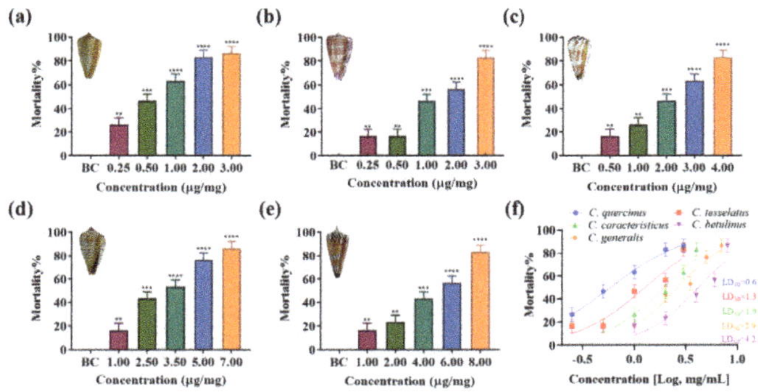

Figure 3. Insecticidal effects and LD_{50} values of crude venom from five species of vermivorous cone snail on *Spodoptera litura* larvae at various concentrations. (**a–e**) The insecticidal effects of crude venoms from *C. quercinus*, *C. tesselatus*, *C. caracteristicus*, *C. betulinus*, and *C. generalis* on *Spodoptera litura* at a specified concentration, $n = 3$. (**f**) The LD_{50} values of these five species of vermivorous cone snail against *Spodoptera litura* larvae at varying concentrations. ** $p < 0.01$, *** $p < 0.001$, **** $p < 0.0001$ versus blank control (BC) group.

2.4. Cytotoxic Activity

The impact of crude venom from various cone snail species on ovarian cancer cell viability was assessed, revealing significant differences in anti-proliferative activity (Figure 4a–e). Increasing doses of crude venom from *C. quercinus, C. tessellatus, C. caracteristicus,* and *C. betulinus* corresponded with lower viability rates in ovarian cancer cells. Notably, the venom of *C. tessellatus* and *C. quercinus* exhibited the strongest toxicity, with an IC_{50} value of 5.2 mg/mL, indicating significant anti-proliferative activity ($p < 0.05$). In contrast, the crude venom of *C. generalis* showed an insignificant impact on ovarian cancer cell inhibition, with no statistically significant difference in activity compared to the control ($p = 0.1673$), and thus its IC_{50} value was not determined in this study. The IC_{50} values for the other species were as follows: *C. caracteristicus* at 5.7 mg/mL, and *C. betulinus* at 8.5 mg/mL (Figure 4f).

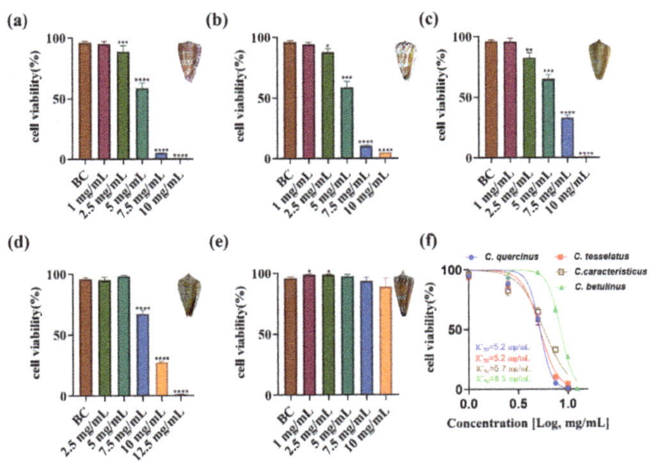

Figure 4. Cytotoxicity assays of crude venom from five species of vermivorous cone snail on ovarian cancer cells at different concentrations. (**a–e**) Cytotoxicity assays of crude venom from *C. quercinus*, *C. tesselatus*, *C. caracteristicus*, *C. betulinus*, and *C. generalis* on ovarian cancer cells at predetermined concentrations, $n = 3$. (**f**) IC_{50} value for the crude venom of four species of vermivorous cone snail on ovarian cancer cells at varying concentrations. * $p < 0.05$, ** $p < 0.01$, *** $p < 0.001$, **** $p < 0.0001$ compared to the blank control (BC) group.

2.5. Toxicity Assessment on Adult Zebrafish

To assess the toxicity of crude venom from five species of vermivorous cone snail on vertebrate models, experiments were conducted using adult zebrafish. Figure 5 highlights significant differences in zebrafish mortality rates when exposed to crude venoms from five species of cone snails, revealing a measurable impact on survival rates compared to a blank control group ($p < 0.05$). Among the venoms tested, *C. tesselatus* exhibited the highest lethal activity, inducing higher mortality rates at lower concentrations (at a concentration of 1.00 µg/mg, *C. tesselatus* venom resulted in a mortality rate of 86% ± 10%), higher than *C. caracteristicus* (at a concentration of 1.00 µg/mg, *C. caracteristicus* venom resulted in a mortality rate of 12% ± 10%), with rapid onset of motility disorders. The LD_{50} value for *C. tesselatus* was determined to be 0.2 µg/mg, indicating its high potency, whereas *C. caracteristicus* had a higher LD_{50} value of 1.9 µg/mg, suggesting lower potency. Mortality rates increased significantly with higher venom concentrations, a trend more pronounced for *C. tesselatus* compared to *C. caracteristicus*, reinforcing the higher lethality of *C. tesselatus* venom. In contrast, the venoms of *C. quercinus*, *C. betulinus*, and *C. generalis* exhibited minimal or negligible toxicity, with low or absent mortality rates across tested concentrations, making it impossible to determine IC_{50} values due to insufficient mortality.

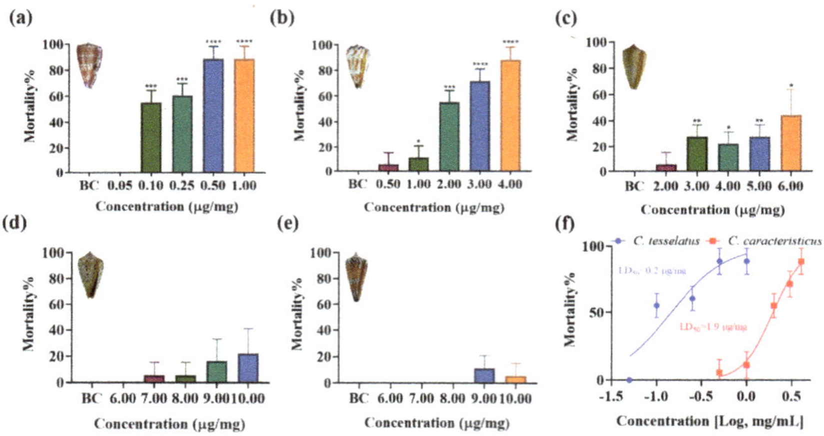

Figure 5. Toxicity analysis of crude venom from five species of vermivorous cone snail on zebrafish at varying concentrations. (**a–e**) The mortality rates of zebrafish exposed to crude venom from *C. tesselatus*, *C. caracteristicus*, *C. quercinus*, *C. betulinus*, and *C. generalis* at specific concentrations, $n = 3$. (**f**) The LD_{50} values for the crude venom of *C. tesselatus* and *C. caracteristicus*. * $p < 0.05$, ** $p < 0.01$, *** $p < 0.001$, **** $p < 0.0001$ versus the blank control (BC) group.

2.6. Assessment of Nociceptive-like Swimming Behavior in Larval Zebrafish

Building on the understanding of the analgesic effects of crude venoms from five species of vermivorous cone snails, behavioral assays were conducted on zebrafish larvae to evaluate these effects. Zebrafish larvae were selected due to their observable locomotor behaviors, which can be effectively tracked. The larvae were exposed to crude venoms from *C. tesselatus*, *C. caracteristicus*, *C. quercinus*, *C. betulinus*, and *C. generalis*, each at a concentration of 5, 10, and 20 µg/mL, respectively, and their locomotor activity was analyzed.

Increased locomotor activity in zebrafish larvae in response to acetic acid may be interpreted as a nocifensive-like escape behavior in response to a noxious stimulus. Lidocaine, a widely recognized non-selective sodium channel blocker, offers rapid pain relief, making it ideal for evaluating immediate analgesic effects., while tramadol, with its dual mechanism as a µ-opioid receptor agonist, provides potent central analgesia [28]. The acetic acid group elicited an acute and rapid swimming behavior immediately upon treatment (exhibited a mean movement of 2750 ± 150 mm over 20 min), while the tramadol (moved a mean distance of 1850 ± 140 mm) and lidocaine groups (moved a mean distance of 2020 ± 100 mm) showed reduced movement, confirming the model's validity. Notably, the *C. caracteristicus* venom-treated group demonstrated a significant reduction in movement distance (mean movement distances of 1240 ± 130 mm), similar to the pure water groups (mean movement distances of 1390 ± 140 mm), suggesting its analgesic efficacy (Figure 6a; One-way ANOVA, $p < 0.0001$).

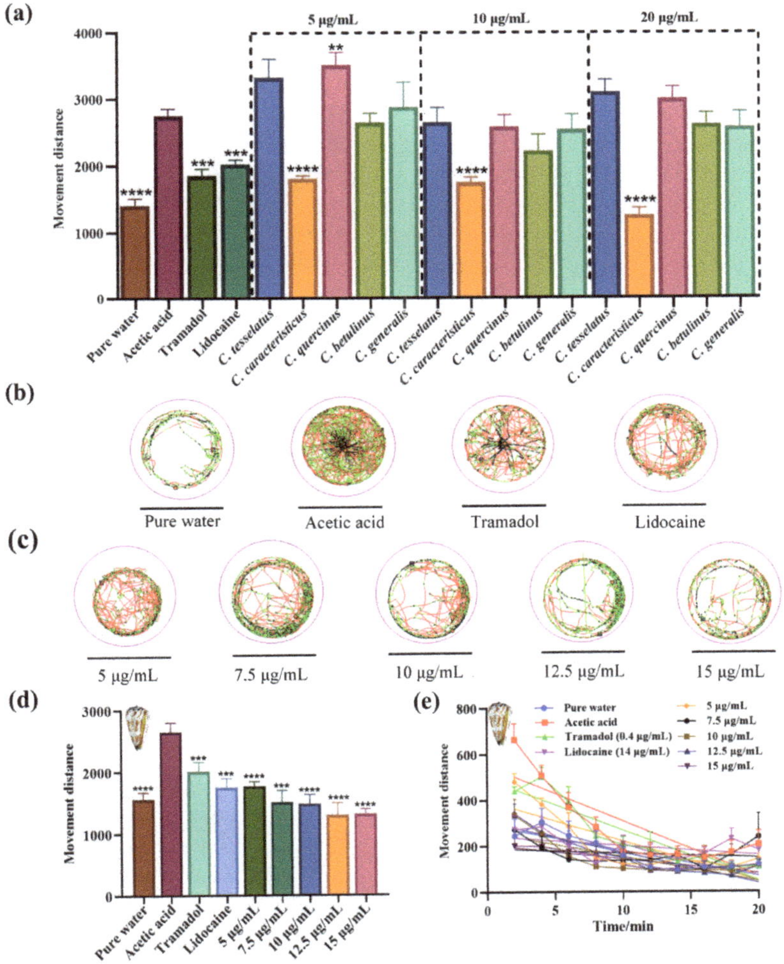

Figure 6. Analgesic effects of crude venoms from five species of vermivorous cone snail on zebrafish larvae. (**a**) Analgesic effects of crude venoms from five species of vermivorous cone snail on zebrafish larvae, evaluated based on their swimming distance, $n = 12$. (**b**) Motion trajectory diagrams for the blank control group, acetic acid model group, and positive drug (Tramadol and Lidocaine) group. Motion trajectories were recorded every 5 s, with instantaneous velocity indicated by color: black (<2 mm/s), green (2–8 mm/s), and red (>8 mm/s). (**c**) Motion trajectory diagram of zebrafish larvae treated with crude venom from *C. caracteristicus*. (**d**) Analgesic effects of *C. caracteristicus* crude venom on zebrafish larvae, $n = 12$. (**e**) Line graph showing movement changes in zebrafish larvae treated with *C. caracteristicus* venom, measured every two minutes. ** $p < 0.01$, *** $p < 0.001$, **** $p < 0.0001$ versus the acetic acid group.

The motion trajectory diagrams (Figure 6b,c) further illustrate these findings, with *C. caracteristicus* venom showing a distinct pattern of reduced locomotion in a dose-dependent manner. The line graph (Figure 6e) highlights the time-dependent changes in movement, reinforcing the analgesic potential of *C. caracteristicus* venom. Statistical analysis confirmed the significance of these observations, with *p*-values indicating strong analgesic effects compared to the acetic acid group.

3. Discussion

The genus Conus, with over 900 known species, exhibits remarkable diversity in its venom composition [4]. Each species has evolved to express a unique cocktail of conopeptides, adapted to the biological targets of their specific prey [29,30]. The venom composition varies not only between species but also within individuals of the same species, due to factors such as predatory or defensive needs [16,31,32]. This intraspecific and interspecific venom variation provides a rich source of bioactive peptides but also presents challenges for venom characterization. The pharmacological activity of conotoxins has been extensively studied, with applications ranging from pain management to cancer treatment [33–36]. Our study aimed to address the question of which pharmacological direction to pursue by examining the venom diversity and identifying potential therapeutic targets.

The COI gene has been widely used in molecular phylogenetic studies of cone snails and other gastropods [3,37,38]. It has helped elucidate evolutionary relationships within the diverse Conus genus [39]. COI, along with other mitochondrial genes including 16S and 12S rRNA, has revealed major clades and subclades within Conus, providing insights into their diet evolution, biogeography, and toxin diversity [40]. Beyond phylogenetics, targeted sequencing of venom genes has shed light on conotoxin evolution, revealing patterns of positive selection, gene expression regulation, and extensive gene turnover across species [41]. In this study, we employed hierarchical cluster analysis to elucidate the relationships among conotoxins based on RP-HPLC fingerprinting data and mitochondrial COI gene sequence homology. Our results demonstrate that the compositional diversity of crude venom is underpinned by genetic variability, with distinct correlations observed among different conotoxin types. These findings significantly contribute to the identification and characterization of novel and potentially more efficacious conotoxins, thereby advancing our understanding of their evolutionary and functional diversity.

Traditionally, vermivorous cone snails are known for their insecticidal activity, with specific conotoxins targeting insect nicotinic acetylcholine receptors (nAChRs) [7–9,42]. Our group has explored the potential of conotoxins from vermivorous cone snails as novel insecticides. We have identified and synthesized various conotoxins from species like *C. quercinus* and *C. betulinus*, demonstrating their insecticidal activities through in vitro and in vivo assays [7–9]. Both α-conotoxins with two disulfide bonds and cysteine-free conopeptides have shown promising results, with some peptides exhibiting high insecticidal efficacy. Specifically, two cysteine-free peptides from *C. betulinus*, Bt010 and Bt016, showed high efficacy with LD_{50} values of 9.07 nM and 10.93 nM, respectively [8,9]. Molecular docking analyses showed that these conotoxins bind to the central pore of nAChRs, forming multiple hydrogen bonds with key residues like Tyr-112 and Lys-116 [8,9]. Our findings on the vermivorous cone snails corroborate the previous literature [7–9], demonstrating their potential as natural insecticides. However, a significant discovery in our study was the substantial toxicity of the crude venom from these cone snails towards zebrafish, as evidenced by the LD_{50} values: *C. tesselatus* venom at 0.2 μg/mg and *C. caracteristicus* venom at 1.9 μg/mg. This toxicity profile contrasts with some previous studies [7–9], suggesting a divergence that may be attributed to the complexity and diversity inherent in the crude venom of cone snails. It is plausible that individual insecticidal peptides within the venom possess high selectivity and specificity, which could account for the observed differences in toxicity. Further research is needed to isolate and characterize specific peptides that possess selective insecticidal activity without harming vertebrates.

Cone snail venoms have shown promising anticancer potential across various studies. Extracts from *C. virgo*, *C. vexillum*, *C. textile*, *C. flavidus*, and *C. geographus* demonstrated cytotoxic effects against multiple cancer cell lines, including breast, liver, lung, and ovarian cancers [33,43–47]. Specifically, *C. virgo* venom extracts showed IC_{50} values between 173.89 and 481.27 μg/mL against breast, liver, and colorectal cancer cells [48]. *C. flavidus* venom exhibited potent effects against liver ($HepG_2$, $IC_{50} = 1.593 \pm 0.05$ μg/mL) and lung (A_{549}, $IC_{50} = 7.836 \pm 0.4$ μg/mL) cancer cells [46]. *C. vexillum* venom from Marsa Alam was effective against lung cancer cells (A_{549}, $IC_{50} = 4.511 \pm 0.03$ μg/mL) [46]. Notably, *C. geographus* extract demonstrated broad antiproliferative activity against ovarian (SKOV-3, $IC_{50} = 22.7 \pm 2.2$ μg/mL), breast (MDA-MB-231, $IC_{50} = 68.7 \pm 6.2$ μg/mL; MCF-7, $IC_{50} = 47 \pm 4.2$ μg/mL), and liver ($HepG_2$, $IC_{50} = 19 \pm 2.1$ μg/mL) cancer cell lines [47]. Individual conotoxins from various species showed anti-ovarian cancer effects, with IC_{50} values ranging from 24.29 to 111.6 μM [33]. Our study finds that crude venom from different cone snail species exhibits varying levels of anti-proliferative activity against ovarian cancer cells. Among the tested species, *C. tessellatus* showed the strongest toxicity, with an IC_{50} value of 5.238 mg/mL ($p < 0.05$). In contrast, *C. generalis* had no significant effect on ovarian cancer cell viability ($p = 0.1673$). The IC_{50} values for *C. quercinus*, *C. caracteristicus*, and *C. betulinus* were 5.226 mg/mL, 5.711 mg/mL, and 8.496 mg/mL, respectively, highlighting species-specific differences in venom potency.

Peptides with potent analgesic properties found in cone snail venoms present promising opportunities for advancing pain management therapies. Studies have identified novel conotoxins from various Conus species, including *C. coronatus*, *C. frigidus* [49], *C. textile* [50], *C. moncuri* [51], *C. lividus* [49], and *C. marmoreus* [52]. These peptides exhibit analgesic effects in animal models of acute and chronic pain, often comparable to morphine [50]. Mechanisms of action include inhibition of voltage-gated calcium channels [51] and interaction with cannabinoid receptors [53]. The ω-conotoxin MVIIA (Prialt®) has been clinically evaluated for severe chronic pain [54]. Combining different conotoxins, such as conantokin-G and ω-conotoxin MVIIA, can produce synergistic antinociceptive effects with reduced side effects [55]. Our study explored the pain-relieving effects of venoms from five cone snail species using zebrafish larvae. The venoms reduced pain-related movement caused by acetic acid in a dose-dependent manner. Among the venoms tested, *C. caracteristicus* showed the strongest analgesic effect, significantly reducing acetic acid-induced hyperlocomotion to levels similar to the Lidocaine group. Statistical analysis validated these results, underscoring the venom's significant potential as a subject for further investigation.

In conclusion, this study provides a comprehensive investigation into the crude venom extracted from five species of vermivorous cone snails native to the South China Sea, uncovering their diverse biological activities and potential for therapeutic applications. Through detailed analyses of venom composition, phylogenetic relationships, and pharmacological properties, we identified significant insecticidal, anticancer, and analgesic activities across different species. Notably, *C. quercinus* demonstrated the strongest insecticidal activity, *C. tessellatus* exhibited potent anticancer effects, and *C. caracteristicus* showed promising analgesic properties, highlighting the species-specific diversity of these bioactive peptides. The integration of RP-HPLC fingerprinting and COI gene sequence analysis revealed a clear correlation between venom composition and genetic variability, offering valuable insights for future studies on the evolution and functional diversity of conotoxins. These findings not only expand our understanding of the pharmacological potential of cone snail venoms but also underscore the untapped value of vermivorous species in the development of innovative marine-derived drugs. Moving forward, the isolation and characterization of individual peptides from these venoms will be essential for advancing their application in therapeutic and agricultural domains, while also promoting sustainable utilization and conservation of marine biodiversity.

4. Materials and Methods

4.1. Venom Extraction

Five species of vermivorous cone snails were collected in the Sanya area of the South China Sea (Figure 1a). Crude venom was extracted by cutting and crushing the shells to extract the snail flesh, followed by removing the venom duct (as illustrated in Figure 1b). The venom duct tissue was then cut into small sections (1–2 mm) and stored at a low temperature for further processing. The venom duct tissue was immersed in a 30% acetonitrile aqueous solution and subjected to low-temperature ultrasonication for 5 s. The solution was vortexed thoroughly and the process was repeated five times. The mixture was then centrifuged at 10,000 rpm and 4 °C for 5 min using a high-speed centrifuge, and the supernatant was collected. The remaining venom duct tissue was re-extracted using the same 30% acetonitrile solution following the same procedure [56]. The supernatants from the two extraction cycles were combined and frozen at −80 °C. Finally, the crude venom solution was concentrated using a freeze-dryer, yielding the crude venom from the vermivorous cone snails.

4.2. Bicinchoninic Acid (BCA) Protein Assay

Bicinchoninic Acid (BCA) Protein Assay was employed to quantify the protein concentration in the extracted crude venom of vermivorous Conus species [57]. Absorbance values of all samples were measured at a wavelength of 562 nm using an enzyme-linked immunosorbent assay (ELISA) reader, and the readings were recorded. A sample without BSA was used as the blank control. A standard curve was generated with absorbance at 562 nm (A_{562}) plotted on the vertical axis and BSA content on the horizontal axis. This standard curve was then used to calculate the protein concentration in each crude venom sample.

4.3. SDS-PAGE

SDS-PAGE was performed essentially as described by Laemmli [58]. A total of 10 mg of each crude venom of vermivorous cone snail was weighed and dissolved in 1 mL of deionized water and then centrifuged at 1000 rpm for 5 min. Then, 16 μL of the supernatant was mixed with 4 μL of the sample loading buffer (0.01 M Tris-HCl, 10% (*w*/*v*) SDS, 10% (*v*/*v*) glycerol, 0.1% (*w*/*v*) bromophenol blue, 2% β-mercaptoethanol) and then boiled for 10 min. Samples were loaded onto gel slab consisting of 12% separating gel (pH 8.8) and 4% stacking gel (pH 6.8), and subjected to electrophoresis at 100 V for 1 h. The electrophoresis gels were stained using Coomassie Brilliant Blue R-250.

4.4. Chromatographic Conditions

The crude venom fractions of vermivorous cone snails were preliminarily analyzed using RP-HPLC with a UV detector and a C_{18} column (Diamonsil, 250 × 4.6 mm, 5 μm). For each sample, 10 mg of crude venom was dissolved in 1 mL of deionized water and centrifuged at 4 °C and 10,000 rpm for 10 min. The supernatant was filtered through a 0.45 μm microporous membrane, and 100 μL of the filtered sample was injected into the RP-HPLC system using an auto-sampler. The mobile phase consisted of solution A (0.1% trifluoroacetic acid [TFA] in acetonitrile) and solution B (0.1% TFA in ultrapure water). Both solvents were filtered through 0.45 μm filter papers and degassed for 15 min using an ultrasonicator before use. The separation was performed at a constant flow rate of

1 mL/min over 65 min using a linear gradient from 5% to 70% solution A. The elution was monitored by measuring absorbance at 214 nm.

4.5. Establishment of Fingerprint and Similarity Analysis

Fingerprint mapping was employed to analyze venom composition differences within and between five species of cone snail (*C. generalis*, *C. betulinus*, *C. caracteristicus*, *C. tesselatus*, and *C. quercinus*) [59]. Crude venom samples were prepared, filtered, and analyzed using RP-HPLC to generate chromatographic fingerprints, with Chromatogram 5 selected as the reference spectrum. The "Similarity Evaluation System for Chromatographic Fingerprint of Traditional Chinese Medicine developed by Chinese Pharmacopoeia Committee (Version 2004A) (Beijing, China)" software was used to calculate fingerprint similarities across 10 batches of venom per species. Cluster analysis, performed in SPSS 26.0 using the relative peak areas of characteristic peaks from 25 samples, produced a dendrogram to assess venom quality, variability, and interspecies relationships.

4.6. Phylogenetic Tree Analysis

COI gene sequences of five species of vermivorous cone snails were obtained from the National Center for Biotechnology Information (NCBI) database (http://www.ncbi.nlm.nih.gov/, accessed on 20 November 2024). Representative gene sequences were selected, processed, and analyzed using MEGA-X (v10.2.6) software. The Maximum Likelihood (ML) method was employed to construct an evolutionary phylogenetic tree, illustrating the genetic relationships among the species [60].

4.7. Insect Toxicity Test

Spodoptera litura eggs were purchased from Guangzhou Zhixin Biotechnology Co., Ltd (GuangZhou, China) and cultured at room temperature to 4–6 days post-fertilization (dpf) for experimental use. The insect toxicity test was conducted using the fourth to sixth instar larvae of *Spodoptera litura*, weighing approximately 180 mg. Prior to the experiment, the larvae were fed with standard feed ad libitum. Five venoms were dissolved in 0.7% saline to the required concentrations. A liquid micro-injector (Shanghai High Pigeon, Shanghai, China) was used to administer 3 μL of the solution into the abdomen of each larva. The control group received an injection of 0.7% saline. Each concentration was tested in triplicate ($n = 3$), with lethality rates determined using 10 larvae per replicate. Mortality of *Spodoptera litura* larvae was observed and recorded over 36 h following the injection of the venom solutions or saline control [7].

4.8. Cytotoxicity Assays

The cytotoxic effects of five vermivorous venom on selected human ovarian cancer cells were evaluated using the Cell Counting Kit-8 (CCK-8) assay [61]. Briefly, ovarian cancer cells were harvested and resuspended to prepare a cell suspension. The cells were seeded into 96-well plates at a density of 5×10^4 cells per well and incubated in a biochemical incubator at 37 °C for 24 h to allow attachment and growth. The five venom samples were dissolved in the culture medium to the desired concentrations and filtered to ensure sterility. Each sample (100 μL) was added to the wells, with three replicates for each concentration. Wells containing only culture medium served as the blank control group. Following a 24 h incubation at 37 °C, 90 μL of culture medium and 10 μL of CCK-8 reagent were added to each well. The plates were then incubated for an additional 10 h. Finally, the optical density (OD) was measured at 450 nm using a Synergy HTX multifunctional mioplate reader (manufactured by Agilent BioTek, Burlington, VT, USA) to assess cell viability.

4.9. Adult Zebrafish Toxicity Assay

Three-month-old zebrafish with a standard body length of 2–2.5 cm were used for the toxicity study. Five distinct conotoxin solutions were prepared by dissolving the toxins in 0.7% physiological saline to achieve the desired concentrations. Using a precision microsyringe, 4 μL of each venom solution was injected into the lower abdominal cavity of the zebrafish. To ensure complete delivery of the conotoxin and minimize residual toxin in the syringe, the syringe was rinsed with 10 μL of 0.7% physiological saline, which was subsequently injected into the zebrafish.

Physiological saline was used as the control. Each concentration was tested in triplicate ($n = 3$), with lethality rates determined using six zebrafish per replicate. The zebrafish were maintained in a stable aquatic environment throughout the experiment, and their physiological status was carefully monitored and recorded over a 24 h observation period [62,63].

4.10. Locomotor Behavior of Zebrafish Larval

AB series wild-type zebrafish eggs were sourced from the Institute of Hydrobiology, Chinese Academy of Sciences, and incubated in our laboratory's zebrafish breeding system. The embryos were maintained at 28 °C in an incubator until they reached 3–5 dpf to facilitate behavioral verification. Each juvenile fish was individually placed into a well of a 24-well plate, with ten replicates per treatment group ($n = 10$). Experimental groups were designed as follows: (1) a control group exposed to pure water, (2) a model group treated with pure water and a 0.01% acetic acid aqueous solution, (3) treatment groups exposed to different concentrations of conotoxin crude toxin aqueous solutions (5, 10, and 20 μg/mL) mixed with 0.01% acetic acid aqueous solution, (4) positive control group 1 treated with a tramadol solution (0.4 μg/mL) mixed with 0.01% acetic acid aqueous solution, and (5) positive control group 2 treated with a lidocaine solution (14 μg/mL) mixed with 0.01% acetic acid aqueous solution. Following the treatments, the locomotor activity of the juvenile zebrafish was evaluated over a 20 min observation period using the zebrafish tracking system from Viewpoint Life Sciences, Montreal, QC, Canada. Each larva was recorded ten times, with each recording lasting 2 min, and the total distances swum by the larva were ultimately documented. The movement of each larva was captured at 1 min intervals, accumulating a total of 20 recordings per individual [64].

4.11. Statistical Analysis

Statistical analyses were performed using GraphPad Prism (v8). Unless stated otherwise, data are expressed as the mean ± standard error of the mean (s.e.m.). For comparisons involving three or more groups, one-way analysis of variance (ANOVA) followed by Tukey's multiple-comparison test was used.

Supplementary Materials: The following supporting information can be downloaded at: https://www.mdpi.com/article/10.3390/md23030102/s1, Figure S1: Fingerprint of crude venom of *C. tesselatus*; Figure S2: Fingerprint of crude venom of *C. generalis*; Figure S3: Fingerprint of crude venom of *C. caracteristicus*; Figure S4: Fingerprint of crude venom of *C. betulinus*; Figure S5: Fingerprint of crude venom of *C. quercinus*; Table S1: Similarity of crude venom fingerprints of 10 batches of *C. caracteristicus*; Table S2: Similarity of crude venom fingerprints of 10 batches of *C. betulinus*; Table S3: Similarity of crude venom fingerprints of 10 batches of *C. generalis*; Table S4: Similarity of crude venom fingerprints of 10 batches of *C. quercinus*; Table S5: Similarity of crude venom fingerprints of 10 batches of *C. tesselatus*.

Author Contributions: Conceptualization, B.G. and K.M.; methodology, S.S.; data curation, Y.L. (Yanling Liao), Y.C. and J.F.; writing—original draft preparation, S.S. and Y.L. (Yanxia Liang); writing—review and editing, B.G. and K.M.; funding acquisition, B.G. All authors have read and agreed to the published version of the manuscript.

Funding: This research was supported by Hainan Provincial Key Point Research and Invention Program (no. ZDYF2022SHFZ309), Hainan Provincial Natural Science Foundation of China (no. 821MS046), and Hainan Provincial Natural Science Foundation of China (no. 424QN258).

Institutional Review Board Statement: Not applicable.

Informed Consent Statement: Not applicable.

Data Availability Statement: The original data presented in the study are included in the article; further inquiries can be directed to the corresponding author.

Conflicts of Interest: The authors declare no conflicts of interest.

References

Gutthavilli, P.R.; Bharne, A.M.; Marimuthu, K.; Thiruchitrambalam, G. Unveiling the enigmatic cone snails along the coastal environments of the South Andaman Islands: Diversity, distribution and their habitat preference. *Front. Mar. Sci.* **2024**, *11*, 1477472. [CrossRef]

Kumar, P.S.; Kumar, D.S.; Umamaheswari, S. A perspective on toxicology of Conus venom peptides. *Asian Pac. J. Trop. Med.* **2015**, *8*, 337–351. [CrossRef] [PubMed]

Laxmilatha, P.; Ameri, S.; Labeeb, K.A.; Ranjith, L.; Kathirvelpandian, A. A new species of cone snail, *Conus laccadivensis* sp. nov.: (Gastropoda: Conidae) from the Islands of Lakshadweep Archipelago. *Reg. Stud. Mar. Sci.* **2021**, *44*, 101783. [CrossRef]

Fiorotti, H.B.; Figueiredo, S.G.; Campos, F.V.; Pimenta, D.C. Cone snail species off the Brazilian coast and their venoms: A review and update. *J. Venom. Anim. Toxins Incl. Trop. Dis.* **2023**, *29*, e20220052. [CrossRef]

Neves, J.L.B.; Imperial, J.S.; Morgenstern, D.; Ueberheide, B.; Gajewiak, J.; Antunes, A.; Robinson, S.D.; Espino, S.; Watkins, M.; Vasconcelos, V.; et al. Characterization of the First Conotoxin from *Conus ateralbus*, a Vermivorous Cone Snail from the Cabo Verde Archipelago. *Mar. Drugs* **2019**, *17*, 432. [CrossRef]

Jin, A.-H.; Dutertre, S.; Dutt, M.; Lavergne, V.; Jones, A.; Lewis, R.J.; Alewood, P.F. Transcriptomic-Proteomic Correlation in the Predation-Evoked Venom of the Cone Snail, Conus imperialis. *Mar. Drugs* **2019**, *17*, 177. [CrossRef] [PubMed]

Gao, B.; Peng, C.; Lin, B.; Chen, Q.; Zhang, J.; Shi, Q. Screening and Validation of Highly-Efficient Insecticidal Conotoxins from a Transcriptome-Based Dataset of Chinese Tubular Cone Snail. *Toxins* **2017**, *9*, 214. [CrossRef]

Lin, C.; Qin, H.; Liao, Y.; Chen, J.; Gao, B. Chemical Synthesis and Insecticidal Activity Research Based on α-Conotoxins. *Molecules* **2024**, *29*, 2846. [CrossRef] [PubMed]

Chen, J.; Zhang, X.; Lin, C.; Gao, B. Synthesis and insecticidal activity of cysteine-free conopeptides from Conus betulinus. *Toxicon* **2023**, *233*, 107253. [CrossRef] [PubMed]

0. Liao, Y.; Fu, J.; Gao, B.; Tang, T. The complete mitochondrial DNA genome of a cone snail, *Conus betulinus* (Neogastropoda: Conidae), from the South China sea. *Mitochondrial DNA Part B* **2021**, *6*, 1696–1698. [CrossRef]

1. Peng, C.; Yao, G.; Gao, B.-M.; Fan, C.-X.; Bian, C.; Wang, J.; Cao, Y.; Wen, B.; Zhu, Y.; Ruan, Z.; et al. High-throughput identification of novel conotoxins from the Chinese tubular cone snail (*Conus betulinus*) by multi-transcriptome sequencing. *GigaScience* **2016**, *5*, 17. [CrossRef]

2. Espiritu, M.J.J.; Taylor, J.K.K.; Sugai, C.K.K.; Thapa, P.; Loening, N.M.M.; Gusman, E.; Baoanan, Z.G.G.; Baumann, M.H.H.; Bingham, J.-P. Characterization of the Native Disulfide Isomers of the Novel χ-Conotoxin PnID: Implications for Further Increasing Conotoxin Diversity. *Mar. Drugs* **2023**, *21*, 61. [CrossRef] [PubMed]

3. Grandal, M.; Hoggard, M.; Neely, B.; Davis, W.C.; Mari, F. Proteogenomic Assessment of Intraspecific Venom Variability: Molecular Adaptations in the Venom Arsenal of *Conus purpurascens*. *Mol. Cell. Proteom.* **2021**, *20*, 100100. [CrossRef]

4. Li, X.; Chen, W.; Zhangsun, D.; Luo, S. Diversity of Conopeptides and Their Precursor Genes of *Conus Litteratus*. *Mar. Drugs* **2020**, *18*, 464. [CrossRef] [PubMed]

5. Mir, R.; Karim, S.; Kamal, M.A.; Wilson, C.M.; Mirza, Z. Conotoxins: Structure, Therapeutic Potential and Pharmacological Applications. *Curr. Pharm. Des.* **2016**, *22*, 582–589. [CrossRef] [PubMed]

6. Ratibou, Z.; Inguimbert, N.; Dutertre, S. Predatory and Defensive Strategies in Cone Snails. *Toxins* **2024**, *16*, 94. [CrossRef] [PubMed]

7. Thapa, P.; Espiritu, M.J.; Cabalteja, C.C.; Bingham, J.-P. Conotoxins and their regulatory considerations. *Regul. Toxicol. Pharmacol.* **2014**, *70*, 197–202. [CrossRef]

8. Giribaldi, J.; Haufe, Y.; Evans, E.R.J.; Wilson, D.T.; Daly, N.L.; Enjalbal, C.; Nicke, A.; Dutertre, S. Synthesis, Structural and Pharmacological Characterizations of CIC, a Novel α-Conotoxin with an Extended N-Terminal Tail. *Mar. Drugs* **2021**, *19*, 141. [CrossRef]

9. Margiotta, F.; Micheli, L.; Ciampi, C.; Ghelardini, C.; McIntosh, J.M.; Di Cesare Mannelli, L. *Conus regius*-Derived Conotoxins: Novel Therapeutic Opportunities from a Marine Organism. *Mar. Drugs* **2022**, *20*, 773. [CrossRef]

20. Green, B.R.; Bulaj, G.; Norton, R.S. Structure and function of μ-conotoxins, peptide-based sodium channel blockers with analgesi activity. *Future Med. Chem.* **2014**, *6*, 1677–1698. [CrossRef]
21. Jimenez, E.C. Pharmacological Classes of *Conus Peptides* Targeted to Calcium, Sodium, and Potassium Channels. *Protein Pept. Let* **2023**, *30*, 913–929. [CrossRef] [PubMed]
22. McMahon, K.L.; Tran, H.N.T.; Deuis, J.R.; Lewis, R.J.; Vetter, I.; Schroeder, C.I. Discovery, Pharmacological Characterisation an NMR Structure of the Novel μ-Conotoxin SxIIIC, a Potent and Irreversible Na_V Channel Inhibitor. *Biomedicines* **2020**, *8*, 39 [CrossRef]
23. McMahon, K.L.; Vetter, I.; Schroeder, C.I. Voltage-Gated Sodium Channel Inhibition by μ-Conotoxins. *Toxins* **2024**, *16*, 5 [CrossRef] [PubMed]
24. Pei, S.; Wang, N.; Mei, Z.; Zhangsun, D.; Craik, D.J.; McIntosh, J.M.; Zhu, X.; Luo, S.; Wahlen, G.E. Conotoxins Targetin Voltage-Gated Sodium Ion Channels. *Pharmacol. Rev.* **2024**, *76*, 828–845. [CrossRef]
25. Wilson, D.T.; Bansal, P.S.; Carter, D.A.; Vetter, I.; Nicke, A.; Dutertre, S.; Daly, N.L. Characterisation of a Novel A-Superfamil Conotoxin. *Biomedicines* **2020**, *8*, 128. [CrossRef]
26. Lin, J.; Chen, S.; Butt, U.D.; Yan, M.; Wu, B. A comprehensive review on ziconotide. *Heliyon* **2024**, *10*, e31105. [CrossRef]
27. Lin, Z.; Torres, J.P.; Watkins, M.; Paguigan, N.; Niu, C.; Imperial, J.S.; Tun, J.; Safavi-Hemami, H.; Finol-Urdaneta, R.K.; Neve J.L.B.; et al. Non-Peptidic Small Molecule Components from Cone Snail Venoms. *Front. Pharmacol.* **2021**, *12*, 655981. [CrossRef]
28. Zaig, S.; Scarpellini, C.d.S.; Montandon, G. Respiratory depression and analgesia by opioid drugs in freely behaving larva zebrafish. *eLife* **2021**, *10*, e63407. [CrossRef]
29. Giribaldi, J.; Wilson, D.; Nicke, A.; El Hamdaoui, Y.; Laconde, G.; Faucherre, A.; Maati, H.M.O.; Daly, N.L.; Enjalbal, C.; Dutertr S. Synthesis, Structure and Biological Activity of CIA and CIB, Two α-Conotoxins from the Predation-Evoked Venom of *Conu catus*. *Toxins* **2018**, *10*, 222. [CrossRef]
30. Jin, A.-H.; Vetter, I.; Himaya, S.W.A.; Alewood, P.F.; Lewis, R.J.; Dutertre, S. Transcriptome and proteome of *Conus planorbi* identify the nicotinic receptors as primary target for the defensive venom. *Proteomics* **2015**, *15*, 4030–4040. [CrossRef] [PubMed
31. Pushpabai, R.R.; Alphonse, C.R.W.; Mani, R.; Apte, D.A.; Franklin, J.B. Diversity of Conopeptides and Conoenzymes from th Venom Duct of the Marine Cone Snail *Conus bayani* as Determined from Transcriptomic and Proteomic Analyses. *Mar. Drug* **2021**, *19*, 202. [CrossRef] [PubMed]
32. Turner, A.H.; Craik, D.J.; Kaas, Q.; Schroeder, C.I. Bioactive Compounds Isolated from Neglected Predatory Marine Gastropods *Mar. Drugs* **2018**, *16*, 118. [CrossRef] [PubMed]
33. Ju, S.; Zhang, Y.; Guo, X.; Yan, Q.; Liu, S.; Ma, B.; Zhang, M.; Bao, J.; Luo, S.; Fu, Y. Anti-Ovarian Cancer Conotoxins Identifie from *Conus Venom*. *Molecules* **2022**, *27*, 6609. [CrossRef]
34. Lopez-Millan, J.M.; Coca-Gamito, C. Intrathecal Ziconotide Infusion Therapy for Oncologic Refractory Neuropathic Pain: Cas Report with High-Dose Intraventricular Ziconotide After 15 Years of Treatment. *Neural Interface* **2024**, *27*, 1251–1253. [CrossRef]
35. Monroe, L.K.; Truong, D.P.; Miner, J.C.; Adikari, S.H.; Sasiene, Z.J.; Fenimore, P.W.; Alexandrov, B.; Williams, R.F.; Nguyen, H.B Conotoxin Prediction: New Features to Increase Prediction Accuracy. *Toxins* **2023**, *15*, 641. [CrossRef] [PubMed]
36. Sukul, V.V. Intrathecal Pain Therapy for the Management of Chronic Noncancer Pain. *Neurosurg. Clin.* **2019**, *30*, 195–201 [CrossRef] [PubMed]
37. Chiappa, G.; Fassio, G.; Smriglio, C.; Mariottini, P.; Albano, P.G.; Modica, M.V.; Zuccon, D.; Puillandre, N.; Oliverio, M The polymorphic top-shell puzzle: Iterative taxonomy of Calliostoma Swainson, 1840 (Gastropoda: Calliostomatidae), in the Mediterranean Sea. *J. Molluscan Stud.* **2024**, *90*, eyae026. [CrossRef]
38. Lawler, A.J.; Duda, T.F., Jr. Molecular and morphometric data suggest the presence of a neglected species in the marine gastropo family Conidae. *Mol. Phylogenetics Evol.* **2017**, *109*, 421–429. [CrossRef] [PubMed]
39. Puillandre, N.; Bouchet, P.; Duda, T.F., Jr.; Kauferstein, S.; Kohn, A.J.; Olivera, B.M.; Watkins, M.; Meyer, C. Molecular phylogen and evolution of the cone snails (Gastropoda, Conoidea). *Mol. Phylogen. Evol.* **2014**, *78*, 290–303. [CrossRef] [PubMed]
40. Puillandre, N.; Stoecklin, R.; Favreau, P.; Bianchi, E.; Perret, F.; Rivasseau, A.; Limpalaer, L.; Monnier, E.; Bouchet, P. Whe everything converges: Integrative taxonomy with shell, DNA and venomic data reveals *Conus conco*, a new species of cone snail (Gastropoda: Conoidea). *Mol. Phylogenetics Evol.* **2014**, *80*, 186–192. [CrossRef] [PubMed]
41. Phuong, M.A.; Mahardika, G.N. Targeted Sequencing of Venom Genes from Cone Snail Genomes Improves Understanding of Conotoxin Molecular Evolution. *Mol. Biol. Evol.* **2018**, *35*, 1210–1224. [CrossRef]
42. Spiezia, M.C.; Chiarabelli, C.; Polticelli, F. Recombinant expression and insecticidal properties of a *Conus ventricosus* conotoxin-GST fusion protein. *Toxicon* **2012**, *60*, 744–751. [CrossRef]
43. Nakmode, D.D.; Singh, B.; Abdella, S.; Song, Y.; Garg, S. Long-acting parenteral formulations of hydrophilic drugs, proteins, and peptide therapeutics: Mechanisms, challenges, and therapeutic benefits with a focus on technologies. *Drug Deliv. Transl. Res* **2024**, 1–25. [CrossRef]

4. Abdel-Rahman, M.A.; Abdel-Nabi, I.M.; El-Naggar, M.S.; Abbas, O.A.; Strong, P.N. Conus vexillum venom induces oxidative stress in Ehrlich's ascites carcinoma cells: An insight into the mechanism of induction. *J. Venom. Anim. Toxins Incl. Trop. Dis.* **2013**, *19*, 10. [CrossRef]
5. Afsharirad, T.; Tahmasvand, R.; Amini, M.; Daraei, B.; Salimi, M. Two novel anticancer compounds with minimum cardiotoxic property. *BMC Pharmacol. Toxicol.* **2020**, *21*, 1–8. [CrossRef]
6. Magdy, N.A.; Nafie, M.S.; El-Naggar, M.S.; Abu El-Regal, M.A.; Azeiz, A.Z.A.; Abdel-Rahman, M.A.; El-Zawahry, M. Cytotoxicity and apoptosis induction of the marine *Conus flavidus* venom in HepG2 cancer cell line. *J. Biomol. Struct. Dyn.* **2023**, *41*, 7786–7793. [CrossRef]
7. Alburae, N.A.; Mohammed, A.E. Antiproliferative effect of the Red Sea cone snail, *Conus geographus*. *Trop. J. Pharm. Res.* **2020**, *19*, 577–581. [CrossRef]
8. Ganesan, D.S.; Mohamed, M.A.R.; Lhanzin, P.; Neelan, K.; Vadakkuvaselvi, L.; Subramani, S.; Chinnasamy, A. Antiproliferative Effect of Crude Venom from *Conus virgo* on Human Lung Cancer Cell Line and Toxicity Assessment on Adult Zebra Fish (*Danio rerio*). *Indian J. Pharm. Educ. Res.* **2020**, *54*, 85–94. [CrossRef]
9. Qiang, Y.; Wu, Y.; Zhao, D.; Zhao, B.; Wang, F.; Ren, S.; Wen, Y.; Gu, J.; Zhang, L.; Liu, K.; et al. Discovery and characterization of the novel conotoxin Lv1d from *Conus lividus* that presents analgesic activity. *Toxicon* **2021**, *194*, 70–78. [CrossRef] [PubMed]
0. Tabaraki, N.; Shahbazzadeh, D.; Moradi, A.M.; Vosughi, G.; Mostafavi, P.G. Analgesic effect of Persian Gulf *Conus textile* venom. *Iran. J. Basic Med. Sci.* **2014**, *17*, 793–797. [CrossRef]
1. Sousa, S.R.; McArthur, J.R.; Brust, A.; Bhola, R.F.; Rosengren, K.J.; Ragnarsson, L.; Dutertre, S.; Alewood, P.F.; Christie, M.J.; Adams, D.J.; et al. Novel analgesic ω-conotoxins from the vermivorous cone snail *Conus moncuri* provide new insights into the evolution of conopeptides. *Sci. Rep.* **2018**, *8*, 13397. [CrossRef]
2. Gao, B.; Zhangsun, D.; Hu, Y.; Wu, Y.; Sheng, L.; Fang, L.; Wu, X.; Yu, J.; Luo, S. Expression and secretion of functional recombinant μO-conotoxin MrVIB-His-tag in *Escherichia coli*. *Toxicon* **2013**, *72*, 81–89. [CrossRef] [PubMed]
3. Jergova, S.; Perez, C.; Imperial, J.S.; Gajavelli, S.; Jain, A.; Abin, A.; Olivera, B.M.; Sagen, J. Cannabinoid receptor agonists from Conus venoms alleviate pain-related behavior in rats. *Pharmacol. Biochem. Behav.* **2021**, *205*, 173182. [CrossRef] [PubMed]
4. Mollica, A.; Costante, R.; Novellino, E.; Stefanucci, A.; Pieretti, S.; Zador, F.; Samavati, R.; Borsodi, A.; Benyhe, S.; Vetter, I.; et al. Design, Synthesis and Biological Evaluation of Two Opioid Agonist and $Cav_{2.2}$ Blocker Multitarget Ligands. *Chem. Biol. Drug Des.* **2015**, *86*, 156–162. [CrossRef]
5. Anonymous. Abstracts for the 21st Annual Meeting of the American Society for Neural Therapy and Repair. *Cell Transplant.* **2014**, *23*, 761–790. [CrossRef] [PubMed]
6. Yu, S.; Yang, B.; Yan, L.; Dai, Q. Sensitive Detection of α-Conotoxin GI in Human Plasma Using a Solid-Phase Extraction Column and LC-MS/MS. *Toxins* **2017**, *9*, 235. [CrossRef] [PubMed]
7. Li, J.; Zhao, Y.; Jiang, X. Quantitative analysis of protein in thermosensitive hydroxypropyl chitin for biomedical applications. *Anal. Biochem.* **2020**, *599*, 113745. [CrossRef]
8. Laemmli, U.K. Cleavage of structural proteins during the assembly of the head of bacteriophage T4. *Nature* **1970**, *227*, 680–685. [CrossRef]
9. Kong, W.-J.; Zhang, S.-S.; Zhao, Y.-L.; Wu, M.-Q.; Chen, P.; Wu, X.-R.; Ma, X.-P.; Guo, W.-Y.; Yang, M.-H. Combination of chemical fingerprint and bioactivity evaluation to explore the antibacterial components of *Salvia miltiorrhizae*. *Sci. Rep.* **2017**, *7*, 8112. [CrossRef]
0. Newman, L.; Duffus, A.L.J.; Lee, C. Using the Free Program MEGA to Build Phylogenetic Trees from Molecular Data. *Am. Biol. Teach.* **2016**, *78*, 608–612. [CrossRef]
1. Cai, L.; Qin, X.; Xu, Z.; Song, Y.; Jiang, H.; Wu, Y.; Ruan, H.; Chen, J. Comparison of Cytotoxicity Evaluation of Anticancer Drugs between Real-Time Cell Analysis and CCK-8 Method. *ACS Omega* **2019**, *4*, 12036–12042. [CrossRef] [PubMed]
2. Horzmann, K.A.; Freeman, J.L. Making Waves: New Developments in Toxicology with the Zebrafish. *Toxicol. Sci.* **2018**, *163*, 5–12. [CrossRef]
3. Lantz-McPeak, S.; Guo, X.; Cuevas, E.; Dumas, M.; Newport, G.D.; Ali, S.F.; Paule, M.G.; Kanungo, J. Developmental toxicity assay using high content screening of zebrafish embryos. *J. Appl. Toxicol.* **2015**, *35*, 261–272. [CrossRef] [PubMed]
4. Steenbergen, P.J.; Bardine, N. Antinociceptive effects of buprenorphine in zebrafish larvae: An alternative for rodent models to study pain and nociception? *Appl. Anim. Behav. Sci.* **2014**, *152*, 92–99. [CrossRef]

marine drugs

MDPI

Article

Bioprospecting Bioactive Peptides in *Halobatrachus didactylus* Body Mucus: From In Silico Insights to Essential In Vitro Validation

Marta Fernandez Cunha [1], Ezequiel R. Coscueta [1,*], María Emilia Brassesco [1], Frederico Almada [2], David Gonçalves [3] and Maria Manuela Pintado [1]

[1] CBQF—Centro de Biotecnologia e Química Fina—Laboratório Associado, Escola Superior de Biotecnologia, Universidade Católica Portuguesa, Rua Diogo Botelho 1327, 4169-005 Porto, Portugal; mfcunha@ucp.pt (M.F.C.); mbrassesco@ucp.pt (M.E.B.); mpintado@ucp.pt (M.M.P.)

[2] MARE—Marine and Environmental Sciences Centre, ISPA Instituto Universitário de Ciências Psicológicas, Sociais e da Vida, Rua Jardim do Tabaco, 34, 1149-041 Lisbon, Portugal

[3] Institute of Science and Environment, University of Saint Joseph, Rua de Londres 106, Macau SAR, China; david.goncalves@usj.edu.mo

* Correspondence: ecoscueta@ucp.pt; Tel.: +351-225-580-001 (ext. 8047)

Abstract: Fish body mucus plays a protective role, especially in *Halobatrachus didactylus*, which inhabits intertidal zones vulnerable to anthropogenic contaminants. In silico predicted bioactive peptides were identified in its body mucus, namely, EDNSELGQETPTLR (HdKTLR), DPPNPKNL (HdKNL), PAPPPPPP (HdPPP), VYPFPGPLPN (HdVLPN), and PFPGPLPN (HdLPN). These peptides were studied in vitro for bioactivities and aggregation behavior under different ionic strengths and pH values. Size exclusion chromatography revealed significant peptide aggregation at 344 mM and 700 mM ionic strengths at pH 7.0, decreasing at pH 3.0 and pH 5.0. Although none exhibited antimicrobial properties, they inhibited *Pseudomonas aeruginosa* biofilm formation. Notably, HdVLPN demonstrated potential antioxidant activity (ORAC: 1.560 μmol TE/μmol of peptide; ABTS: 1.755 μmol TE/μmol of peptide) as well as HdLPN (ORAC: 0.195 μmol TE/μmol of peptide; ABTS: 0.128 μmol TE/μmol of peptide). Antioxidant activity decreased at pH 5.0 and pH 3.0. Interactions between the peptides and mucus synergistically enhanced antioxidant effects. HdVLPN and HdLPN were non-toxic to Caco-2 and HaCaT cells at 100 μg of peptide/mL. HdPPP showed potential antihypertensive and antidiabetic effects, with IC_{50} values of 557 μg of peptide/mL for ACE inhibition and 1700 μg of peptide/mL for α-glucosidase inhibition. This study highlights the importance of validating peptide bioactivities in vitro, considering their native environment (mucus), and bioprospecting novel bioactive molecules while promoting species conservation.

Keywords: bioactive peptides; peptide aggregation; fish mucus; marine bioprospection

1. Introduction

Bioprospection is the exploration of natural bioresources, namely, animals, plants, and microorganisms, whose potential may lead to the development of products that benefit society [1]. As a largely untapped reservoir of bioresources, the marine environment has become attractive for bioprospection due to the promising properties of new marine chemical compounds [2,3]. *Halobatrachus didactylus*, also known as Lusitanian Toadfish, though understudied as a potential source of bioactive compounds, possesses unique characteristics that highlight its potential for further investigation. This species secretes abundant mucus, and based on our observations during the collection of mucus samples in aquariums, fish kept in captivity alongside other species often caused a rapid death of these cohabiting fish. This suggests a potential toxic effect of substances released into the water, potentially originating from the mucus or other sources, such as axial glands [4].

In a previous publication, we unveiled the bioactive properties present in the body mucus of *H. didactylus*, including potential antioxidant, antimicrobial, and antihypertensive activities [4]. Additionally, we conducted size exclusion chromatography to obtain the protein–peptide profile and isolate the most prominent peak which contained the peptide fraction presumed to be responsible for the observed bioactivities. Subsequently, nano-liquid chromatography–tandem mass spectrometry (Nano-LC-MS/MS) determined the most probable peptide sequences associated with the peptide fraction. Bioactivity prediction databases, specifically, PeptideRanker (http://distilldeep.ucd.ie/PeptideRanker/), CAMPR3 (http://www.camp3.bicnirrh.res.in), and BIOPEP–UWM (https://biochemia.uwm.edu.pl/biopep/start_biopep.php), accessed on 15 January 2025, were employed to screen bioactive peptides. A deep analysis successfully identified five peptides with putative bioactivity: HdKTLR (EDNSELGQETPTLR), HdKNL (DPPNPKNL), HdPPP (PAPPPPPP), HdVLPN (VYPFPGPLPN), and HdLPN (PFPGPLPN). These bioactivities include inhibition activity against the following enzymes: dipeptidyl peptidase III (DPP-III), related to cancer and inflammation; dipeptidyl peptidase IV (DPP-IV), associated with type 2 diabetes; the angiotensin-converting enzyme (ACE), related to hypertension; and α-glucosidase, related to diabetes. Additionally, the peptides were predicted with promising antioxidant and antimicrobial properties.

The employment of an in silico approach involves computational methods that manipulate, interpret, and organize information within biological systems [5]. An in silico analysis significantly accelerates the discovery of new bioactive peptides by facilitating the prediction of bioactivities through computational modelling and machine learning. Databases such as BIOPEP gather essential information, including peptide sequences, sources of origin, reference articles, and documented bioactivities [6], providing a comprehensive resource for researchers. This approach proves highly effective in uncovering the bioactivities of peptides, offering high precision and efficiency while saving time, costs, and resources [7]. However, it is crucial to establish a strong correlation between experimental analyses (in vitro assays) and in silico methods [8,9]. This integrated approach enables substantial advancements in the discovery of bioactive peptides.

This study focused on exploring new bioactive molecules, particularly peptides, from the mucus of *H. didactylus*. This involved validating in silico predicted bioactivities through in vitro assays, which is crucial in determining the peptide bioactive potential. Additionally, it was important to study the conditions under which these peptides can achieve optimal bioactivity. Issues such as aggregation may lead to the reduced efficacy and stability of peptides, making it essential to address these challenges during the development process.

2. Results and Discussion

2.1. Peptide Aggregation and Stability Analysis

2.1.1. Molecular Characterization of Aggregated Peptides

A high-performance size exclusion chromatography (HPSEC) analysis was employed to validate the molecular weight of the selected peptides, HdKTLR, HdKNL, HdVLPN, and HdLPN, synthesized by GenScript Biotech (Rijswijk, The Netherlands), following the preparation instructions provided in the technical datasheet. The decision to exclude HdPPP was based on its solubility just in pure dimethyl sulfoxide (DMSO), which presents challenges due to the high viscosity of the solution. At higher concentrations, viscosity may increase due to molecular interactions between DMSO, buffer ions, and water. In isocratic systems, such as HPSEC, an increase in the mobile phase viscosity can reduce flow efficiency, affecting resolution and increasing the system pressure. Additionally, DMSO can alter interactions between the mobile phase and the column matrix.

The HPSEC column was eluted with 344 mM of a phosphate buffer at pH 7.0, replicating the conditions used to analyze the mucus peptide fraction in our previous article [4]. During the analysis, we observed the formation of peptide aggregates, specifically oligomers, including dimers, trimers, and tetramers. Various buffer conditions were tested to determine if a particular pH and ionic strength would enable the peptides to expose their theoretical molecular weights, as specified in the GenScript technical datasheet. Since seawater typically has an ionic strength of approximately 700 mM [10], buffers with this ionic strength were tested at pH 3.0, 5.0, and 7.0, as well as at 344 mM of ionic strength.

An analysis of the chromatograms of the HdKTLR peptide under various buffer conditions revealed that the theoretical molecular weight of 1588.4 Da was not validated due to the formation of aggregates into oligomers. Notably, the buffer conditions most closely approximating the theoretical molecular weight was pH 3.0, with ionic strengths of 344 mM and 700 mM (Figure 1A). These findings also applied to the chromatograms of the HdKNL peptide, which has a theoretical molecular weight of 893.8 Da, with the observation that aggregate formation increased at higher pH conditions (Figure 1B). For the HdLPN and HdVLPN peptides with a theoretical molecular weight of 837.5 Da and 1099.6 Da, respectively, the degree of polymerization was less prominent than in the other peptides (Figure 1C,D), with HdLPN forming only dimers (Figure 1D).

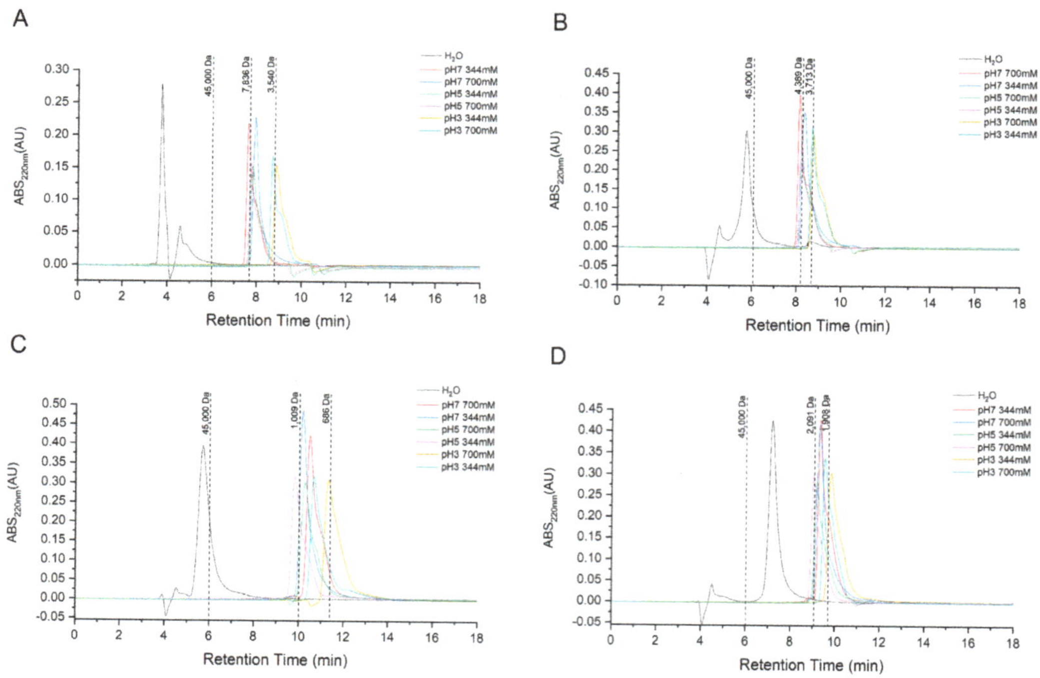

Figure 1. SEC chromatograms of the peptides at different ionic strength and pH conditions: HdKTLR (**A**), HdKNL (**B**), HdVLPN (**C**), and HdLPN (**D**).

Peptides prone to aggregation typically exhibit broader peaks in chromatograms, which is more prominently observed in the HdKTLR peptide. In size exclusion chromatography, increasing the salt concentration to enhance ionic strength reduces electrostatic interactions [11]. A review by Fekete et al. points out that at very high concentrations, these ions can promote hydrophobic interactions or ion-exclusion effects, leading to undesired secondary ionic interactions between peptides and the stationary phase, resulting in further aggregation. Additionally, adjusting the pH closer to the isoelectric point of the peptide can minimize secondary interactions. In our study, peptide aggregation occurs at both ionic strengths, with only pH adjustments causing the peptide's molecular weight to approach the theoretical value.

Understanding that seawater typically has an ionic strength of around 700 mM is crucial, as fish are inherently adapted to their environmental conditions. Peptides exhibit conformational sensitivity due to environmental factors, such as seawater composition, specific cations, and anions in solution, and interactions with polyelectrolytes [12]. This highlights the sensitivity of peptide behavior to specific physicochemical parameters, underscoring the requirement for a comprehensive understanding of their conformational dynamics. Critical factors, including the molar extinction coefficient, water solubility, net charges at physiological pH, isoelectric point, and molecular weight, significantly influence the actual biological effects of molecules [12].

The outcomes of an in silico bioactivity analysis may diverge from in vitro results, necessitating a thorough exploration of diverse molecular parameters. Therefore, in databases such as BIOPEP and $CAMPR_3$, which analyze a compiled literature of reported bioactive peptides, a comprehensive understanding of these molecular characteristics is essential for accurate predictions and interpretations of in silico findings in the context of experimental results.

2.1.2. Computational Assessment of Peptide Aggregation Propensity

We utilized an in silico approach to definitively evaluate the aggregation propensities of peptides. The online software AGGRESCAN was utilized to calculate the number of "hot spots", considering specific amino acids that could form aggregation-prone regions. Hydrophobic amino acids are commonly associated with inducing aggregation [12]. An in silico analysis of the peptides revealed zero hot spots, indicating low sequence aggregation propensity values (Table 1). This finding demonstrates that the peptides have a minimal tendency to form aggregates. However, Groot et al. [13] used β-amyloid peptide as a model to assess the individual aggregation tendencies of natural amino acids. Their research findings indicate the participation of hydrophobic amino acids, including isoleucine, phenylalanine, valine, and leucine, in the aggregation [13]. These amino acids are found in the sequences of peptides, specifically HdKNL (DPPNPKNL), HdVLPN (VYPFPGPLPN), and HdLPN (PFPGPLPN); however, they did not form aggregation-prone regions. Notably, HdVLPN and HdLPN exhibit low solubility (Table 2), which can be attributed to their composition being rich in hydrophobic amino acids.

Table 1. Determination of aggregation propensities by AGGRESCAN [14].

Peptides	Sequence Average Aggregation Propensity Value (a^3vSA)	Hot Spot Number
HdKTLR	−0.612	0
HdKNL	−0.624	0
HdPPP	−0.297	0
HdVLPN	0.271	0
HdLPN	−0.0005	0

Table 2. Physicochemical parameters determined by INNOVAGEN [15].

Peptides	Extinction Coefficient $M^{-1}cm^{-1}$	Isoelectric Point	Net Charge at pH 7	Estimated Solubility
HdKTLR	0	pH 3.54	−3	+
HdKNL	0	pH 6.66	0	+
HdPPP	0	pH 4.26	0	−
HdVLPN	1280	pH 3.63	0	−
HdLPN	0	pH 4.16	0	−

+: good water solubility; −: poor water solubility. Water solubility was estimated according to the isoelectric point, the number of charged residues, and the peptide length.

In a molecular characterization, it was found that the HdKTLR peptide exhibited pronounced aggregation. The theoretical molecular weight of the peptide is 1588.4 Da; however, it increased to approximately 7836 Da under specific buffers with ionic strengths of 344 mM and 700 mM at pH 7.0 and pH 5.0, indicating a high degree of aggregation. An evaluation of the physicochemical parameters in Table 2 indicates that at pH 7.0, the peptide carries a net charge of −3. This charge may not be sufficiently high to cause repulsion between molecules, potentially leading to aggregation. Despite the HdKNL, HdLPN, and HdVLPN peptides having a net charge of 0 at pH 7.0, aggregation was still observed under conditions of high ionic strengths (344 mM and 700 mM) and at both pH levels of 7.0 and 5.0. This suggests that factors other than charge, such as ionic strength and hydrophobic interactions, may also significantly contribute to aggregation under these conditions.

Notably, peptide aggregation not only compromises their stability, resulting in a loss of activity, but also introduces other significant concerns, such as toxicity and immunogenicity [12]. However, several solutions exist to address this issue, such as designing analogs that incorporate a small number of amino acid substitutions, which could normalize or improve biological activity and reduce aggregation propensity [16].

2.2. *In Vitro Validation of Peptide Bioactivities*

2.2.1. Antioxidant Activity

Antioxidant Assays: ORAC and ABTS

The antioxidant potential of *H. didactylus* peptides was assessed through oxygen radical absorbance capacity (ORAC) and 2,2′-Azinobis-(3-Ethylbenzthiazolin-6-Sulfonic Acid (ABTS) assays. The ORAC and ABTS assays assess distinct interactions between antioxidants and generated radicals. In the ORAC assay, antioxidants donate hydrogen to neutralize radicals, whereas in the ABTS assay, the reaction involves the transfer of electrons by the antioxidant [17]. The peptides HdVLPN and HdLPN exhibited antioxidant potential in both ABTS and ORAC assays (Table 3), establishing them as the two most promising candidates. In contrast, the HdKNL and HdPPP peptides demonstrated antioxidant activity exclusively in the ORAC assay, while the HdKTLR peptide did not show antioxidant potential in either the ABTS or ORAC assay. The HdVLPN and HdLPN peptides possess distinct amino acid compositions, where variations in the quantity and arrangement of hydrophobic amino acids (Val, Leu, and Pro for HdVLPN; Leu and Pro for HdLPN) and aromatic amino acids (Tyr and Phe for HdVLPN; Phe for HdLPN) may significantly influence their antioxidant potential [18].

Table 3. Bioactivities: ORAC and ABTS.

Peptides	Sequence	Molecular Weight (Da)	ORAC (μmol TE/μmol Peptide)	ABTS (μmol TE/μmol Peptide)
HdKTLR	EDNSELGQETPTLR	1588.4	nd	nd
HdKNL	DPPNPKNL	893.8	0.048 ± 0.002 [a]	nd
HdPPP	PAPPPPPP	768.8	0.054 ± 0.006 [a,b]	nd
HdVLPN	VYPFPGPLPN	1099.6	1.560 ± 0.110 [c]	1.755 ± 0.138 [a]
HdLPN	PFPGPLPN	837.5	0.195 ± 0.030 [b,c]	0.128 ± 0.017 [b]

nd: not detected. All values are means ($n \geq 3$) ± SD. [a,b,c] The different letter superscripts in the same column indicate that the values differed significantly ($p < 0.05$). ORAC data were analyzed with the Kruskal–Wallis test and ABTS data with the Mann–Whitney U test.

Upon comparing our results with peptides previously identified in the ark shell (*Scapharca subcrenata*) by Jin et al. [19], it became evident that after converting their values to the same units as ours, our peptides exhibited superior antioxidant activity in both the ABTS and ORAC assays. Specifically, in the ABTS assay, P1 (MCLDSCLL) from the ark shell demonstrated an antioxidant activity of 0.678 μmol Trolox equivalents (TE)/μmol of peptide, and P2 (HPLDSLCL) exhibited an activity of 0.005 μmol TE/μmol of peptide. In the ORAC assay, the values were 0.334 μmol TE/μmol of peptide for P1 and 0.239 μmol TE/μmol of peptide for P2. Compared to the HdVLPN peptide, the ark shell peptides exhibited inferior antioxidant activity, achieving 1.560 μmol TE/μmol of peptide in the ORAC assay and 1.755 μmol TE/μmol of peptide in the ABTS assay. Additionally, the HdLPN peptide demonstrated antioxidant activity with 0.128 μmol TE/μmol in the ABTS assay, surpassing the activity of the P2 peptide.

In the food industry, the overproduction of free radicals, which leads to the generation of oxidants, can compromise food quality and shorten shelf-life [18]. Consequently, antioxidant peptides have proven to be a viable alternative to synthetic antioxidants as food preservatives, offering the advantage of mitigating the severe toxicity associated with synthetic antioxidants in food formulations [20]. Thus, the identified antioxidant bioactivity of *H. didactylus* peptides may have a potential application in this industry. Additionally, incorporating a balanced diet rich in antioxidants is believed to help mitigate the risk of chronic diseases associated with oxidative stress, such as Alzheimer's disease, arthritis, heart disease, cancer, and others [18].

Influence of pH and Ionic Strength on Antioxidant Activity (ABTS Assay)

As revealed by the HPSEC analysis, the aggregation of peptides demonstrated sensitivity to varying ionic strength and pH conditions, which were factors further evaluated in relation to antioxidant activity using the ABTS assay. Our results demonstrated that HdKTLR exhibited no observable antioxidant activity. The antioxidant activity of the HdKNL peptide remained unaffected by a decrease in pH, whereas no antioxidant activity was observed in H_2O, only in the presence of buffers (Figure 2A). However, for HdVLPN and HdLPN peptides, a reduction in pH resulted in a significant decrease in antioxidant activity (Figure 2B,C).

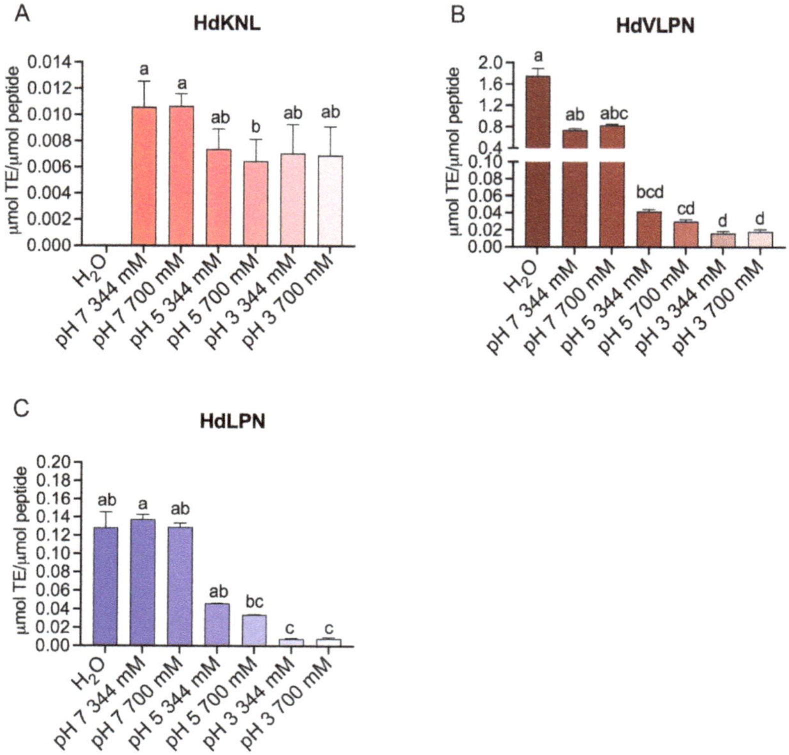

Figure 2. Evaluation of antioxidant activity through the ABTS scavenging assay at different ionic strengths and pH conditions of the peptides: HdKNL (**A**) (no bar is shown for H_2O, as no antioxidant activity was detected); HdVLPN (**B**); and HdLPN (**C**). All values are means (n $\geq$ 3) $\pm$ standard deviation (SD). Different lower-case letters (a–d) at each condition indicate that the values differed significantly ($p < 0.05$). Data (**B**,**C**) were analyzed using the Kruskal–Wallis test and (**A**) a one-way ANOVA with Tukey's test.

Interestingly, studies have unveiled that the reaction rates of specific amino acids—namely, Cys, Tyr, Trp, and His—in the ABTS scavenging assay exhibit a significant increase in ABTS radical scavenging values within a pH range of 5.0–7.0 [21]. These results reveal a strong pH dependence in the ABTS radical scavenging capacity of amino acids; namely, the values of Tyr increased from 2.89 μmol TE/μmol at pH 6.0 to 6.55 μmol TE/μmol at pH 8.0. In addition, this study suggests a minimum incubation time of 30 min for reliable Trolox equivalent values in the ABTS assay at pH 7.4 [21].

Our peptides, HdVLPN and HdLPN, demonstrated a significant correlation with pH variation, reflecting decreased ABTS radical scavenging values from pH 7.0 to 3.0. Specifically, at pH 7.0 and 700 mM, HdVLPN demonstrates values of 0.827 $\pm$ 0.021 μmol TE/μmol of peptide, whereas HdLPN shows 0.283 $\pm$ 0.177 μmol TE/μmol of peptide. Contrastingly, at pH 3 and 344 mM, HdVLPN records 0.163 $\pm$ 0.002 μmol TE/μmol of peptide, while HdLPN displays 0.012 $\pm$ 0.004 μmol TE/μmol of peptide (Figure 2B,C). The HdVLPN peptide includes the amino acid Tyr in its sequence, a characteristic that appears to align with the fact that the HdVLPN peptide and the amino acid Tyr exhibit greater antioxidant activity at pH 7.0. Interestingly, despite the peptide aggregates observed in the HPSEC chromatograms at pH 7.0, their antioxidant activity remained unaffected. On the contrary, they exhibited the greatest antioxidant activity, as demonstrated through the ABTS scavenging assay. However, additional studies are necessary to deepen our understanding of the behavior of peptides and their constituent amino acids at different pH levels. This is particularly important, as pH variation substantially impacts peptides' antioxidant potential.

Mucus Interaction and Antioxidant Capacity (ORAC Assay)

We employed an experimental approach by mixing the peptides with the mucus to thoroughly investigate potential synergies. Therefore, we tested the combined effect of peptides and mucus on antioxidant activity using the ORAC assay (Figures 3 and 4). This approach aimed to uncover any positive interactions, considering the diverse array of mucus molecules that could interact with the peptides.

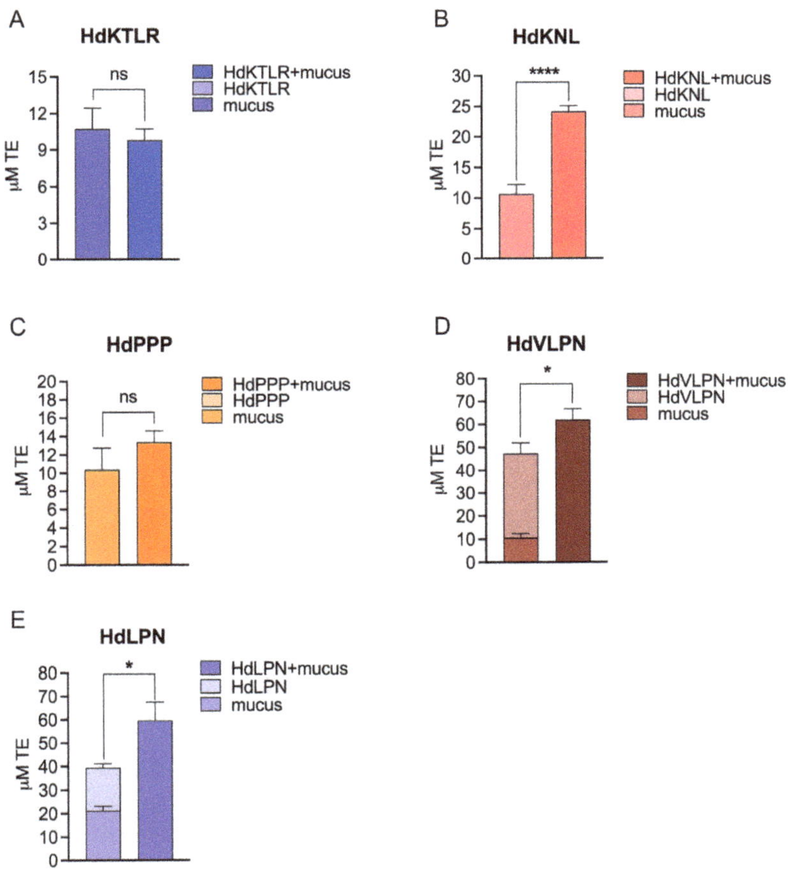

Figure 3. Evaluation of the synergistic effect between peptides and the captive mucus sample through the ORAC assay. HdKTLR (**A**), HdKNL (**B**), HdPPP (**C**), HdVLPN (**D**), and HdLPN (**E**). In graphs (**A–C**), no bar is shown for the peptides, as no antioxidant activity was detected. All values are means (n ≥ 3) ± standard deviation. ns: $p > 0.05$, * $p < 0.05$, and **** $p < 0.0001$. Data were analyzed using a *t*-test.

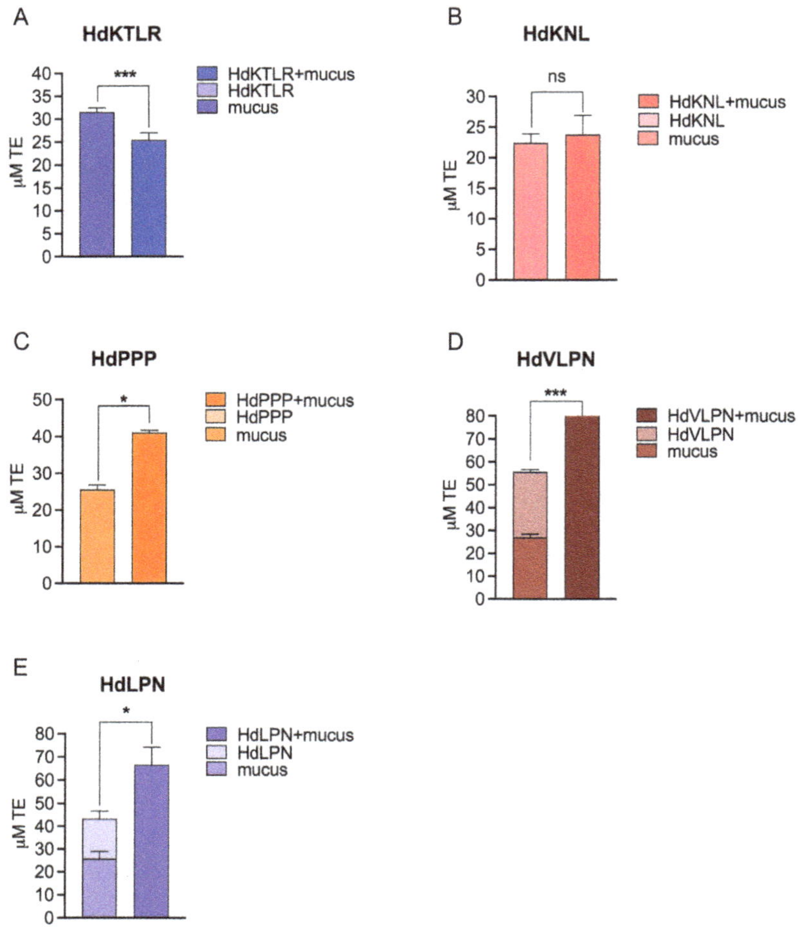

Figure 4. Evaluation of the synergistic effect between peptides and the wild mucus sample through the ORAC assay. HdKTLR (**A**), HdKNL (**B**), HdPPP (**C**), HdVLPN (**D**), and HdLPN (**E**). In graphs (**A–C**), no bar is shown for the peptides, as no antioxidant activity was detected. All values are means ($n \geq 3$) ± SD. ns: $p > 0.05$, * $p < 0.05$, and *** $p < 0.001$. Data were analyzed using a *t*-test.

Two mucus samples, obtained from wild and captive individuals, were used to evaluate the synergistic effect of the peptides. We pooled the collected mucus of several *H. didactylus* individuals, resulting in a protein value of 5888 ± 750 µg BSA/ mL for the captive sample and 111 ± 5 µg BSA/mL for the wild sample. Specifically, the HdVLPN peptide, when combined with mucus from captive individuals, exhibited a notable 31% increase in antioxidant activity (Figure 3D) compared with the sum of the bioactivity of the individual systems (mucus and peptide). The increment was even more pronounced at 44% for the wild mucus sample (Figure 4D). Likewise, HdLPN demonstrated significant enhancements, showing a substantial 51% increase in antioxidant activity when combined with mucus from captive individuals (Figure 3E) and a 54% increase for the wild mucus sample (Figure 4E). Furthermore, the HdKNL peptide displayed a remarkable 126% increase in antioxidant activity for the captive mucus sample (Figure 3B), and the HdPPP peptide showcased a 61% increase for the wild mucus sample (Figure 4C). The findings reveal a notable enhancement in antioxidant activity upon the incorporation of peptides into mucus.

The mucus comprises ca. 95% water and mucins, which are high molecular weight glycoproteins that are crucial in determining mucus viscosity [22,23]. Additionally, mucins encompass a diverse array of molecules with significant bioactivities. These include glycoproteins, glycosaminoglycans, pheromones, proteolytic enzymes, lectins, galectins, lysozymes, calmodulins, immunoglobulins, complements, C-reactive proteins, and antimicrobial peptides [24].

The diverse and abundant range of molecules within the mucus composition potentially contributes to interactions with peptides, namely, HdKNL, HdPPP, HdVLPN, and HdLPN, resulting in enhanced antioxidant activity through a synergistic effect. This synergistic combination of antioxidants could mirror a dynamic interplay, as highlighted in a study by Aklakur [25]. According to this study, pairing two antioxidants demonstrated a unique mechanism: one antioxidant reacted with the peroxy radical, being consumed in the process, while the second antioxidant could regenerate the first, preserving its functionality.

2.2.2. Assessment of Antihypertensive Activity

The antihypertensive potential of bioactive peptides was tested through the inhibition of an angiotensin-converting enzyme (iACE) assay. The mucus of *H. didactylus* demonstrated an inhibitory effect, with an IC_{50} of 60 $\pm$ 7 μg of mucus protein/mL [4]. To identify the molecules responsible for this bioactivity, the peptides identified from the fraction were tested for their inhibitory properties. The iACE activity was only identified for the HdPPP peptide, which inhibited 50% of ACE at a concentration of 577 $\pm$ 90 μg of peptide/mL, representing the mean $\pm$ standard deviation (n = 3). This aligns with our previous article, which describes the peptides in in silico predictions using the BIOPEP database and identifies the HdPPP peptide as having the highest potential as an ACE inhibitor. The observed inhibitory potentials using BIOPEP were as follows: HdKTLR (A = 0.375), HdKNL (A = 0.125), HdVLPN (A = 1.200), HdLPN (A = 1.000), and HdPPP (A = 1.375). "A" represents the frequency of bioactive fragments in a protein or peptide [26].

When compared to other bioactive peptides derived from fish protein hydrolysates, such as those from *Syngnathus schlegeli*—specifically Thr-Phe-Pro-His-Gly-Pro (IC_{50} of 600 μg/mL) and His-Trp-Thr-Thr-Gln-Arg (IC_{50} of 1440 μg/mL)—and peptides from *Salmo salar*, including Ile-Trp, Ile-Tyr, Thr-Val-Tyr, Val-Trp, Val-Pro-Trp, and Val-Tyr, which exhibit IC_{50} values between 190 and 1040 μg/mL [27], the HdPPP peptide demonstrates comparable antihypertensive activity. Although the methodology differs, the HdPPP peptide also displays antihypertensive activity, aligning with the effects observed in these peptides from other fish sources.

The most widely used ACE blocker is the commercial drug Captopril®, which functions through the proline binding component to ACE at the following sites: Trp67, Asn68, Thr71, Asn72, Met340, and Arg348 [28]. The peptide HdPPP, rich in proline amino acids, emerges as our most promising candidate for exhibiting antihypertensive effects. Furthermore, the potential of this activity could be enhanced through gastrointestinal digestion. A study by Abachi et al. [29] provides examples of synthesized peptides where fragmentation and splitting during digestion resulted in di- and/or tri-peptides, enhancing their activity. For instance, the transformation of IKPLNY (IC_{50} 32 μg/mL) into IKP (IC_{50} 0.6 μg/mL) led to a remarkable 96% improvement in activity [28]. Likewise, for other peptides (HdKTLR, HdKNL, HdVLPN, and HdLPN) from which antihypertensive activity was undetermined, gastrointestinal digestion could have the potential to amplify their bioactivity via enzymatic hydrolysis. The increasing focus on peptides identified from natural sources is driven by concerns regarding common antihypertensive drugs' adverse effects, including symptoms like coughing, taste disturbances, renal complications, and angioedema [27]. This shift towards natural peptides highlights the necessity for safer and potentially more effective alternatives in the treatment of hypertension.

2.2.3. Evaluation of Antidiabetic Activity

The peptides were tested for antidiabetic activity based on the capacity to inhibit 50% of α-glucoside. The HdPPP was the only peptide that could inhibit 50% of α-glucoside activity at 1700 ± 40 μg of peptide/mL (mean ± standard deviation; n = 3). As predicted in silico by BIOPEP, HdPPP was the peptide with the highest probability of inhibiting α-glucoside (HdKNL A = 0.125, HdVLPN A = 0.100, and HdPPP A = 0.625). Although, compared to acarbose, which has an IC_{50} of 757 ± 75 μg/mL (mean ± standard deviation; n = 3), HdPPP proved less effective. Likewise, compared to other peptides hydrolyzed from food sources, namely, RVPSLM (23.05 μM) and TPSPR (40.02 μM) [29], the HdPPP peptide showed a higher IC_{50} of 2211 μM. As previously discussed, regarding antihypertensive activity, there is a potential for a similar augmentation of antidiabetic activity during digestion. The enzymatic hydrolysis during digestion results in the fractionation of peptides into di-peptides and tri-peptides, potentially generating active molecules that could inhibit α-glycoside.

2.2.4. Antimicrobial Activity: Growth Inhibition Curves

In our prior investigation, the mucus collected from *H. didactylus* exhibited a notable inhibitory effect, with an approximately 76% inhibition against *Escherichia coli* ATCC 25922 at a concentration of 442 μg of mucus protein/mL and around 66% inhibition against *Listeria monocytogenes* NCTC 10357 at 221 μg of mucus protein/mL [4]. Considering the promising antimicrobial activity exhibited by the mucus of *H. didactylus*, growth inhibition curves were determined for strains of *E. coli* ATCC 25922, *Salmonella enterica* serovar Enteritidis ATCC 13076, *Pseudomonas aeruginosa* ATCC 27853, and *Staphylococcus aureus* ATCC 6538 using the peptides HdKTLR (262 μg/mL), HdKNL (262 μg/mL), HdPPP (130 μg/mL), HdVLPN (204 μg/mL), and HdLPN (227 μg/mL). The peptides did not demonstrate antimicrobial potential for the tested strains, except for the peptide HdPPP, which showed slight inhibition for the strain *S. aureus* (Figure 5). The HdPPP peptide at the tested concentration of 130 μg/mL has 5% DMSO, and to confirm that DMSO did not cause the inhibition of *S. aureus*, a control was made with only 5% DMSO, where no inhibition of the strain was observed.

In our previous study, we employed an in silico approach through the database $CAMPR_3$ classifiers (SVM, RF, ANN, and DA) to predict the antimicrobial potential of peptides [4]. The HdKNL peptide notably surpassed the identified peptides, resulting in potential antimicrobial activity in all four classifiers. However, in vitro antimicrobial activity was not demonstrated for the HdKNL or the other peptides.

Generally, AMPs are cationic, with a net charge of +2 and up to +8 [30]. This allows the AMPs to interact with the negatively charged head groups of the microbial cytoplasmic membrane, adopting an amphipathic helical conformation that enables them to insert the hydrophobic face into the bilayer [31]. However, the peptides HdKNL, HdPPP, HdVLPN, and HdLPN have a net charge of 0 at pH 7, and the peptide HdKTLR has a net charge of −3 at pH 7, an atypical characteristic for AMPs, which aligns with their observed lack of antimicrobial activity. However, these peptides possess other characteristics that could contribute to antimicrobial activity, namely, the presence of proline in the sequence being particularly notable in some identified peptides. Studies have reported that proline hinges play a crucial role as a structural factor in facilitating the penetration of the molecule into the cell [32].

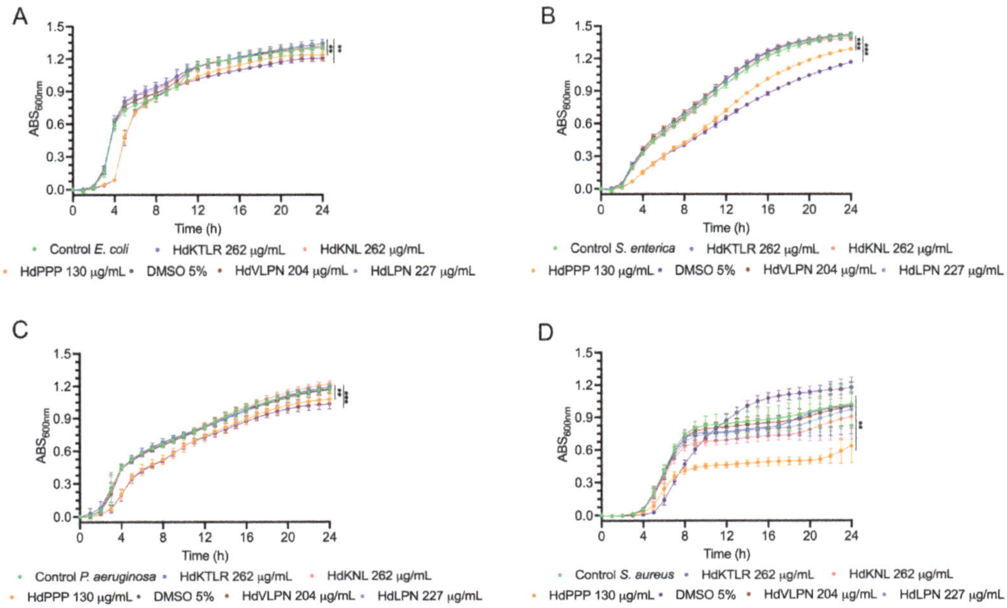

Figure 5. Inhibition growth curves of peptides HdKTLR, HdKNL, HdVLPN, and HdPPP against pathogenic bacteria: *E. coli* (**A**), *S. enterica* (**B**), *P. aeruginosa* (**C**), and *S. aureus* (**D**). ** $p < 0.01$, *** $p < 0.001$. Data were analyzed using repeated measures with a Bonferroni post hoc test.

2.2.5. Antibiofilm Activity Assessment

The peptides HdKTLR (262 µg/mL), HdKNL (262 µg/mL), HdPPP (130 µg/mL), HdVLPN (204 µg/mL), and HdLPN (227 µg/mL) were assessed for their efficacy in reducing biofilm formation after 24 h of incubation with strains of *E. coli*, *S. enterica*, *P. aeruginosa*, and *S. aureus* (Figure 6). When exposed to the *P. aeruginosa* strain, it was observed that the peptides exhibited significant differences compared to the control, resulting in substantially reduced biofilm production. However, this outcome did not occur with the other strains (*E. coli*, *S. enterica*, and *S. aureus*), as no significant differences were observed between the peptides and the control.

The reduction in biofilm production by *P. aeruginosa* through the action of peptides represents a promising outcome, particularly given the strain's association with serious chronic lung diseases [33]. *P. aeruginosa* is particularly proficient at biofilm production [34]. Inhibiting biofilm formation during the mature phase offers substantial benefits, as microorganisms in biofilms exhibit heightened resistance to antimicrobial agents. Consequently, peptides could disrupt biofilm formation, transitioning bacteria from a biofilm state to a free-living one, thereby facilitating bacterial elimination [33]. Exploring additional approaches to address peptides' relatively limited antimicrobial potential involves considering their combination with other antimicrobial agents. This strategy could enhance their efficacy, offering a promising pathway to strengthen antimicrobial activity.

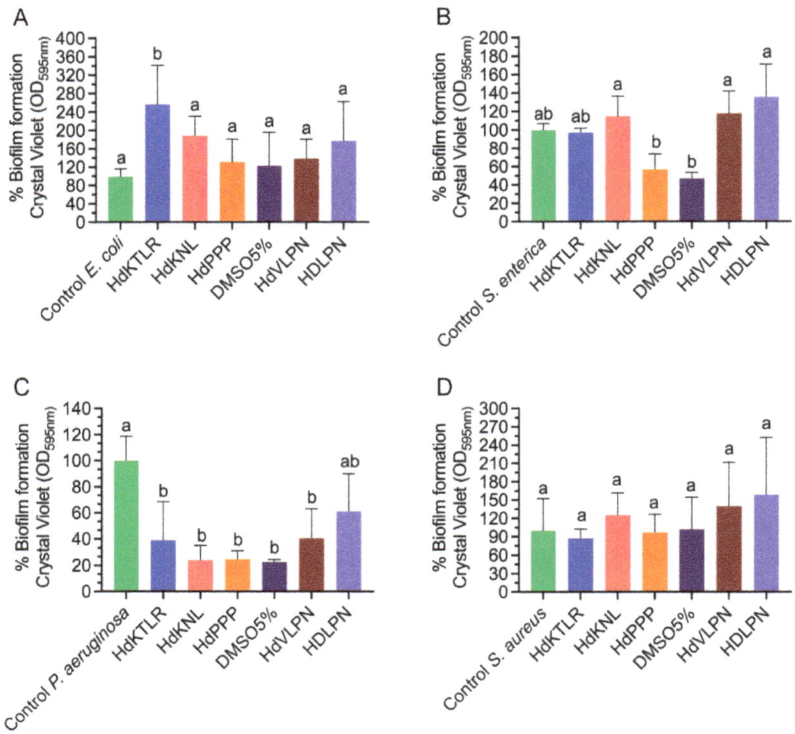

Figure 6. Percentage of biofilm formation of the strains *E. coli* (**A**), *S. enterica* (**B**), *P. aeruginosa* (**C**), and *S. aureus* (**D**) in the presence of peptides. All values are means ($n \geq 3$) ± standard deviation (SD). Different lower-case letters (a,b) at each condition indicate that the values differed significantly ($p < 0.05$). Data were analyzed using one-way ANOVA for *E. coli* and *P. aeruginosa* and Kruskal–Wallis for *S. aureus* and *S. enterica*.

2.2.6. Evaluation of Peptides on Probiotic Growth and Viability

The effect of the peptides on the probiotics, particularly regarding their positive and negative interactions, was assessed for *Bifidobacterium* and *Lactobacillus* species. The *Bifidobacterium* tested included *B. longum* BG3, *B. animalis* Bo, *B. animalis* BB-12, *B. breve*, and *B. animalis* BLC. The *Lactobacillus* tested included *L. acidophilus* Ki, *L. plantarum* 299V, *L. casei*, *L. acidophilus* La5, and *L. rhamnosus* RII. The peptides (HdKTLR 262 μg/mL, HdKNL 262 μg/mL, HdVLPN 204 μg/mL, and HdLPN 227 μg/mL) did not inhibit the growth of probiotics and could instead promote growth (Figures 7 and 8). The peptide HdPPP was not tested due to its solubility in DMSO, as this solvent is unsuitable for further applications in the food industry. Probiotics can benefit not only from the essential amino acids provided by the peptides, which could play a crucial role in modulating the integrity of bacterial cell membranes and walls [35].

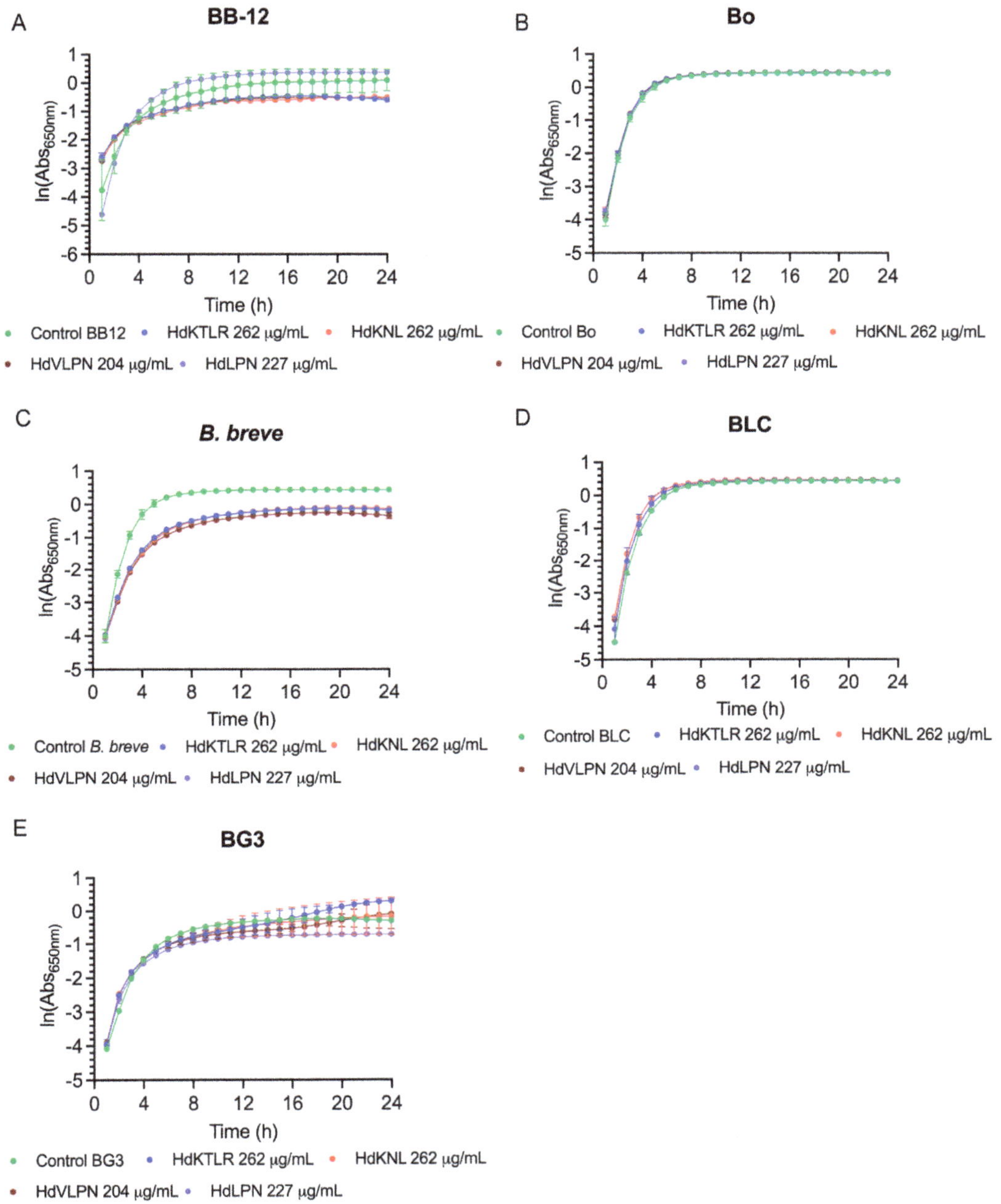

Figure 7. Growth curves of peptides HdKTLR, HdKNL, HdVLPN, and HdLPN with the probiotics of *Bifidobacterium* species: *B. animalis* BB-12 (**A**), *B. animalis* Bo (**B**), *B. breve* (**C**), *B. animalis* BLC (**D**), and *B. longum* BG3 (**E**).

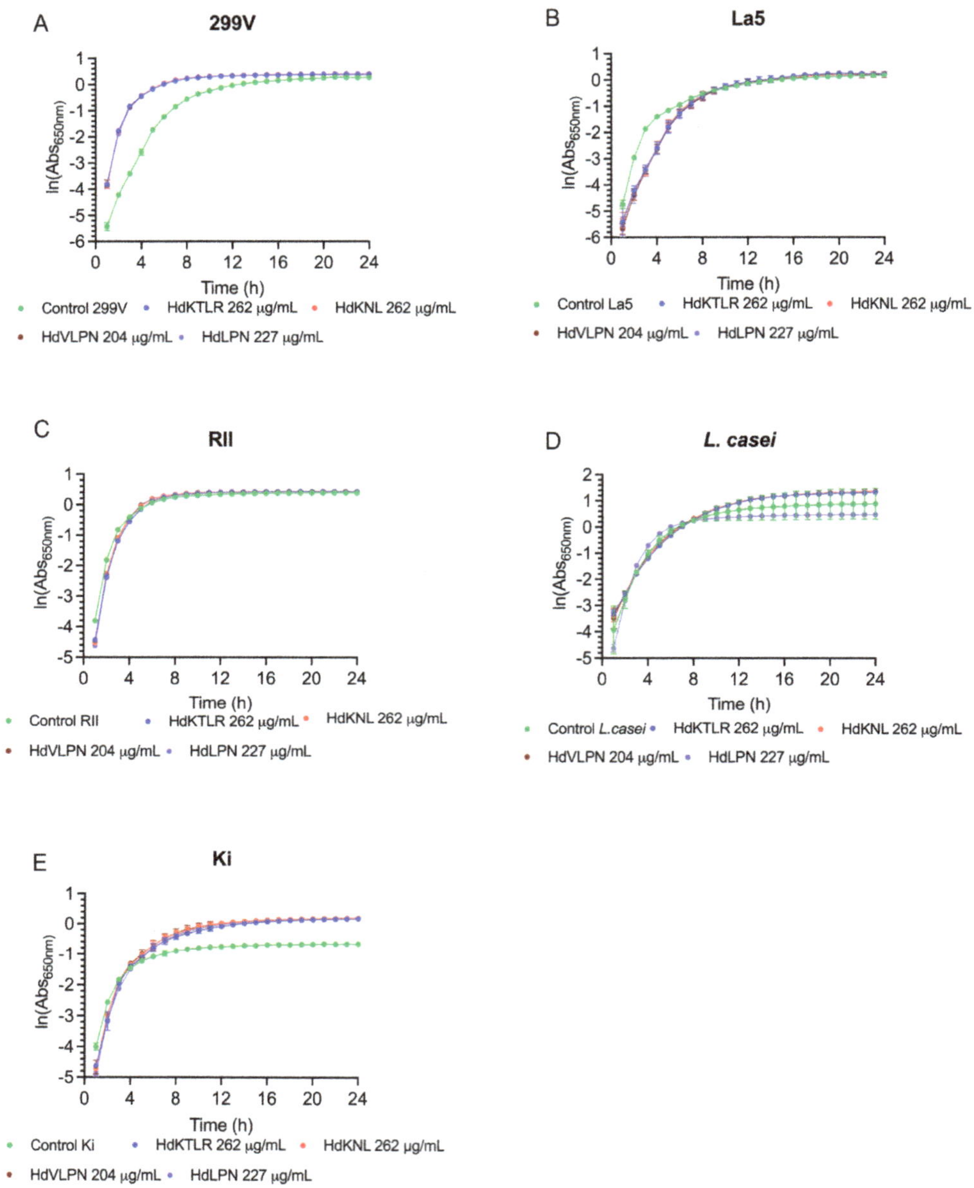

Figure 8. Growth curves of peptides HdKTLR, HdKNL, HdVLPN, and HdLPN with the probiotics of *Lactobacillus* species: *L. plantarum* 299V (**A**), *L. acidophilus* La5 (**B**), *L. rhamnosus* RII (**C**), *L. casei* (**D**), and *L. acidophilus* Ki (**E**).

Moreover, probiotics like *Lactobacillus lactis* depend on oligopeptides as a nitrogen source. Additionally, other interactions enhance the strain's resistance to acidic conditions and promote probiotic growth by increasing protease or peptidase activity. An illustrative example is the Leu-Pro peptide, which boosts the protease activity of *Lactobacillus helveticus* [36]. Furthermore, the antioxidant properties of HdVLPN and HdLPN peptides can neutralize reactive oxygen species, which may help protecting probiotics from oxidative stress.

These results indicate a promising potential for applications in food products, such as acting as preservatives, since the peptides demonstrated no detrimental effects on the probiotics. Future research could focus on exploring the potential of bioactive peptides in modulating and preserving a healthy microbiome [36].

2.2.7. Cytotoxicity Assessment on Caco-2 and HaCaT Cell Lines

Considering the prospective industrial applications of peptides, especially in the food industry, it is essential to assess the cytotoxicity of these peptides on the proliferation of human Caco-2 cells, which serve as a representative model for the intestine. Consequently, we conducted cytotoxicity tests on the two most promising peptides, HdVLPN and HdLPN, which were previously discussed regarding their significant antioxidant activity. These peptides were also subjected to a gastrointestinal simulation (see Section 2.2.8), which will also be necessary for investigating whether they are safe for humans to consume. The results demonstrate that both peptides (HdVLPN and HdLPN) exhibited no cytotoxic effects on human Caco-2 cells at tested concentrations from 25 to 100 μg/mL, with a metabolic inhibition below 30% (Figure 9A). This evaluation was crucial in determining the safety of these peptides for human consumption, particularly since they underwent a gastrointestinal simulation to ascertain potential bioaccessibility. Examining their impact on Caco-2 cells contributes to understanding their safety profile and provides valuable insights into their suitability for incorporation into human dietary applications.

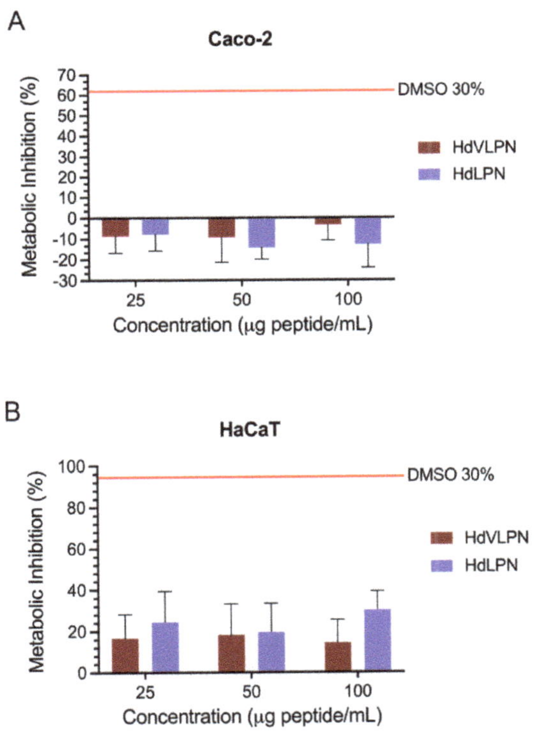

Figure 9. Cytotoxicity evaluation of HdVLPN and HdLPN peptides of human Caco-2 cell line (**A**) and HaCaT human keratinocyte cell line (**B**). All values are means (n ≥ 3) ± standard deviation (SD). There was no significant difference ($p > 0.05$) in metabolic inhibition between the two peptides at the range of concentrations tested (25 to 100 μg of peptide/mL) for Caco-2 and HaCaT cells, though these differed significantly from DMSO 30% ($p < 0.001$). Data were analyzed using one-way ANOVA, followed by Tukey's test.

To further investigate the potential benefits of the antioxidant peptides HdVLPN and HdLPN for skin health, we also utilized the HaCaT keratinocyte human cell line as an in vitro model to assess the cytotoxicity of the peptides. This cell line is extensively employed by researchers in skin biology studies as keratinocytes, constituting 95% of epidermal cells in the cutaneous layer [37,38], which play a fundamental role in skin function. The results demonstrate that both peptides (HdVLPN and HdLPN) exhibited no cytotoxic effects on human HaCaT cells at tested concentrations from 25 to 100 μg peptide/mL, with a metabolic inhibition below 30% (Figure 9B). These findings are promising, suggesting favorable applications for HdVLPN and HdLPN peptides as antioxidants for the skin.

Bioactive peptides derived from marine sources remain relatively underexplored for use in industrial products, such as functional foods, dietary supplements, medicines, and cosmetics [39]. Consequently, our finding holds promise for advancing the safe utilization of antioxidant peptides in these domains, offering a positive contribution towards their practical and beneficial implementation.

2.2.8. Bioactivity Evaluation Post-Gastrointestinal Digestion: Antioxidant and Antihypertensive Effects

It was essential to assess the capacity of our bioactive peptides to resist the gastrointestinal tract and cross the intestinal epithelial barrier, thereby confirming their possible applicability as health potentiators [40]. The two most promising peptides, HdLPN and HdVLPN, exhibited significant antioxidant activity and were submitted to a gastrointestinal simulation at a concentration of 1.15 mg/mL for HdLPN and 1 mg/mL for HdVLPN. The digested content from the dialysis process, both the permeate and retentate, using 3.5 kDa of membranes, were evaluated in vitro for their antioxidant activity through the ORAC assay and only the permeate for antihypertensive potential through the inhibition of the angiotensin-I-converting enzyme (iACE).

The antioxidant activity of the two digested peptides, in both permeate and retentate forms, was comparable to the control (which utilized water in place of peptides); similar results were obtained for their antihypertensive activity in the permeate form. The results for both bioactivities show no significant differences when comparing the digested peptides' retentate and permeate forms with the control (Figure 10). This suggests that the enzymatic hydrolysis occurring during digestion reduces the bioactivity of these peptides, as they exhibited antioxidant activity before digestion. Alternatively, the peptide concentration used may have been insufficient to observe the impact of digestion. Additionally, there was no subsequent activation or potentiation related to antihypertensive activity, which was also absent before digestion.

Therefore, a comprehensive analysis, including mass spectrometry, is essential to fully understand the impact of digestion on the hydrolysis of these peptides. Moreover, the exploration of biocompatible materials for safe delivery methods of peptides, e.g., encapsulation, to enable them to withstand gastrointestinal digestion and effectively reach target organs to exert their intended bioactivity should be conducted [40].

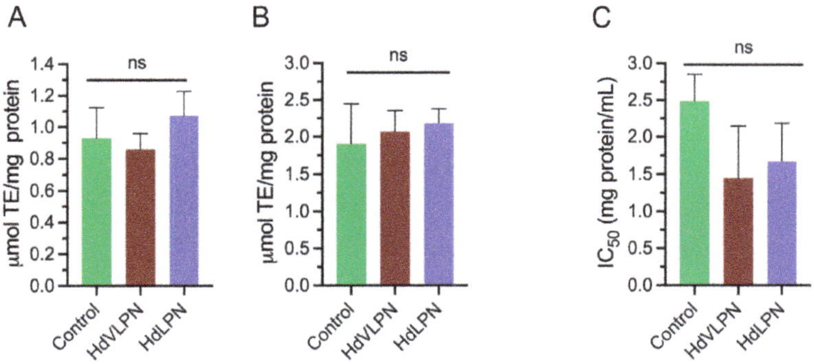

Figure 10. Antioxidant and inhibitory activities during the final phase of peptide digestion and dialysis: ORAC antioxidant activity in the retentate (**A**) and permeate forms (**B**). No significant differences were observed between peptides and the control in both the retentate and permeate forms ($p > 0.05$). Data for the retentate were analyzed using the Kruskal–Wallis test, while a one-way ANOVA, followed by Tukey's post hoc test was applied to the permeate. An iACE 50% inhibition concentration from dialysis permeates during the final digestion phase (**C**). Similarly, no significant differences were observed between peptides and the control ($p > 0.05$), with data analyzed using the Kruskal–Wallis test.

3. Materials and Methods

3.1. Materials and Reagents

The angiotensin-I-converting enzyme (peptidyl-di-peptidase A, EC 3.4.15.1, 5.1 U/mg); Trolox (6-hydroxy-2,5,7,8-tetramethyl-chroman-2-carboxylic acid); AAPH [2,2′-azo-bis-(2-methylpropionamidine)-dihydrochloride] chloramphenicol; pepsin, from porcine stomach mucosa (P7012, 500 U/mg); pancreatin, from porcine pancreas (P7545, 8 trypsin U/mg); bile salts (B863); and and DMSO (≥99.9%) were obtained from Sigma-Aldrich (St. Louis, MO, USA) and were used without further purification. α-glucosidase from *S. cerevisiae*, acarbose, and ρ-nitrophenyl-α-D-Glucopyranoside were obtained from Sigma-Aldrich (Sintra, Portugal). Fluorescein [3′,6′-dihydroxyspiro (isobenzofuran-1 [3H], 9′ [9H]-xanthen)-3-one] was purchased from Fisher Scientific (Hanover Park, IL, USA). The tri-peptide Abz-Gly-Phe (NO_2)-Pro was obtained from Bachem Feinchemikalien (Bubendorf, Switzerland). Tris [tris (hydroxymethyl) aminomethane] was obtained from Honeywell Fluka (Charlotte, NC, USA). Mueller–Hinton (MH) and Man-Rogosa-Sharpe (MRS) broths were purchased from Biokar Diagnostics (Beauvais, France). The Pierce BCA Protein Assay Kit was purchased from Thermo Scientific (Vantaa, Finland). All remaining chemicals and reagents employed were of analytical grade or suitable for chromatography.

3.2. Sourcing and Preparing Peptides

The peptides HdKTLR (EDNSELGQETPTLR), HdKNL (DPPNPKNL), HdPPP (PAPPPPPP), HdVLPN (VYPFPGPLPN), and HdLPN (PFPGPLPN) were synthesized by GenScript, Piscataway, NJ, USA, with the purity ≥ 90%. The peptides were prepared following the guidelines provided in the technical datasheet, ensuring proper solubility and concentration, as detailed in Table 4.

Table 4. Summary of peptide solubility and concentration values.

Peptides	Sequences	Molecular Weight (Da)	Concentration (mg/mL)	Solubility
HdKTLR	EDNSELGQETPTLR	1588.4	2.6	H_2O
HdKNL	DPPNPKNL	893.8	2.6	H_2O
HdPPP	PAPPPPPP	768.8	2.6	DMSO
HdVLPN	VYPFPGPLPN	1099.6	2.0	H_2O
HdLPN	PFPGPLPN	837.5	2.3	H_2O

3.3. *Mucus Collection and Fish Source*

The mucus of the body *H. didactylus* was collected according to the method described by [4]. Mucus was collected from *Halobatrachus didactylus* specimens caught by fishermen in central-west Portugal and were subsequently released by our team. Two pooled mucus samples were analyzed: one from five captive individuals at the Vasco da Gama Aquarium and the other from four wild individuals provided by the Portuguese Institute for Sea and Atmosphere (IPMA).

3.4. *Soluble Protein Content*

The Bicinchoninic Acid (BCA) methodology, using the Pierce BCA Protein Assay Kit, was employed to determine the soluble protein concentration in both the permeate and retentate forms of the samples obtained from digestion and dialysis. This methodology was also utilized to determine protein concentrations in the pooled mucus samples.

3.5. *Antioxidant Activity*

3.5.1. ABTS Assay

ABTS radical activity is based on an antioxidant's capacity to scavenge the ABTS's oxidized state. For this, ABTS radical scavenging activity was performed in a 96-well microplate, following the method described by [4]. The absorbance was measured with a multidetection plate reader (Synergy H1, Winooski, VT, USA). For each sample, the standard and control analyses were duplicated. All measurements were performed in duplicate. The results are expressed in µmol of Trolox equivalents/µmol of peptide. To test the antioxidant activity under different pH and ionic strength conditions, the ABTS radical solution was prepared using the same buffers employed in the HPSEC analysis.

3.5.2. ORAC Assay

The ORAC assay followed the protocol outlined by Coscueta et al. [41]. A black polystyrene 96-well microplate (Nunc, Roskilde, Denmark) was utilized, and measurements were taken using a multidetection plate reader (Synergy H1, Winooski, VT, USA) controlled by the Gen5 Biotek software version 3.04. Fluorescence was monitored for 80 min at 1 min intervals. Each sample, standard, blank, or control, analysis was duplicated. The final ORAC values were expressed as µmol TE/µmol of peptide for the peptides and µmol TE /mg of protein for the digestion retentate and permeate contents.

The synergistic interaction between the mucus and peptides was performed by mixing the peptides with mucus at a ratio of one part peptide to four parts mucus. In addition, control analyses were performed by mixing one part of the phosphate buffer with four parts of the mucus and another with one part of the peptide to four parts of the phosphate buffer. The mixtures were then tested in the ORAC assay. The final ORAC values were expressed as µM TE.

3.6. Antihypertensive Activity

The iACE assay was conducted following the procedure outlined by Coscueta et al. [41]. The assay was performed using a black polystyrene 96-well microplate (Nunc, Denmark), and fluorescence was monitored at 1 min intervals for 80 min using a multidetection plate reader (Synergy H1, Winooski, VT, USA) controlled by the Gen5 Biotek software version 3.04. Each sample, blank, and control, analysis was carried out in triplicate. The iACE was quantified as the concentration required to inhibit 50% of the enzymatic activity (IC_{50}). The IC_{50} values were determined using non-linear modelling of the data obtained.

3.7. Antidiabetic Activity

The α-glucosidase inhibitory activity was evaluated through the following method, outlined by Vilas-Boas et al. [42]. The assay was performed in a 96-well microplate (Stansted) and incubated at 25 °C for 5 min, and the absorbance was monitored at 405 nm with a microplate reader (Synergy H1, BioTek Instruments, Winooski, VT, USA). A negative control was performed, with 50 µL of a buffer solution that replaced the peptides. Acarbose was used as a positive control at a 10 mg/mL concentration. Each sample and control analysis was triplicated. The inhibition of α-glucosidase was quantified as the concentration required to inhibit 50% of the enzymatic activity (IC_{50}). The IC_{50} values were determined using non-linear modelling of the data obtained. The IC_{50} value for acarbose was determined to use as a comparative reference.

3.8. Growth Inhibition Curves

The antimicrobial activity of the mucus samples was evaluated following the method by [4] against the following pathogenic bacteria: Gram-negative *Escherichia coli* ATCC 25922, *Salmonella enterica* serovar Enteritidis ATCC 13076, and *Pseudomonas aeruginosa* ATCC 27853 and Gram-positive *Staphylococcus aureus* ATCC 6538. The mucus samples were combined with a 1/10 ratio of inoculum containing 1% (*v*/*v*) of bacteria cultured for 24 h in MH broth. For the positive control, deionized water was used instead of mucus samples with a 1% inoculum, while deionized water with MH served as the negative control. A control using chloramphenicol was made at a concentration of 16 µg/mL for *E. coli*, *S. enterica*, and *P. aeruginosa* and 100 µg/mL for *S. aureus*. The mixture was then transferred to a 96-well microplate (Sarstedt, Germany), and the optical density (OD) at 600 nm was measured every hour for a 24 h period at 37 °C using a microplate reader (Multiskan GO, Thermo Scientific, Finland). An increase in OD indicated bacterial growth.

3.9. Antibiofilm Activity

The antibiofilm activity was determined according to the following method, with some modifications [43]. Taking the microplate from 24 h inhibition growth curves, the contents of the plate were discarded, and each well was carefully washed three times with distilled water to remove non-adherent cells. The adhered cells were fixed with 150 µL of methanol for 10 min. Then, the methanol was discarded, the plates were left to dry, and then, the fixed biofilm was stained with 150 mL of crystal violet (CV) for 10 min. Excess stain was rinsed out by placing the plate under slow-running tap water. Once the plates were air-dried, the dye attached to the adherent cells was resolubilized by adding 150 µL of 33% (*v*/*v*) glacial acetic acid and agitated for 15 min. The optical density of the solution obtained was measured at 595 nm using a multidetection plate reader (Synergy H1, Winooski, VT, USA). All tests were carried out in duplicate, with a positive control using an inoculated culture in MH broth and a negative control using MH broth only. The results are presented as the percentage of biofilm formation, calculated according to the following formula, where the OD of the sample is the measured OD of the inoculated medium mixed with the peptides, and the OD of the positive control is the OD measured for the positive control.

$$\% \text{ Biofilm formation} = (OD_{sample}/OD_{positive\ control}) \times 100 \quad (1)$$

3.10. Probiotic Growth Curves

The effect of the peptides on probiotics was evaluated using the following probiotic *Lactobacillus* and *Bifidobacterium* species: *L. acidophilus* Ki (LAS), *L. acidophilus* La5, *L. casei* L26, *L. plantarum* 299V, *L. rhamnosus* RII, *B. longum* BG3 *B. animalis* spp. *lactis* BB12, *B. animalis* BLC, and *B. animalis* Bo. All the assays were performed using MRS broth supplemented with 0.5 g/L of L-cysteine [44]. The peptides were combined with a 1/10 ratio of inoculum, containing 2% (*v*/*v*) of the probiotic and cultured for 48 h in MRS broth. For the positive control, deionized water was used instead of the peptides, with a 2% inoculum, while deionized water with MRS served as the negative control. Peptide wells with *Bifidobacterium* were sealed with paraffin to assure anaerobiosis conditions. The bacterial growth was monitored at 650 nm for 24 h at 37 °C in a microplate reader (Multiskan GO, Thermo Scientific, Finland), and the optical density (OD) was measured hourly. Growth curves were created by transforming the absorbance values in natural logarithms ln (Abs).

3.11. Cytotoxicity Assay

To evaluate the cytotoxicity of the peptides HdVLPN and HdLPN, two cell lines were used, namely, caucasian colon adenocarcinoma cells—Caco-2 (ECACC 86010202) and 5—and human keratinocytes—HaCaT (300493, CLS, Eppelheim, Germany). The assay was carried out as described by [45]. Afterward, the culture media were carefully removed and replaced with 25, 50, and 100 μg/mL of peptide (sterile filtered). The cytotoxicity was evaluated using the PrestoBlue HS Cell Viability assay (Thermo Scientific, Waltham, MA, USA), following the protocol described by the manufacturer. Fluorescence was measured using a fluorescence excitation wavelength of 560 nm and an emission of 590 nm by the multidetection plate reader Synergy H1 (BioTek Instruments, Winooski, VT, USA) controlled by the Gen5 BioTek software version 3.04. The metabolic inhibition was determined as previously described by Coscueta et al. [46]

3.12. Gastrointestinal Simulation

The gastrointestinal simulation was based on the standardized static digestion model INFOGEST 2.0 protocol [46]. The procedure is divided into three phases, the oral, gastric, and intestinal, with the corresponding fluids mimicking the in vivo conditions. Therefore, the oral phase consisted in the dilution of the peptides HdVLPN (1 mg/mL) and HdLPN (1.15 mg/mL) at 1:1 (*v*/*v*) with simulated salivary fluid and agitated in the orbital shaker MaxQ 6000 at 200 rpm and at 37 °C for 2 min for mastication simulation. After this, the oral phase was diluted at 1:1 (*v*/*v*) with simulated gastric fluid and gastric enzyme, namely, pepsin that was prepared according to INFOGEST protocol and agitated at 130 rpm and at 37 °C, at pH 3.0 for 2 h. The resultant gastric phase was subsequently diluted at 1:1 (*v*/*v*) with simulated intestinal fluid, bile salts, and pancreatic enzymes (pancreatin from the porcine pancreas) and incubated at pH 7.0 for 2 h, at 45 rpm and at 37 °C. For each sample two replicas were performed, and a negative control without a sample, using ultrapure water, was used.

After digestion, to assess the possible absorption of the peptides, a dialysis process was performed using 3.5 kDa of membranes (cellulose membranes in dialysis tubes). Therefore, the final content of the digestion was poured inside the membrane (8 mL of the sample), and a significant volume of water was added to the outside, covering the sample inside the membrane (70 mL of ultrapure water). The dialysis samples were placed in an orbital shaker at 50 rpm and at 37 °C overnight. After the incubation time, the contents of the membrane (retentate) and the water (permeate) were collected and lyophilized. The lyophilized contents from the retentate were reconstituted in 4 mL of ultrapure water and from the permeate were reconstituted in 8 mL of ultrapure water.

3.13. Molecular Size Validation

The peptides' molecular size validation was determined using high-performance size exclusion chromatography (HPSEC). An Agilent AdvanceBio SEC column (Agilent Technologies, London, UK), 2.7 µm in particle size, 130 Å in pore size, and 7.8 in inner diameter × 300 mm in length, was used. The column was eluted isocratically using various buffers with different pH and ionic strengths, namely, 344 mM at pH 7 (0.15 M of a phosphate buffer with NaH_2PO_4), 700 mM at pH 7 (0.15 M of a phosphate buffer with NaH_2PO_4 and NaCl), 344 mM at pH 5 (0. 15 M of an acetate buffer with acetic acid and NaCl), 700 mM at pH 5 (0.15 M of an acetate buffer with acetic acid and NaCl), 344 mM at pH 3 (0.05 M of a formate buffer with fomic acid and NaCl), and 700 mM at pH 3 (0.05 M of a formate buffer with formic acid and NaCl) at a flow rate of 1 mL/min. The sample injection volume was 10 µL. The instrument used was Waters 2690 with a photodiode array detector (PDA 190–600 nm). The software Empower 3 was used for data collection. To determine the molecular weights of the resulting chromatogram peaks, a calibration curve was made with the following protein standards: Ovalbumin (44,300 Da), Myoglobin (17,600 Da), Cytochrome C (12,327 Da), Aprotinin (6511 Da), Neurotensin (1672 Da), Angiotensin-II (1040 Da), Tyr-Phe dipeptide (328.4 Da), and L-tryptophan (204 Da).

3.14. Statistical Analysis

Different tests were applied for analyzing the data. Kruskal–Wallis was used for the ORAC assay, Mann–Whitney U for the ABTS assay, and repeated measures with Bonferroni post hoc for antimicrobial growth curves. For the antibiofilm assay, a one-way ANOVA was applied for *E. coli* and *P. aeruginosa* and Kruskal–Wallis for *S. aureus* and *S. enterica*. Cytotoxicity was assessed with a one-way ANOVA, followed by Tukey's test. The antioxidant activity of the digested peptides was evaluated using the Kruskal–Wallis test for the retentate form and a one-way ANOVA, followed by Tukey's test, for the permeate form. The iACE activity of the digested peptides was evaluated using the Kruskal–Wallis test for the permeate form. The effects of pH and ionic strength on ABTS were analyzed using Kruskal–Wallis for the peptides HdVLPN and HdLPN and a one-way ANOVA with Tukey's test for the peptide HdKNL. Lastly, the mucus–peptide interaction was analyzed with *t*-tests. All analyses were performed using SPSS 28.0.0.0 (SPSS Inc., Chicago, IL, USA).

4. Conclusions

Our study delves into using in silico methodologies to predict peptide bioactivities, followed by their validation through in vitro assays. Employing in silico techniques for the bioactivity prediction of novel peptides offers a promising and efficient path for a preliminary exploration, although validation through in vitro assays remains imperative.

This study identifies five bioactive peptides through in silico prediction: HdPPP with the PAPPPPPP sequence, HdVLPN with the VYPFPGPLPN sequence, and HdLPN with the PFPGPLPN sequence. In vitro studies have revealed that HdPPP exhibits antihypertensive and antidiabetic properties, while HdLPN and HdVLPN demonstrate significant antioxidant activity. However, the bioactivities of these peptides could face limitations and not reach their full potential due to their tendency to aggregate under the physiological conditions of pH 7 with ionic strengths of 700 mM and 344 mM.

Consequently, a comprehensive study was conducted to uncover this issue, discovering a valuable interaction between the peptides and mucus. It appears that the molecules present in mucus can augment the bioactivity of peptides through synergistic interactions, thereby optimizing their conformation for enhanced functionality and/or in neutralization oxidants by facilitating electron and proton donation to stabilize the peptides in antioxidant activity.

Further investigations are imperative to delineate the optimal conditions for peptide bioactivity. Subsequent research should explore encapsulation techniques and the utilization of matrices mimicking mucins to enhance peptide bioactivity.

Overall, this study highlights the potential of the marine environment, and in particular marine fish, to contribute to the bioprospection of novel bioactive molecules with wide applications in food, pharmaceutical, nutraceutical, and cosmeceutical products.

5. Patents

New peptides and uses thereof: Instituto Nacional da Propriedade Industrial (patent N°.: PT118365; 5 December 2022).

Author Contributions: Conceptualization, M.F.C. and E.R.C.; methodology, M.F.C. and E.R.C.; validation, M.E.B., F.A., D.G. and M.M.P.; investigation, M.F.C. and E.R.C.; formal analysis, M.F.C. and E.R.C.; data curation, M.F.C. and E.R.C.; writing—original draft preparation, M.F.C., M.E.B. and E.R.C.; writing—review and editing, M.F.C., E.R.C., M.E.B., F.A., D.G. and M.M.P.; supervision, F.A., D.G. and M.M.P.; project administration, F.A., D.G. and M.M.P.; funding acquisition, F.A., D.G. and M.M.P. All authors have read and agreed to the published version of the manuscript.

Funding: This research received financial support from Fundação para a Ciência e a Tecnologia—Fundo para o Desenvolvimento das Ciências e da Tecnologia (FCT-FDCT) through the "FISHMUC—Bioactive properties of external mucus isolated from coastal fish of Macao and Portugal" project, reference MACAU/0003/2019; the FDCT (Macao) project with reference 0005/2019/APJ; and the FCT (Portugal) fund with reference UID/04292/MARE-Centro de Ciências do Mar e do Ambiente.

Institutional Review Board Statement: No ethical concerns regarding the animal study protocol were raised by Fundação para a Ciência e a Tecnologia-Fundo para o Desenvolvimento das Ciências e da Tecnologia (FCT-FDCT) during the assessment and approval of project FISHMUC. The animal study has been conducted at ISPA-MARE by staff and facilities authorized in 2021 by the Portuguese National Authority for Animal Health—Direção Geral de Alimentação e Veterinária (DGAV reference 0421/000/000/2021). All procedures for animal experiments were performed in strict accordance with the EU Directive 2010/63/EU.

Data Availability Statement: The data presented in this study are available on request from the corresponding author. The data are not publicly available, as the patenting process is still ongoing at the time this article was submitted.

Acknowledgments: The authors would like to thank Aquário Vasco da Gama and Fátima Gil for all the assistance with the collection of epidermal mucus from captive fish. Access to this fish facility was crucial for all preliminary tests that allowed us to fine tune the methodologies applied in this paper. We would also like to extend our gratitude to the researchers Rita Marques and José Neto from MARE-ISPA for their invaluable assistance with the collection of fish epidermal mucus.

Conflicts of Interest: The authors declare no conflicts of interest.

Statement: During the preparation of this work, the authors used GPT-4 to improve readability and language for the final version of the manuscript. After using this tool/service, the authors reviewed and edited the content as needed and take full responsibility for the content of this publication.

eferences

Oyemitan, I.A. Chapter 27—African Medicinal Spices of Genus Piper. In *Medicinal Spices and Vegetables from Africa*; Kuete, V.B.T., Ed.; Academic Press: Cambridge, MA, USA, 2017; pp. 581–597, ISBN 978-0-12-809286-6.

Montaser, R.; Luesch, H. Marine Natural Products: A New Wave of Drugs? *Future Med. Chem.* **2011**, *3*, 1475–1489. [CrossRef]

Bhatia, P.; Chugh, A. Role of Marine Bioprospecting Contracts in Developing Access and Benefit Sharing Mechanism for Marine Traditional Knowledge Holders in the Pharmaceutical Industry. *Glob. Ecol. Conserv.* **2015**, *3*, 176–187. [CrossRef]

Fernandez Cunha, M.; Coscueta, E.R.; Brassesco, M.E.; Marques, R.; Neto, J.; Almada, F.; Gonçalves, D.; Pintado, M. Exploring Bioactivities and Peptide Content of Body Mucus from the Lusitanian Toadfish Halobatrachus Didactylus. *Molecules* **2023**, *28*, 6458. [CrossRef] [PubMed]

Tu, M.; Cheng, S.; Lu, W.; Du, M. Advancement and Prospects of Bioinformatics Analysis for Studying Bioactive Peptides from Food-Derived Protein: Sequence, Structure, and Functions. *TrAC Trends Anal. Chem.* **2018**, *105*, 7–17. [CrossRef]

Du, Z.; Comer, J.; Li, Y. Bioinformatics Approaches to Discovering Food-Derived Bioactive Peptides: Reviews and Perspectives. *TrAC Trends Anal. Chem.* **2023**, *162*, 117051. [CrossRef]

Zhou, L.; Mendez, R.L.; Kwon, J.Y. In Silico Prospecting for Novel Bioactive Peptides from Seafoods: A Case Study on Pacific Oyster (*Crassostrea gigas*). *Molecules* **2023**, *28*, 651. [CrossRef]

Devita, L.; Lioe, H.N.; Nurilmala, M.; Suhartono, M.T. The Bioactivity Prediction of Peptides from Tuna Skin Collagen Using Integrated Method Combining In Vitro and In Silico. *Foods* **2021**, *10*, 2739. [CrossRef] [PubMed]

Coscueta, E.R.; Batista, P.; Gomes, J.E.G.; da Silva, R.; Pintado, M.M. Screening of Novel Bioactive Peptides from Goat Casein: In Silico to In Vitro Validation. *Int. J. Mol. Sci.* **2022**, *23*, 2439. [CrossRef] [PubMed]

10. Böni, L.J.; Zurflüh, R.; Baumgartner, M.E.; Windhab, E.J.; Fischer, P.; Kuster, S.; Rühs, P.A. Effect of Ionic Strength and Seawater Cations on Hagfish Slime Formation. *Sci. Rep.* **2018**, *8*, 9867. [CrossRef] [PubMed]
11. Fekete, S.; Beck, A.; Veuthey, J.L.; Guillarme, D. Theory and Practice of Size Exclusion Chromatography for the Analysis of Protein Aggregates. *J. Pharm. Biomed. Anal.* **2014**, *101*, 161–173. [CrossRef] [PubMed]
12. Zapadka, K.L.; Becher, F.J.; Gomes dos Santos, A.L.; Jackson, S.E. Factors Affecting the Physical Stability (Aggregation) of Peptide Therapeutics. *Interface Focus* **2017**, *7*, 20170030. [CrossRef] [PubMed]
13. De Groot, N.S.; Pallarés, I.; Avilés, F.X.; Vendrell, J.; Ventura, S. Prediction of "Hot Spots" of Aggregation in Disease-Linked Polypeptides. *BMC Struct. Biol.* **2005**, *15*, 18. [CrossRef]
14. AGGRESCAN. Available online: http://bioinf.uab.es/aggrescan/ (accessed on 7 February 2024).
15. INNOVAGEN. Available online: https://pepcalc.com/peptide-solubility-calculator.php (accessed on 29 January 2024).
16. Fowler, S.B.; Poon, S.; Muff, R.; Chiti, F.; Dobson, C.M.; Zurdo, J. Rational Design of Aggregation-Resistant Bioactive Peptides: Reengineering Human Calcitonin. *Proc. Natl. Acad. Sci. USA* **2005**, *102*, 10105–10110. [CrossRef] [PubMed]
17. Huang, D.; Boxin, O.U.; Prior, R.L. The Chemistry behind Antioxidant Capacity Assays. *J. Agric. Food Chem.* **2005**, *53*, 1841–1856. [CrossRef] [PubMed]
18. Wen, C.; Zhang, J.; Zhang, H.; Duan, Y.; Ma, H. Plant Protein-Derived Antioxidant Peptides: Isolation, Identification, Mechanism of Action and Application in Food Systems: A Review. *Trends Food Sci. Technol.* **2020**, *105*, 308–322. [CrossRef]
19. Jin, J.; Ahn, C.; Je, J. Puri Fi Cation and Characterization of Antioxidant Peptides from Enzymatically Hydrolyzed Ark Shell (*Scapharca subcrenata*). *Process Biochem.* **2018**, *72*, 170–176. [CrossRef]
20. Zhu, Y.; Lao, F.; Pan, X.; Wu, J. Food Protein-Derived Antioxidant Peptides: Molecular Mechanism, Stability and Bioavailability. *Biomolecules* **2022**, *12*, 1622. [CrossRef] [PubMed]
21. Zheng, L.; Zhao, M.; Xiao, C.; Zhao, Q.; Su, G. Practical Problems When Using ABTS Assay to Assess the Radical-Scavenging Activity of Peptides: Importance of Controlling Reaction PH and Time. *Food Chem.* **2016**, *192*, 288–294. [CrossRef] [PubMed]
22. Fernández-Montero, Á.; Torrecillas, S.; Montero, D.; Acosta, F.; Prieto-Álamo, M.-J.; Abril, N.; Jurado, J. Proteomic Profile and Protease Activity in the Skin Mucus of Greater Amberjack (*Seriola dumerili*) Infected with the Ectoparasite *Neobenedenia girellae*—An Immunological Approach. *Fish Shellfish Immunol.* **2021**, *110*, 100–115. [CrossRef] [PubMed]
23. Roberts, S.D.; Powell, M.D. The Viscosity and Glycoprotein Biochemistry of Salmonid Mucus Varies with Species, Salinity and the Presence of Amoebic Gill Disease. *J. Comp. Physiol. B* **2005**, *175*, 1–11. [CrossRef] [PubMed]
24. Guardiola, F.A.; Cuartero, M.; del Mar Collado-González, M.; Díaz Baños, F.G.; Cuesta, A.; Moriñigo, M.Á.; Esteban, M.Á. Terminal Carbohydrates Abundance, Immune Related Enzymes, Bactericidal Activity and Physico-Chemical Parameters of the Senegalese Sole (*Solea senegalensis*, Kaup) Skin Mucus. *Fish Shellfish Immunol.* **2017**, *60*, 483–491. [CrossRef]
25. Aklakur, M. Natural Antioxidants from Sea: A Potential Industrial Perspective in Aquafeed Formulation. *Rev. Aquac.* **2018**, *10*, 385–399. [CrossRef]
26. Minkiewicz, P.; Dziuba, J.; Iwaniak, A.; Dziuba, M.; Darewicz, M. BIOPEP Database and Other Programs for Processing Bioactive Peptide Sequences. *J. AOAC Int.* **2008**, *91*, 965–980. [CrossRef]
27. Yathisha, U.G.; Bhat, I.; Karunasagar, I.; Mamatha, B.S. Antihypertensive Activity of Fish Protein Hydrolysates and Its Peptides. *Crit. Rev. Food Sci. Nutr.* **2019**, *59*, 2363–2374. [CrossRef]

28. Abachi, S.; Bazinet, L.; Beaulieu, L. Antihypertensive and Angiotensin-i-Converting Enzyme (ACE)-Inhibitory Peptides from Fis as Potential Cardioprotective Compounds. *Mar. Drugs* **2019**, *17*, 613. [CrossRef] [PubMed]
29. Antony, P.; Vijayan, R. Bioactive Peptides as Potential Nutraceuticals for Diabetes Therapy: A Comprehensive Review. *Int. J. Mc Sci.* **2021**, *22*, 9059. [CrossRef] [PubMed]
30. Subramanian, S.; Ross, N.W.; MacKinnon, S.L. Myxinidin, a Novel Antimicrobial Peptide from the Epidermal Mucus of Hagfis *Myxine glutinosa* L. *Mar. Biotechnol.* **2009**, *11*, 748–757. [CrossRef] [PubMed]
31. Cantisani, M.; Leone, M.; Mignogna, E.; Kampanaraki, K.; Falanga, A.; Morelli, G.; Galdiero, M.; Galdiero, S. Structure-Activit Relations of Myxinidin, an Antibacterial Peptide Derived from the Epidermal Mucus of Hagfish. *Antimicrob. Agents Chemothe* **2013**, *57*, 5665–5673. [CrossRef] [PubMed]
32. Li, T.; Liu, Q.; Wang, D.; Li, J. Characterization and Antimicrobial Mechanism of CF-14, a New Antimicrobial Peptide from th Epidermal Mucus of Catfish. *Fish Shellfish Immunol.* **2019**, *92*, 881–888. [CrossRef]
33. Akbarian, M.; Chen, S.-H.; Kianpour, M.; Farjadian, F.; Tayebi, L.; Uversky, V.N. A Review on Biofilms and the Currently Availabl Antibiofilm Approaches: Matrix-Destabilizing Hydrolases and Anti-Bacterial Peptides as Promising Candidates for the Foo Industries. *Int. J. Biol. Macromol.* **2022**, *219*, 1163–1179. [CrossRef]
34. Alves, M.J.; Ferreira, I.C.F.R.; Lourenço, I.; Costa, E.; Martins, A.; Pintado, M. Wild Mushroom Extracts as Inhibitors of Bacteria Biofilm Formation. *Pathogens* **2014**, *3*, 667–679. [CrossRef] [PubMed]
35. Gao, P.P.; Liu, H.Q.; Ye, Z.W.; Zheng, Q.W.; Zou, Y.; Wei, T.; Guo, L.Q.; Lin, J.F. The Beneficial Potential of Protein Hydrolysate as Prebiotic for Probiotics and Its Biological Activity: A Review. *Crit. Rev. Food Sci. Nutr.* **2023**, *64*, 13045–13057. [CrossRef [PubMed]
36. Zhang, C.; Zhang, Y.; Li, H.; Liu, X. The Potential of Proteins, Hydrolysates and Peptides as Growth Factors for: Lactobacillus an Bifidobacterium: Current Research and Future Perspectives. *Food Funct.* **2020**, *11*, 1946–1957. [CrossRef] [PubMed]
37. Casciaro, B.; Loffredo, M.R.; Cappiello, F.; Verrusio, W.; Corleto, V.D.; Mangoni, M.L. Frog Skin-Derived Peptides Agains *Corynebacterium jeikeium*: Correlation between Antibacterial and Cytotoxic Activities. *Antibiotics* **2020**, *9*, 448. [CrossRef] [PubMed
38. Wang, S.; Yang, M.; Yin, S.; Zhang, Y.; Zhang, Y.; Sun, H.; Shu, L.; Liu, Y.; Kang, Z.; Liu, N.; et al. A New Peptide Originated from Amphibian Skin Alleviates the Ultraviolet B-Induced Skin Photodamage. *Biomed. Pharmacother.* **2022**, *150*, 112987. [CrossRef [PubMed]
39. Fernando, I.P.S.; Jayawardena, T.U.; Wu, J. Marine Proteins and Peptides: Production, Biological Activities, and Potentia Applications. *Food Innov. Adv.* **2023**, *2*, 69–84. [CrossRef]
40. Amigo, L.; Hernández-Ledesma, B. Current Evidence on the Bioavailability of Food. *Molecules* **2020**, *25*, 4479. [CrossRef]
41. Coscueta, E.R.; Campos, D.A.; Osório, H.; Nerli, B.B.; Pintado, M. Enzymatic Soy Protein Hydrolysis: A Tool for Biofunctiona Food Ingredient Production. *Food Chem. X* **2019**, *1*, 100006. [CrossRef] [PubMed]
42. Vilas-Boas, A.A.; Goméz-García, R.; Machado, M.; Nunes, C.; Ribeiro, S.; Nunes, J.; Oliveira, A.L.S.; Pintado, M. Lavandula Pedunculata Polyphenol-Rich Extracts Obtained by Conventional, MAE and UAE Methods: Exploring the Bioactive Potentia and Safety for Use a Medicine Plant as Food and Nutraceutical Ingredient. *Foods* **2023**, *12*, 4462. [CrossRef]
43. Costa, E.; Silva, S.; Tavaria, F.; Pintado, M. Antimicrobial and Antibiofilm Activity of Chitosan on the Oral Pathogen *Candida albicans*. *Pathogens* **2014**, *3*, 908–919. [CrossRef] [PubMed]
44. Araújo-Rodrigues, H.; Coscueta, E.R.; Pereira, M.F.; Cunha, S.A.; Almeida, A.; Rosa, A.; Martins, R.; Pereira, C.D.; Pintado, M.E Membrane Fractionation of Cynara Cardunculus Swine Blood Hydrolysate: Ingredients of High Nutritional and Nutraceutica Value. *Food Res. Int.* **2022**, *158*, 111549. [CrossRef] [PubMed]
45. Coscueta, E.R.; Sousa, A.S.; Reis, C.A.; Pintado, M. Chitosan-Olive Oil Microparticles for Phenylethyl Isothiocyanate Delivery Optimal Formulation. *PLoS ONE* **2021**, *16*, e0248257. [CrossRef] [PubMed]
46. Brodkorb, A.; Egger, L.; Alminger, M.; Alvito, P.; Assunção, R.; Ballance, S.; Bohn, T.; Bourlieu-Lacanal, C.; Boutrou, R.; Carrière F.; et al. INFOGEST Static In Vitro Simulation of Gastrointestinal Food Digestion. *Nat. Protoc.* **2019**, *14*, 991–1014. [CrossRef]

marine drugs

rticle

Revealing the Diversity of Sequences, Structures, and Targets of Peptides from South China Sea *Macrodactyla doreensis* Based on Transcriptomics

iqiang Hua †, Yanling Liao †, Jinxing Fu, Xinru Li, Qianxia Xu, Limin Lin, Meiling Huang * and Bingmiao Gao *

Engineering Research Center of Tropical Medicine Innovation and Transformation of Ministry of Education, Hainan Key Laboratory for Research and Development of Tropical Herbs, International Joint Research Center of Human-Machine Intelligent Collaborative for Tumor Precision Diagnosis and Treatment of Hainan Province, School of Pharmacy, Hainan Medical University, Haikou 571199, China; huazzq@icloud.com (Z.H.); liaoyanling@hainmc.edu.cn (Y.L.); fjx1900201062@163.com (J.F.); 13588261944@163.com (X.L.); 17757940986@163.com (Q.X.); liminlin239@gmail.com (L.L.)

* Correspondence: mgling213@163.com (M.H.); gaobingmiao@hainmc.edu.cn (B.G.)

† These authors have contributed equally to this work.

Abstract: The South China Sea is rich in sea anemone resources, and the protein and peptide components from sea anemone toxins comprise an important treasure trove for researchers to search for leading compounds. This study conducted a comprehensive transcriptomic analysis of the tentacles and column of *Macrodactyla doreensis* and explored the distribution and diversity of proteins and peptides in depth using bioinformatics, initially constructing a putative protein and peptide database. In this database, typical peptide families are identified through amino acid sequence analysis, and their 3D structures and potential biological activities are revealed through AlphaFold2 modeling and molecular docking. A total of 4239 transcripts were identified, of which the putative protein accounted for 81.53%. The highest content comprised immunoglobulin and a variety of proteases, mainly distributed in the column and related to biological functions. Importantly, the putative peptide accounted for 18.47%, containing ShK domain and Kunitz-type peptides, mainly distributed in the tentacles and related to offensive predatory behavior. Interestingly, 40 putative peptides belonging to eight typical peptide families were identified, and their structures and targets were predicted. This study reveals the diversity and complexity of *Macrodactyla doreensis* toxins and predicts their structure and targets based on amino acid sequences, providing a feasible approach for research regarding the discovery of peptides with potentially high activity.

Keywords: *Macrodactyla doreensis*; sea anemone toxins; transcriptomics; AlphaFold2 modeling; molecular docking

1. Introduction

Sea anemones are ancient sessile invertebrates, named for their resemblance to sunflowers. They inhabit shallow coral reefs across various marine regions, including the Indian Ocean, the Pacific Ocean, the Red Sea, and the Samoan Islands. Among these regions, the South China Sea, part of the Pacific Ocean, is noted for its particularly high abundance of sea anemones. Sea anemones have traversed an evolutionary path spanning approximately 700 million years, evolving sophisticated sea anemone toxins [1]. Sea anemone toxins are a complex mixture of peptides, proteins, purines, biogenic amines, and assorted compounds, with cysteine-rich peptides being of particular interest due to their unique properties [2]. Sea anemones possess cnidocytes with nematocysts specialized for producing and delivering toxins crucial for their survival in the marine ecosystem [3]. Activation of these nematocysts leads to the discharge of venom in response to mechanical or chemical stimuli, which is essential for both predation and defense [4], and which are regarded as a pharmacological tool or a reservoir for natural drugs [5]. Marine biotoxins exhibit enormous potential for the development of new drugs. So far, more than 20 drugs derived from the ocean have been approved for the market worldwide. For example, ω-Conotoxin MVIIA (Ziconotide, Prialt), which is a peptide derived from *Conus magus*, is used to relieve pain. The ShK peptide from the sea anemone *Stichodactyla helianthus* [6] is a potent blocker of voltage-gated potassium channels. Its derivative, ShK-186, is currently undergoing early clinical trials to treat T-cell autoimmune diseases such as multiple sclerosis by inhibiting the Kv1.3 channel [7]. Therefore, research on marine biotoxins is currently receiving great attention.

Sea anemone toxins produced by prickly cells are typically classified into multiple categories, including phospholipase A2; cytolysins; insulin-like peptides; protease inhibitors, such as serine protease inhibitors; and neurotoxic peptides affecting a range of ion channels, like voltage-gated sodium (Nav), calcium (Cav), and potassium (Kv) channels [8–12]. Sea anemone toxins notably engage in inflammatory and nociceptive processes [13,14], cellular membrane lysis [15], and the modulation of ionic channel dynamics, including the immunosuppressive effects of ShK domain peptides on the Kv1.3 channel [16].

The South China Sea boasts abundant sea anemone resources, including *Heteractis crispa* (*H. crispa*), *Exaiptasia diaphana* (*E. diaphana*), *Heteractis magnifica* (*H. magnifica*), *Macrodactyla doreensis* (*M. doreensis*), etc. At present, a large number of studies have reported the diversity of venom from *H. crispa*, *E. diaphana*, and *H. magnifica*, all of which have been systematically explored through transcriptomics owing to their rich peptides [17–19]. However, there are currently almost no reports on *M. doreensis*, and this study is the first to conduct a systematic transcriptomic analysis of this resource. *M. doreensis* is widespread in the Indo-Pacific region, particularly in the South China Sea, where it is very common. The preliminary research on the venom of *M. doreensis* proved its pharmacological anti-cancer, insecticidal, and analgesic activities. Therefore, it is particularly warranted to conduct in-depth research on the protein and peptide components in *M. doreensis*.

In the early study of sea anemone toxins, traditional methods such as the fractionation and purification of crude venom, as well as gene cloning through PCR primer design, were predominantly employed to discover novel sea anemone toxins. These separation and purification techniques were both laborious and time-consuming, with significant limitations. With the rapid development of high-throughput sequencing (HTS) technology, the limitations of traditional research techniques have been overcome. It has the advantages of high sequencing throughput, high sensitivity, low cost, and the ability to discover novel protein and peptide gene sequences. Therefore, this study used HTS to perform comprehensive transcriptome analysis on the tentacles and column of *M. doreensis* and constructed a putative protein and peptide database. Compared with emphasis of previous studies on the transcriptome of sea anemones, this study focuses on typical peptide families, predicts their 3D structures through AlphaFold2 modeling, and evaluates their potential activities through molecular docking, laying a solid foundation for the development of new marine drugs.

2. Results

2.1. Transcriptome Sequence Assembly

HTS has revealed that 49,077,330 and 45,623,592 paired-end short reads (150 bp in length) were obtained from the tentacles and column, respectively, with Q30 quality scores surpassing 91%. Subsequent to the removal of inferior reads and adapters, the refined dataset yielded a total of 93,660,778 clean reads, with clean data ratios averaging 98.90%. The de novo assembly process resulted in the generation of 287,880 unique sequences across both tissue samples, exhibiting a mean N50 length of 1499 bp (Table S1).

Analyses of the GC content in the assembled contigs and unigenes revealed percentages of 41.18% in tentacle samples and 38.90% in column samples. Functional annotation against multiple databases, including the Non-Redundant Protein Sequence Database (Nr), the Kyoto Encyclopedia of Genes and Genomes (KEGG), Clusters of Orthologous Groups for Eukaryotic Complete Genomes (KOG), and the Universal Protein Resource (UniProt), revealed a substantial portion of unigenes with functionalities yet to be ascribed. The most prominent categories within the KEGG pathways were metabolism and organismal systems, providing valuable insights into the metabolic capabilities of sea anemones and their interactions within the organismal systems.

GO categorization highlighted 85,897 unigenes across three domains: biological process, cellular component, and molecular function, with notable enrichment in cellular anatomical entities and cellular processes. KOG database annotations further distributed 19,922 and 17,159 unigenes across 25 functional classes, with general function prediction and signal transduction mechanisms being predominant. This comprehensive transcriptomic profiling dramatically enhances our understanding of *M. doreensis* at the molecular level, providing a broader biological and functional genomic context (Figure S1).

2.2. *Family Classification of Putative Proteins and Peptides in the M. doreensis Transcriptome*

In this study, the extensive UniProt database has been utilized to identify these transcripts. The analysis revealed that putative proteins represent a predominant 81.63% (3456) of the transcripts, while putative peptides accounted for the remaining 18.37% (783), which were divided into 216 known families (Figure 1). The putative proteins were distributed across numerous families, the most prominent being the immunoglobulins and EGF proteins (Figure 1A). A total of 606 peptides were derived from the transcriptomic profiles of *M. doreensis'* tentacles and its column, cataloged as MD-001 to MD-606 (Table S2). Based on the framework of cysteine, 40 peptides have been further screened and divided into eight families: ShK domain, Kunitz-type, β-defensins, ICK, BBH, Kazal-like, EGF, and waprin (Figure 1B).

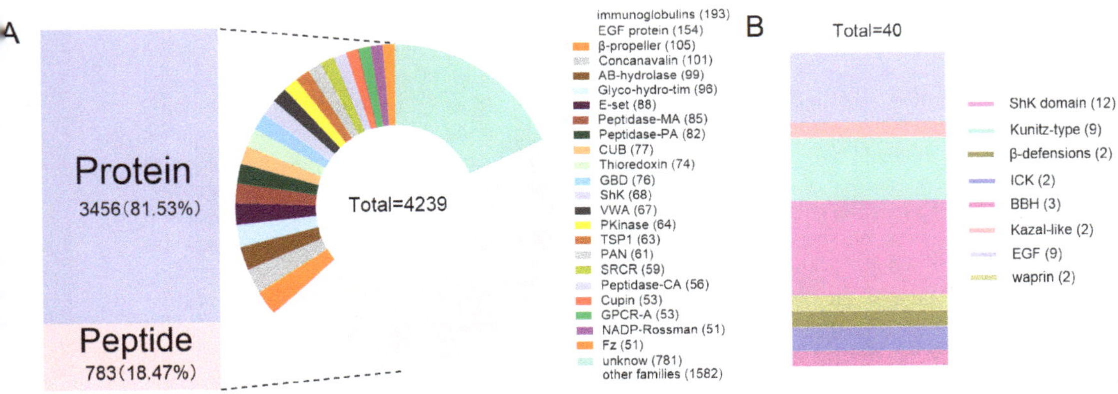

Figure 1. Putative protein and peptide families identified in the transcriptome of *M. doreensis*. (**A**) The proportion of putative proteins to peptides and classification diagram of families. (**B**) The number of putative peptides in representative families.

2.3. *Comparative Analysis of Putative Proteins and Peptides in Different Tissues of M. doreensis*

While exploring the transcriptomic landscapes of *M. doreensis*, distinct regional profiles reveal a diverse array of bioactive molecules. Cysteine-rich peptides are prevalent, suggesting a role in defensive or predatory behaviors. Comprehensive transcriptomic analysis of the tentacle and column regions of *M. doreensis* has revealed the relative classification of these putative proteins and peptides. The transcriptome in the tentacles of *M. doreensis* is divided into 133 families, while the transcriptome in the column is divided into 178 families, of which 95 families are shared (Figure 2A).

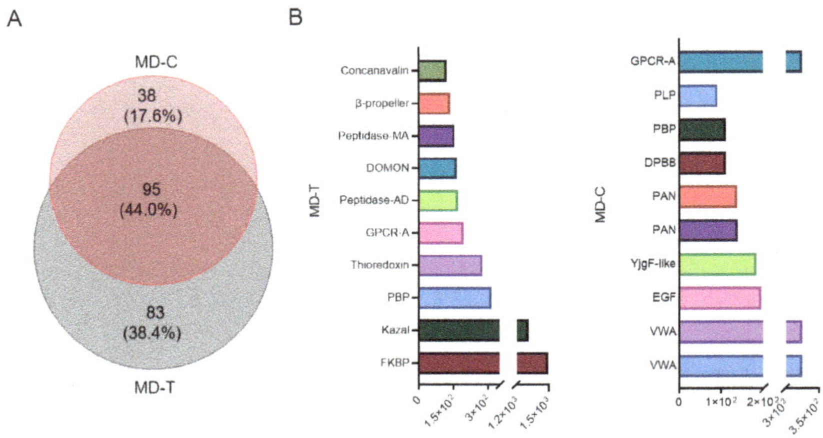

Figure 2. Comparison of transcripts from different parts of *M. doreensis*. (**A**) The Venn diagram of family classification relationship between MD-T and MD-C. (**B**) The ten most highly expressed transcripts from different tissues.

Data were converted to FPKM (fragments per kilobase of exon model per million mapped fragments) to quantify differentially expressed transcripts. The results revealed that FK506 binding protein (FKBP) and Kazal-like peptides exhibited the highest transcript levels in the tentacular region, whereas von Willebrand factor type A domain (VWA) was most prominent in the basal region (Figure 2B).

2.4. Sequence Analysis, Structure, and Target Prediction of Typical Putative Peptides

The peptides derived from sea anemone venom serve as survival tools for sea anemones, exhibiting vast diversity and mostly classified based on cysteine patterns or structural features. They can act on a variety of ion channels or other receptors, such as Kv channels, Nav channels, and serine protease receptors [11,12,20]. This study conducted a comprehensive analysis of 40 putative peptides belonging to eight peptide families, including Kazal-like, EGF, waprin, and anemone types 1–5 potassium channel toxins, namely ShK domain, Kunitz-type, β-defensins, BBH, and ICK. Initially, a phylogenetic analysis was performed on 40 putative peptides, accompanied by a presentation of their corresponding FPKM (Figure 3). The results indicate that all putative peptides can be categorized into four major classes. While some of these classifications are consistent with cysteine pattern-based family classifications, there are significant deviations for the majority. Specifically, the β-defensins, ICK, Kazal-like, and Kunitz-type peptides (excluding MD-368), each form distinct branches, demonstrating strong evolutionary and affinity relationships consistent with classifications based on cysteine patterns. In contrast, ShK domain, EGF, waprin, and BBH peptides cluster into two or more branches. From the perspective of FPKM, Kunitz-type peptides exhibit the highest expression levels, followed by EGF peptides. Although a large number of putative peptides belong to the ShK domain, their expression levels are relatively low. The ShK domain peptides exhibit low amino acid sequence similarity and low FPKM values, indicating a high diversity of peptides in this family, but low individual expression levels. This suggests that the ShK domain peptides may possess a wide range of targets and activities.

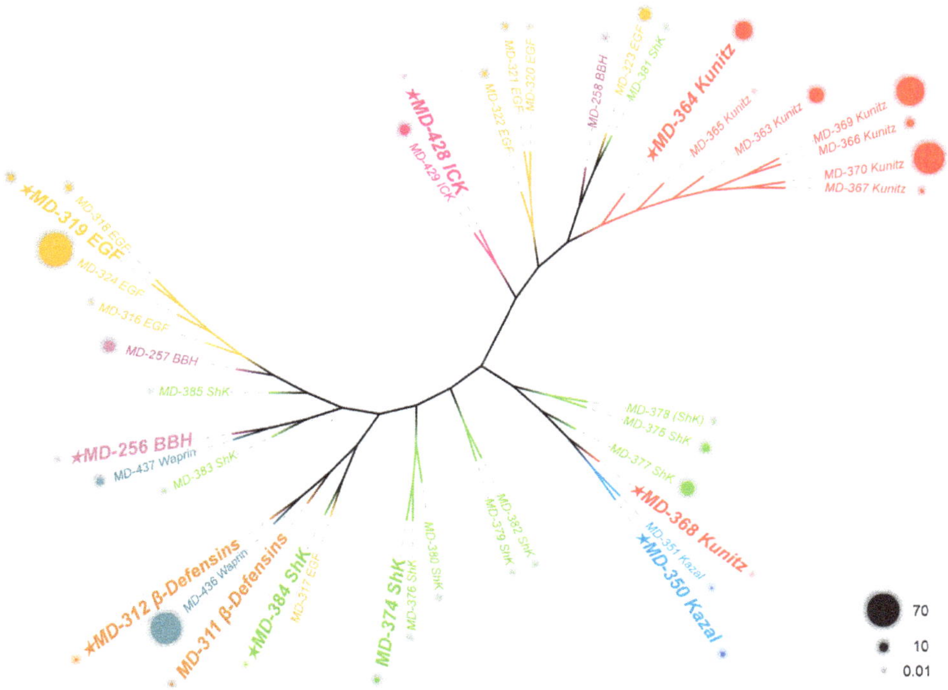

Figure 3. Phylogenetic and FPKM analysis of 40 putative peptides derived from *M. doreensis*. The tree was established using the ML approach. Sequences with the same color indicate peptides from the same family. The size of the outer circle of the sequence shows the FPKM.

2.4.1. ShK Domain

ShK domain peptides also belong to the sea anemone type 1 potassium channel toxin family, and their representative peptide ShK consists of 35 amino acid residues intertwined with three disulfide bridges (Cys1-Cys6, Cys2-Cys4, Cys3-Cys5). These peptides are significant Kv channel inhibitors, predominantly targeting the type 1 channels [21,22]. Drugs used to treat autoimmune diseases typically produce systemic immune suppression, and ShK domain peptides can target the Kv1.3 channel to selectively inhibit CCR7-TEM lymphocytes and reduce this side effect [23].

In this study, eight representative peptides from known ShK structural domain peptides were selected for comparative analysis. ShK mainly inhibits Kv channels [24] and exerts some antibacterial effects [22]. BgK primarily targets the Kv channels (type 1), exhibiting potent inhibition of the Kv1.1 channel [25]. Its inhibitory effect on the remaining subtype channels is relatively weak; AsKs and HmK predominantly interact with the Kv1.2 channel [26,27], with the latter also influencing neurotransmitter release in animal models. AeK acts on the Kv channels (type 1) [28], while BCsTx1 and BCsTx2 focus their inhibitory actions on the Kv1.1 and Kv1.2 channels [29].

Through amino acid sequence analysis of 12 homologous putative peptides (MD-374 to MD-385), it was determined that these peptides showed the highest sequence similarity with seven known peptides of the ShK domain: ShK corresponding to MD-380 (36.59%), BgK to MD-374 (44.19%), AsKs to MD-380 (47.37%), HmK to MD-382 (39.02%), AeK to MD-377 (44.74%), BCsTx1 to MD-384 (46.67%), and BCsTx2 to MD-377 (46.15%) (Figure 4A). Based on the amino acid sequence of each putative peptide, AlphaFold2 was used for modeling to predict their respective 3D structures. The predictive models uniformly adhere to the archetypal architecture of the ShK domain, characterized by intricate arrays of α-helix bundles and helix-turn (loop)-helix motifs. Notably, the α-helices, situated proximate to the N-terminus and C-terminus, display an almost perpendicular, parallel arrangement. Specifically, the model aligned based on eight databases, i.e., AlphaFold/Proteome, AlphaFold/Swiss-Prot, AlphaFold/UniProt50, BFMD, CATH50, GMGCL, MGnify-ESM30, and PDB100. The protein or peptide exhibiting the most structurally similar fragment to each of the putative peptides was identified. Notably, MD-374 structurally aligns most closely with U-actidoxin-Avd8d, derived from the snakelocks anemone *Anemonia viridis*, with a TM-score of 0.94897 and an RMSD of 0.51. Additional information pertaining to other putative peptides is appended in Table S4.

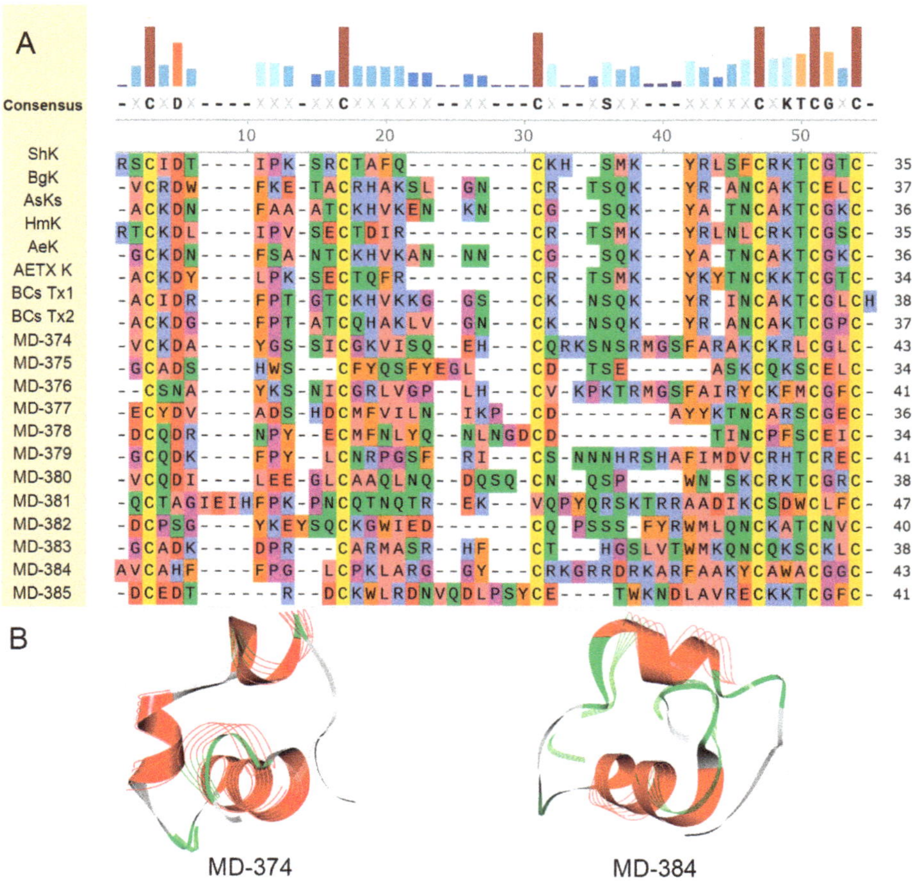

Figure 4. Deduced ShK domain transcripts from *M. doreensis*. (**A**) Multiple sequence alignment of known ShK domain peptides and corresponding putative mature peptides. Residues are shown by physicochemical properties, following their Zappo colors. The representative ShK domain peptides shown are ShK (Uniport: P29187; PDB: 1BEI), from the sea anemone *Stichodactyla helianthus*; BgK (Uniport: P29186; PDB: 1BGK), from the red warty sea anemone *Bunodosoma granuliferum*; AsKs (Uniport: Q9TWG1; AlphaFold: AF-Q9TWG1-F1), from the Mediterranean snakelocks sea anemone *Anemonia sulcata*; HmK (Uniport: O16846; AlphaFold: AF-O16846-F1) from the magnificent sea anemone *Heteractis magnifica*; AeK (Uniport: P81897; AlphaFold: AF-P81897-F1) from the beadlet anemone *Actinia equina*; BCsTx1/2 (Uniport: C0HJC2/C0HJC3; AlphaFold: AF-C0HJC2-F1/AF-C0HJC3-F1) from the sea anemone *Bunodosoma caissarum*. (**B**) The putative peptide models (flat representation) of the ShK domain predicted by AlphaFold2 (the line model that overlaps with them is the best peptide obtained in the model alignment).

The Kv1.3 channel plays an important role in the pathogenesis of autoimmune diseases and is also associated with analgesia and anti-cancer effects [23,24]. A large-scale molecular docking of 20 ShK domain peptides with the Kv1.3 channel was conducted to demonstrate their potential interactions and biological effects on the potassium channels [30]. MD-374 was found to form the most stable peptide–protein complex, exhibiting the potential to form extensive non-covalent bonds with the target protein (Figure 5A). According to the prediction result, the complex has a total of 27 favorable non-covalent bonds, of which 18 are hydrogen bonds, such as Asp161-Arg27, Cys248-Gln21, and Lys276-Leu39 (Figure 5B). The contact surface area between the ligand and the receptor is 526.64 and 552.76, respectively. Compared with the molecular docking results of other known ShK domain peptides, MD-374 may exhibit more effective inhibitory activity on the Kv1.3 channel.

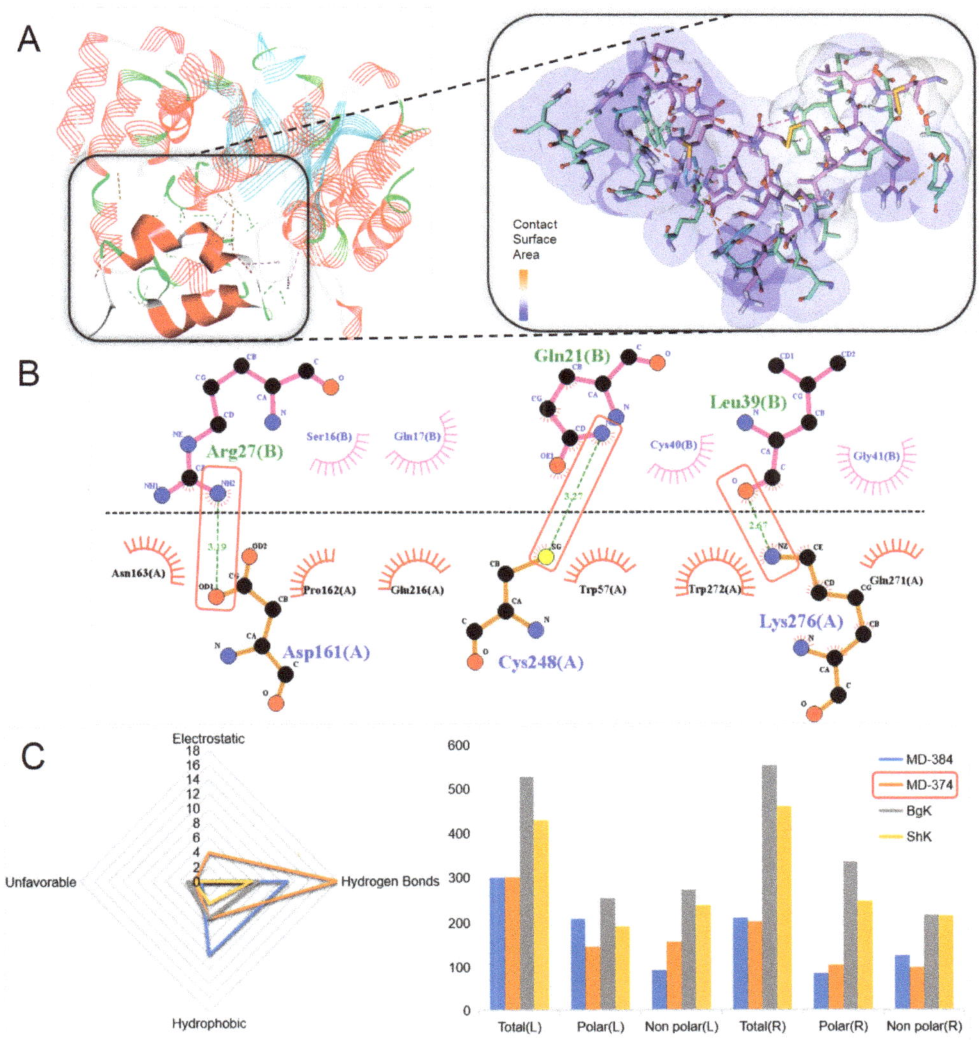

Figure 5. Molecular docking analyses revealed the binding interaction between the ShK domain peptides and the Kv1.3 channel. The green dotted lines indicate hydrogen bond interactions, the red dotted lines represent unfavorable interactions, the yellow dotted lines denote electrostatic interactions,

and the pink dotted lines highlight hydrophobic interactions. (**A**) A docking pattern diagram illustrating MD-374 (flat representation) and the Kv1.3 channel (PDB: 7EJ1) (line representation), along with a more comprehensive 3D diagram displaying the docking locations and contact surface area of MD-374 (colored purple) and the Kv1.3 channel (colored green). (**B**) An eyelash (2D) diagram depicting the key docking sites of MD-374 (colored blue) and the Kv1.3 channel (colored green). (**C**) A radar chart illustrating the number of non-bond interactions between ShK domain peptides and the Kv1.3 channel, along with a 3D surface plot depicting the contact surface area. Ligand (L); receptor (R).

2.4.2. Kunitz-Type

Kunitz-type peptides derived from sea anemones belong to the sea anemone type 2 potassium channel toxin family and possess dual functional capacities. They not only inhibit the Kv1.2 channel but also act as serine protease inhibitors. The configuration of the three disulfide bonds within the sea anemone Kunitz-type peptides follows a specific arrangement, i.e., Cys1-Cys6/Cys2-Cys4 and Cys3-Cys5, as indicate by the cysteine pattern C-C-C-C-CX3C. The first two disulfide bridges play a crucial role in maintaining the conserved nature of the tertiary structure of the domain, while the third bond is critical for stabilizing the active binding site [31].

This study selected 10 representative peptides from known Kunitz-type peptides for comparative analysis. The peptides designated as AsKC1/2/3 are documented to exhibit inhibitory actions against serine proteases, as well as suppression of the Kv channels, primarily the Kv1.2 channel [26]. AXPI-I is characterized by its potent inhibition of trypsin, with a moderate level of inhibition exhibited towards plasmin and only a weak effect on chymotrypsin, elastase, and the metalloprotease thermolysin. Notably, AXPI-I does not display inhibitory activity against the Kv channels [32]. The peptide DTX-α, although homologous to serine proteases, does not confer inhibitory effects upon serine proteases; instead, it is capable of obstructing the Kv channels, specifically the Kv1.1, Kv1.2, and Kv1.6 channels [33,34]. SHTXIII, another peptide in this class, can inhibit both serine proteases and the Kv channels [35].

Through amino acid sequence analysis, a suite of nine putative peptides, MD-363 to MD-371, was identified as homologous to AsKC-1, DTX1, and SHTXIII, with a conserved cysteine pattern of CX8CX15CX7CX12CX3C. The results of the homology comparison indicated that MD-364 exhibited the highest similarity and identity among the aforementioned peptides, with an overall similarity and identity of approximately 56.26% and 49.20%, respectively. Notably, its similarity to AsKC1/2 and AXPI-I exceeded 70.00%, with the identity surpassing 65.00% (Figure 6A).

Based on the amino acid sequence of each putative peptide, AlphaFold2 was used for modeling to predict their respective 3D structures (Figure 6B). All predictive models possess the typical structure of Kunitz-type peptides, which is composed of a helix-loop-sheet-turn-sheet-loop-helix configuration extending from the N-terminus to the C-terminus. This architecture includes a β-hairpin and features a preserved, compact hydrophobic core inherent to the α-β-β motif. Specifically, the protein or peptide with the most similar structural fragment to each putative peptide was discovered. MD-364 is structurally most similar to U-actitoxin-Avd3j from the Mediterranean snakelocks sea anemone *Anemonia sulcata*, with a TM-score of 0.95746, and an RMSD of 0.74. Other putative peptide-related information is attached in Table S4.

Due to the dual inhibitory properties of Kunitz-type peptides on the Kv channels and serine proteases, a large-scale molecular docking study was conducted to elucidate the potential interactions and biological efficacy of 16 Kunitz-type peptides with the Kv1.2 channel and serine proteases. The Kv1.2 channel is primarily implicated in neurological disorders such as epilepsy, intellectual disabilities (ID), attention deficit/hyperactivity disorder (ADHD), autism spectrum disorder (ASD), pain, and autoimmune conditions [36]. Serine proteases are associated with inflammation and cancer. In the screenings of putative peptides in this family, MD-364 emerged as the peptide that forms the most robust peptide–protein complex with serine proteases, showing a potential capacity for extensive non-covalent bonds with target proteins (Figure 7A). Based on predictive analyses, this complex boasts a total of 19 favorable non-covalent bonds, 12 of which are hydrogen bonds, including those between Gly78 and Thr1, His40 and Gly42, as well as Gly38 and Asn43 (Figure 7B). The contact surface area between the ligand and the receptor is 549.64 and 549.64, respectively. MD-364 outperformed all other putative peptides within the same family and even surpassed known peptides, hinting at its superiority. Consequently, MD-364 may demonstrate a stronger inhibitory influence against serine proteases compared to that of other Kunitz-type peptides. With respect to the docking outcomes involving the Kv1.2 channel, within its family, MD-364 remains the peptide predicted as the most likely to form a stable complex with the receptor, thereby implying a possible inhibitory influence on the Kv1.2 channel (Figure 7C).

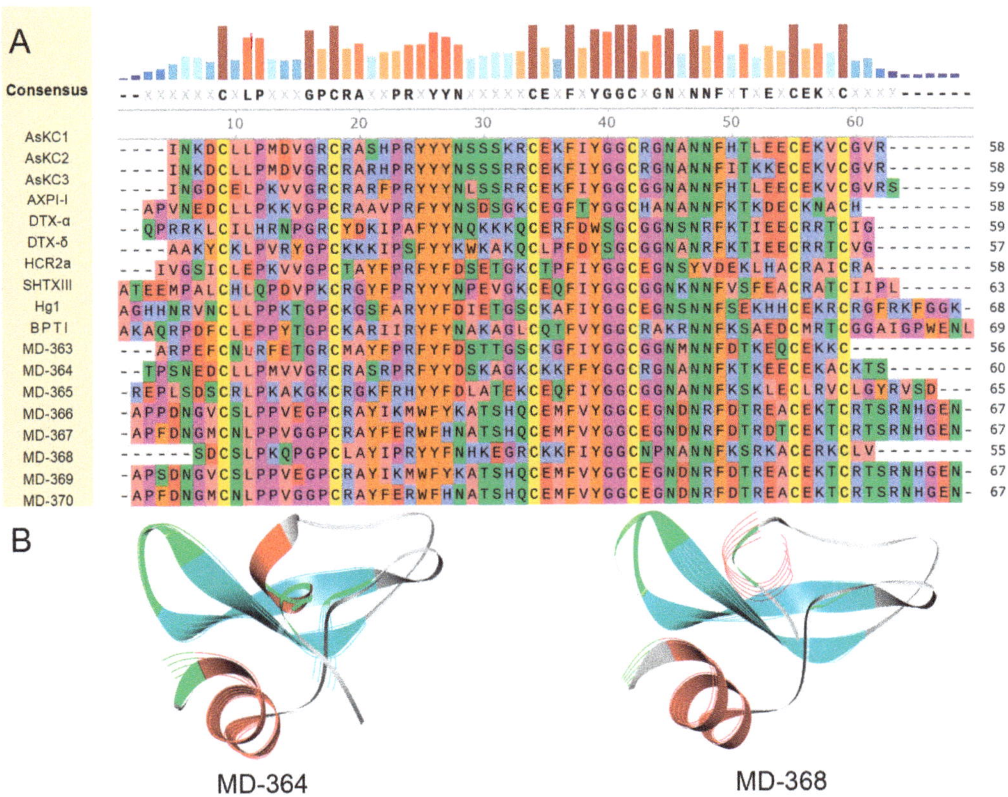

Figure 6. Deduced Kunitz-type transcripts from *M. doreensis*. (**A**) Multiple sequence alignment of known Kunitz-type peptides and corresponding putative mature peptides. The representative Kunitz-type peptides shown are BPT1 (Uniport: P00974), from the Bovine *Bos taurus*; AsKC1/2/3 (Uniport: Q9TWG0/Q9TWF9/Q9TWF8; AlphaFold: AF-Q9TWG0-F1/AF-Q9TWF9-F1/AF-Q9TWF8-F1), from the Mediterranean snakelocks sea anemone *Anemonia sulcata*; AXPI-I (Uniport: P81547; AlphaFold: AF-P81547-F1), from the sea anemone *Anthopleura aff. Xanthogrammica*; DTX-α/δ (Uniport: P00980/P00982; PDB: 1DTX/AlphaFold: AF-P00982-F1), from the Eastern green mamba or Naja angusticeps *Dendroaspis angusticeps*; SHTXIII (Uniport: B1B5I8; AlphaFold: AF-B1B5I8-F1), from the saddle carpet anemone, or Haddon's sea anemone *Stichodactyla haddoni*. (**B**) The Kunitz-type putative peptide models (flat representation) predicted by AlphaFold2 (the line model that overlaps with them is the best peptide obtained in the model alignment).

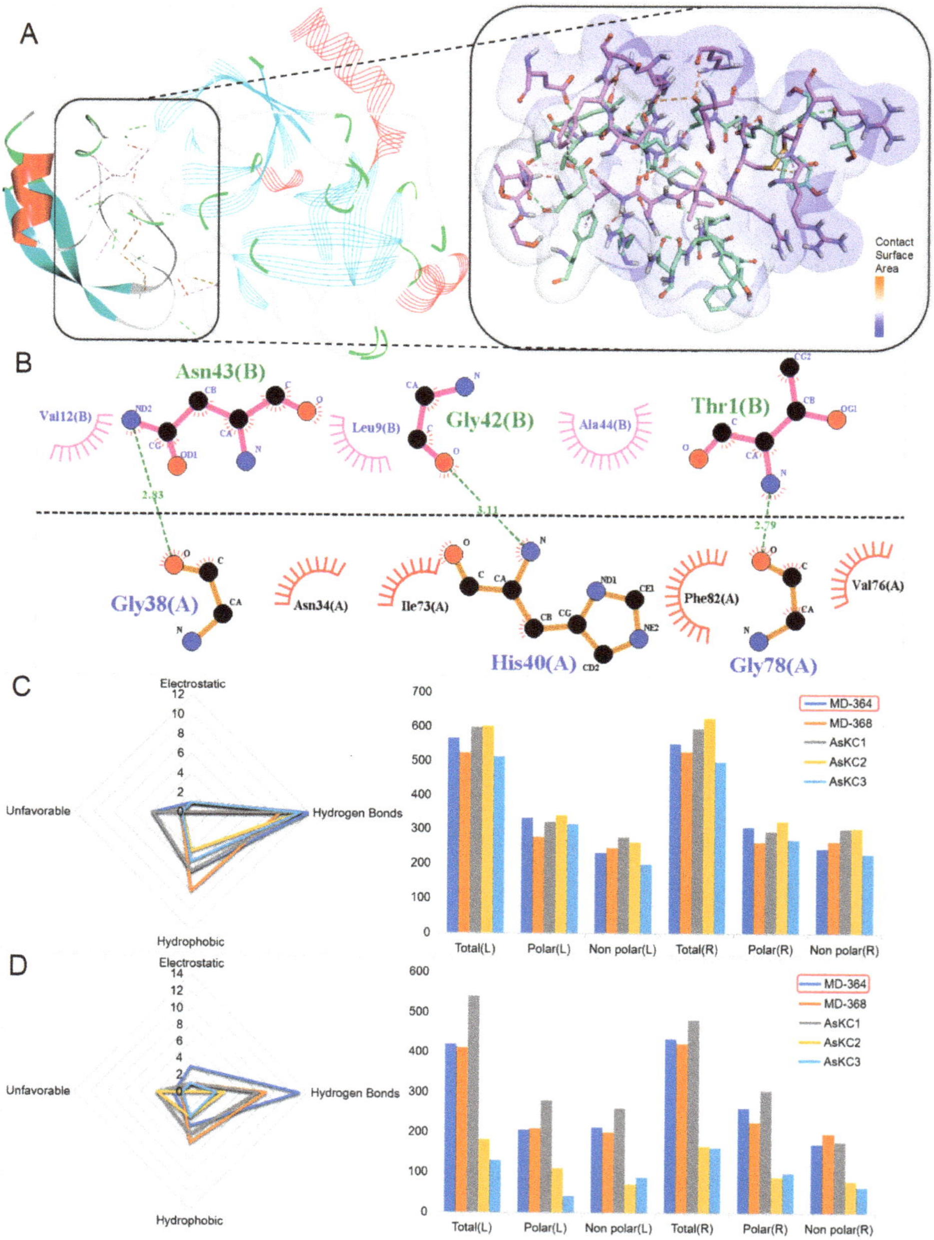

Figure 7. Molecular docking analyses revealed the binding interaction between Kunitz-type peptides and the serine proteinase receptor. (**A**) A docking pattern diagram illustrating MD-364 (flat representation) and the serine proteinase receptor (PDB: 5PTP) (line representation), along with a more comprehensive 3D diagram displaying the docking locations and contact surface area of MD-364 (colored purple) and the serine proteinase receptor (colored green). (**B**) An eyelash diagram depicting the key docking sites of MD-364 (colored blue) and the serine proteinase receptor (colored green). (**C**,**D**) A radar chart illustrating the number of non-bond interactions between Kunitz-type peptides and the serine proteinase receptor/Kv1.2 channel, along with a 3D surface plot depicting the contact surface area.

2.4.3. β-Defensins

The β-defensins discovered in sea anemones are also classified as comprising the sea anemone type 3 potassium channel toxin family. They typically consist of approximately 40 amino acids and display an essential pattern of disulfide linkage (Cys1-Cys5, Cys2-Cys4, and Cys3-Cys6) crucial for structural stability [37]. β-defensins display selective activities against Nav channels (types 1, 2, and 4), the Kv1.3 channel, ASICs, and TRPV1, modulating physiological processes pivotal to host defense [38,39].

This study selected five representative peptides from known β-defensins peptides for comparative analysis. Am II is a low molecular weight neurotoxin that causes paralysis when injected into crabs, suggesting a potential inhibitory or blocking effect on the Nav channels [40]. BDSI and BDSII are inhibitors of the Kv channels, predominantly affecting type 3 [41,42]. BDSI prolongs the opening duration of the Nav channels, explicitly targeting the Nav1.7 channel, resulting in a hypotensive effect [40]. BCIV is another peptide neurotoxin that affects the activity of the Nav channels without impacting the Kv channels [39]. APETx1 primarily affects the ERG potassium ion channels and the sodium ion channels. Its mechanism of action involves upregulating the voltage required to activate the channels and limiting the amplitude of their action potentials [43,44].

Through amino acid sequence analysis, two putative peptides affiliated with the β-defensins family were identified, namely MD-311 and MD-312, exhibiting a cysteine pattern characteristic of CXC-C-C-CC. A subsequent homology comparison revealed that MD-311 showed the highest similarity with BCIV, at 38.10%, and the highest sequence identity with APETx1, at 33.33%. Additionally, MD-312 displayed the highest similarity with BCIV, at 35.29%, and the highest sequence identity with BDSI, at 26.42% (Figure 8A).

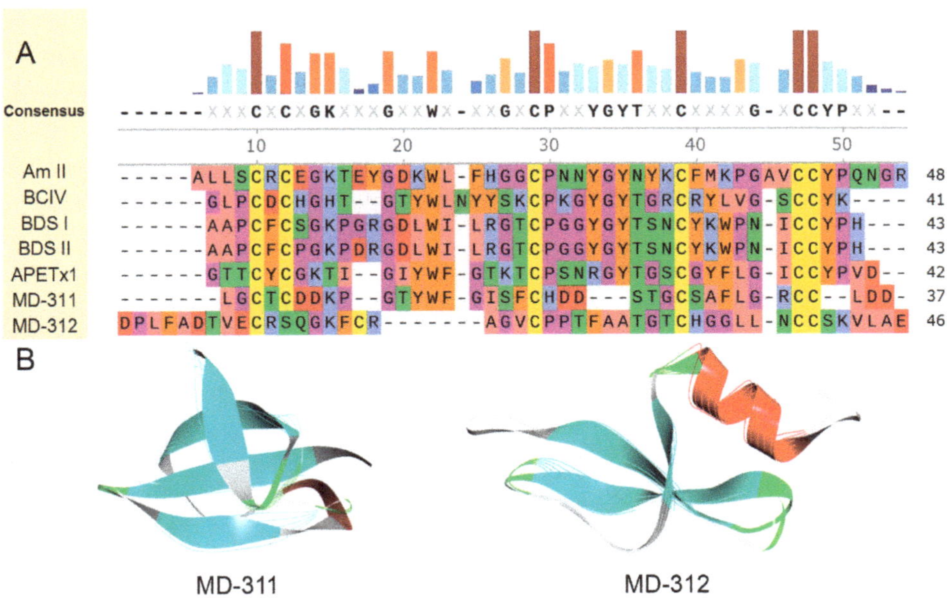

Figure 8. Deduced β-defensins transcripts from *M. doreensis*. (**A**) Multiple sequence alignment of known β-defensins peptides and corresponding putative mature peptides. The representative β-defensins peptides shown are Am II (Uniport: P69930; AlphaFold: AF-P69930-F1), from the sea anemone *Antheopsis maculata*; BCIV (Uniport: P84919; AlphaFold: AF-P84919-F1), from the sea anemone *Bunodosoma caissarum*; BDSI/II (Uniport: P11494/P59084; PDB:1BDS/AlphaFold: AF-P59084-F1), from the Mediterranean snakelocks sea anemone *Anemonia sulcata*; APETx1 (Uniport: P61541;

PDB: 1WQK), from the green aggregating anemone, or Actinia elegantissima *Anthopleura elegantissima*. (**B**) The putative peptide models (flat representation) of β-defensins predicted by AlphaFold2 (the line model that overlaps with them is the best peptide obtained in the model alignment).

Based on the amino acid sequence of each putative peptide, AlphaFold2 was used for modeling to predict their respective 3D structures. The prediction results show that all models exhibit typical structural features, predominantly displaying multiple β-fold composition. All predictive models prominently featured a β-barrel, emphasizing the presence of a highly conserved structural core β-fold, which is characteristic of the β-defensins family. Specifically, the protein or peptide with the most similar structural fragment to each putative peptide was discovered. MD-312 is structurally most similar to gallinacin-10, from chicken *Gallus gallus*, with a TM-score of 0.90943, and an RMSD of 0.97. Other putative peptides-related information is attached in Table S4.

Due to the role of β-defensins peptides as Kv channels blockers, along with their effects on inhibiting or prolonging the opening times of the Nav channels, a comprehensive molecular docking study was conducted involving seven β-defensins peptides with Kv1.2 and Nav1.7 channels to explore their potential interactions and biological efficacy. Both the Kv1.2 and Nav1.7 channels are involved in the mechanisms underlying pain responses, with the Nav1.7 channel encoded by the SCN9A gene and primarily located in the dorsal root ganglion neurons, making it a crucial target for pain management. Loss of function in the Nav1.7 channel results in insensitivity to pain [45]. Within the β-defensins peptides identified from our transcriptomic screening, MD-312 was found to form the most stable peptide–protein complex with both the Kv1.2 and Nav1.7 channels, exhibiting substantial potential for extensive non-covalent bonds with these target proteins (Figure 9A). Specifically, MD-312 was predicted to form nine hydrogen bonds with the Kv1.2 channel, e.g., between Asp86 and Cys22, Gln93 and Ala28, and Val102 and Leu36 (Figure 9B). In total, there were 17 favorable non-covalent bonds, which surpassed the number for the comparative peptides in this aspect. In terms of interactions with the Nav1.7 channel, MD-312 also demonstrated optimal performance. It is expected to form 16 favorable non-covalent bonds, including 8 hydrogen bonds, surpassing other peptides in terms of ligand–receptor contact surface area, without any unfavorable interactions (Figure 9C). These results suggest that MD-312 has exceptionally high biological potential for targeting both the Kv1.2 and Nav1.7 channels.

2.4.4. BBH

The BBH peptides discovered in sea anemones are also classified as comprising the sea anemone type 3 potassium channel toxin family. These peptides, commonly found in sea anemones, are composed of cysteine-rich chains, typically ranging from 25 to 40 amino acid residues in length. They are distinguished by the formation of two discrete disulfide bonds between four cysteine residues, manifesting an intramolecular bonding configuration typified as Cys1-Cys3 and Cys2-Cys4. Such a disulfide bond framework is identified by the cysteine pattern CX2C-C [46].

This study selected six representative peptides from the known BBH peptides for comparative analysis. The peptide SHTX-I/2 has been identified as an inhibitor of the Kv channels [45]. In contrast, Ugr 9a-1 is known to exert an influence on the ASIC3 channel, imbuing it with anti-inflammatory and analgesic characteristics that are attributed to its distinctive spatial structure and mechanism of action [47]. The peptide BCg III 23.41 is speculated to inhibit the function of the Kv channels [48]. Furthermore, Tau-AnmTX Ms 9a-1 has been shown to act as an agonist to mammalian TRPA1 and to demonstrate anti-inflammatory and analgesic activities in vitro [49].

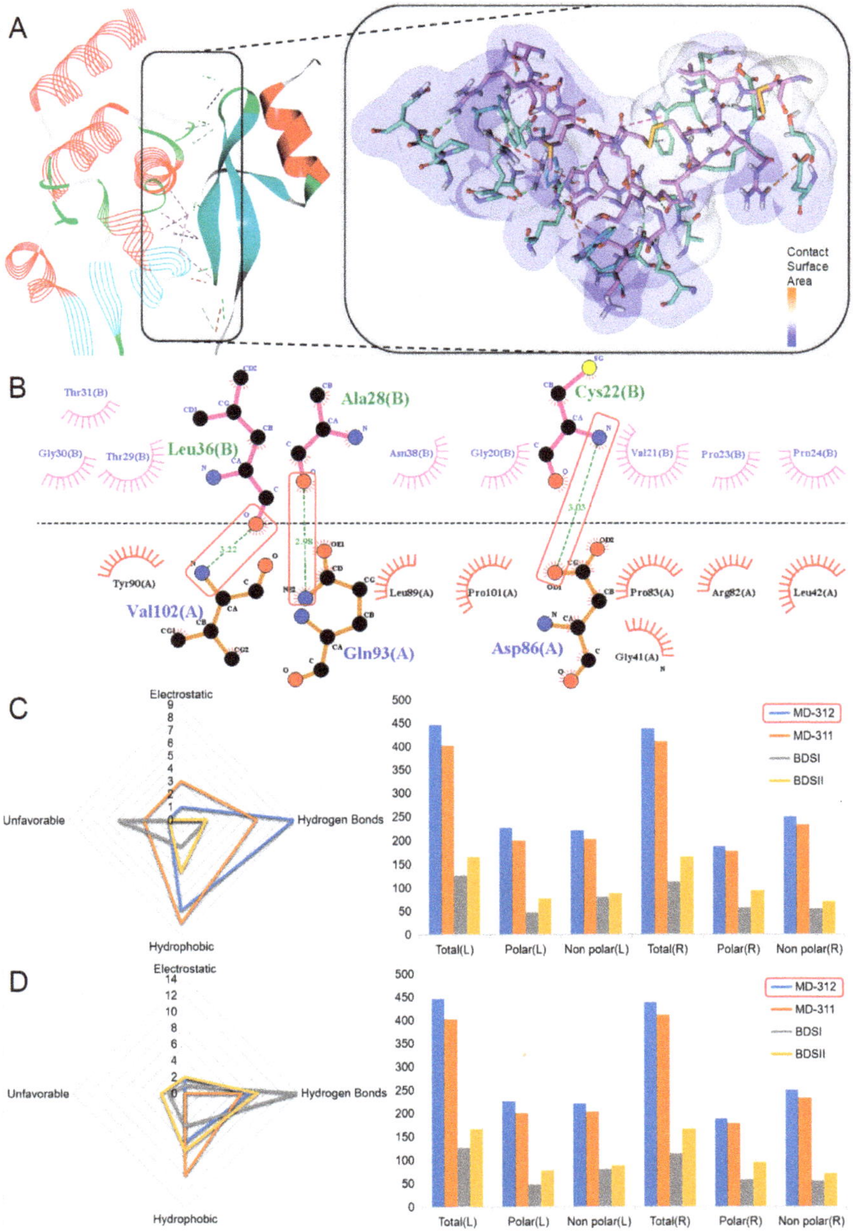

Figure 9. Molecular docking analyses revealed the binding interaction between β-defensins peptides and the Kv1.2 channel. (**A**) A docking pattern diagram illustrating MD-312 (flat representation) and the Kv1.2 channel (PDB: 1QDV; A chain) (line representation), along with a more comprehensive 3D diagram displaying the docking locations and contact surface area of MD-312 (colored purple) and the Kv1.2 channel (colored green). (**B**) An eyelash diagram depicting the key docking sites of MD-312 (colored blue) and the Kv1.2 channel (colored green). (**C**,**D**) A radar chart illustrating the number of non-bond interactions between the β-defensins peptides and the Kv1.2/Nav1.7 channels, along with a 3D surface plot depicting the contact surface area.

Through amino acid sequence analysis, three putative peptides affiliated with the BBH, denoted as MD-256, MD-257, and MD-258, were identified. It was determined that MD-256 shows the highest sequence similarity and identity with SHTX-I/2, at 35.00% and 22.50%, respectively. The similarity levels with the other four peptides are approximately 30%, while the similarity between MD-257 and MD-258 and these peptides is relatively low (Figure S2A).

Based on the amino acid sequence of each putative peptide, AlphaFold2 was used for modeling to predict their respective 3D structures. All predictive models inherently possess the β-hairpin motif, emphasizing the consistent structural essence that characterizes the BBH family.

Since BBH peptides are Kv channel blockers, a large-scale molecular docking of a total of nine BBH peptides with the Kv1.1 channel was conducted to demonstrate the potential interactions and biological potential of these peptides with the Kv1.1 channel. The Kv1.1 channel has been associated with epilepsy, multiple sclerosis, episodic ataxia, and cardiorespiratory dysregulation [50]. Among the BBH peptides identified in our transcriptome screening, MD-256 was found to form the most stable peptide–protein complex, exhibiting the potential to form numerous hydrogen bonds (Thr83-Cys10, etc.) and non-covalent (Arg67-Asp9 etc.) interactions with the target protein (Figure S3A,B). Specifically, it showed similar counts of anticipated favorable non-covalent bonds when compared with SHTX-I/2 (14; 16) but presented fewer unfavorable interactions (1 vs. 2). Additionally, the ligand-receptor interface areas were found to be comparable between the two peptides (Figure S3C). These results suggest that MD-258 may have an inhibitory impact on the Kv1.1 channel similar to that of SHTX-I/2.

Among the BBH peptides identified in our transcriptome screening, MD-256 has been identified as forming the most stable peptide–protein complex, exhibiting the potential to form extensive non-covalent bonds with the target protein (Figure S3A,B). Specifically, it showed a slightly lower number of hydrogen bonds than did SHTX-I/2 (10 vs. 12), while displaying a reduced number of unfavorable interactions (0 vs. 2). Furthermore, an analysis revealed a similarity in the ligand and receptor contact surface area between the two peptides (Figure S3C). These findings imply that MD-256 may possess an inhibitory effect on the Kv1.1 channel, akin to that exhibited by SHTX-I/2.

2.4.5. ICK

The ICK peptides discovered in sea anemones are also classified as comprising the sea anemone type 5 potassium channel toxin family. They are predominantly characterized by their approximately 50 amino acids. These sequences are stabilized through a network of internal pairings among eight cysteine residues, resulting in four discrete intramolecular disulfide bridges. It is posited that a distinct disulfide bonding arrangement prevails within these peptides, involving the linkages Cys1-Cys4, Cys2-Cys5, Cys3-Cys8, and Cys6-Cys7, which establish a framework as C-C-CC-CX3C-C-C [51,52].

This study selected four representative peptides from known ICK peptides for comparative analysis. PP III is recognized as an inhibitor of the Nav and Cav channels. PP III is characterized by a voltage-independent mechanism that obstructs the conductivity of ion channels by blocking the extracellular vestibular pores, which can lead to fatal paralysis of the target insect. The efficacy of PP III in binding to and inhibiting sodium and calcium channels is likely integral to its formidable insecticidal properties [53]. BcsTx3 is known to target the Kv channels, predominantly interacting with the Kv1 subtypes, an intervention that provokes paralytic effects in crabs [51]. Similarly, NvePTx1 and MsePTx1, peptides that closely mirror the primary amino acid composition of BcsTx3 and PP III, are identified as neurotoxins. They are speculated to potentially inhibit the Kv channels [54,55].

Through amino acid sequence analysis, two putative peptides, designated MD-428 and MD-429, were identified. They exhibit a cysteine pattern characteristic of C-C-CC-CX3C-C-C. It has been determined that MD-428 shares notable similarity and identity with Bcs4a, with values of 70.59% and 56.86%, respectively. Furthermore, MD-429 showed the highest sequence similarity and identity with NvePTx1, with values of 72.00% and 60.00%, respectively (Figure 10A).

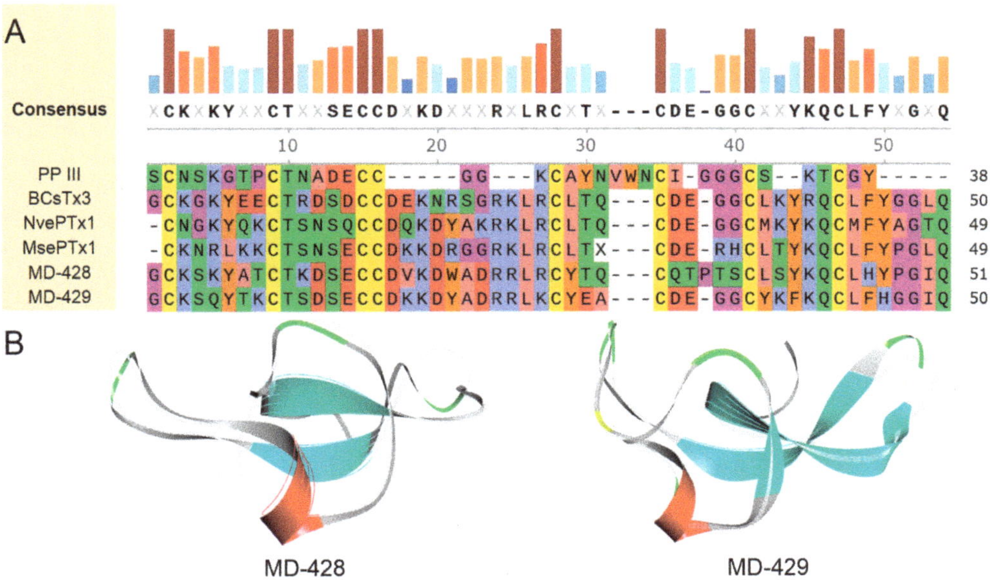

Figure 10. Deduced ICK family transcripts from *M. doreensis*. (**A**) Multiple sequence alignment of known ICK peptides and corresponding putative mature peptides. The representative ICK peptides shown are PP III (Uniport: P49268; PDB:2M36), from the trap-door spider, or Aptostichus schlingeri *Apomastus schlingeri*; BcsTx3 (Uniport: C0HJC4; AlphaFold: AF-C0HJC4-F1), from the saddle carpet anemone, or Haddon's sea anemone *Bunodosoma caissarum*; NvePTx1 (Uniport: A7RMN1; AlphaFold: AF-A7RMN1-F1), from the starlet sea anemone *Nematostella vectensis*; MsePTx1 (Uniport: P0DMD7), from the brown or frilled sea anemone *Metridium senile*. (**B**) The putative peptide models (flat representation) of ICK predicted by AlphaFold2 (the line model that overlaps with them is the best peptide obtained in the model alignment).

Based on the amino acid sequence of each putative peptide, AlphaFold2 was used for modeling to predict their respective 3D structures (Figure 10B). All prediction models include the typical structure of ICK peptides, which is composed of a helix-loop-sheet-loop-sheet configuration extending from the N-terminus to the C-terminus. Central to this conformation is a conserved hydrophobic core, characterized by an α-β-β motif. This conservation in structure suggests the potential for similar functional properties among these peptides. Specifically, the protein or peptide with the most similar structural fragment to each putative peptide was discovered. MD-429 is structurally the most similar to NvePTx1, from the starlet sea anemone *Nematostella vectensis*, with a TM-score of 0.84807 and an RMSD of 2.30. Other putative peptides-related information is attached in Table S4.

Since ICK peptides are Nav1.7 channel inhibitors, a molecular docking of a total of six ICK peptides with the Nav1.7 channel was conducted to demonstrate the potential interactions and biological potential of these peptides with the Nav1.7 channel. Among the ICK peptides identified in our transcriptome screening, MD-429 was found to form the most stable peptide–protein complex, exhibiting the potential to form extensive non-covalent bonds with the target protein (Figure 11A). Specifically, MD-429 is predicted to form 17 favorable non-covalent bonds with the receptor, including 11 hydrogen bonds, e.g., Gly327-Thr7; Cys330-Gln5; Ser321-Try6 (Figure 11B). The contact surface area between the ligand and the receptor was measured to be 583.35 and 605.31, respectively (Figure 11C). Compared with the known peptides PPIII and BCsTx3 within the same family, MD-429 exhibits a higher number of hydrogen bonds than BCsTx3, yet a lower number compared to PPIII. Furthermore, the contact surface area of the ligand and the receptor for MD-429 is greater than for either PPIII and BCsTx3. Consequently, it is speculated that MD-429 may possess a certain inhibitory effect on the Nav1.7 channel.

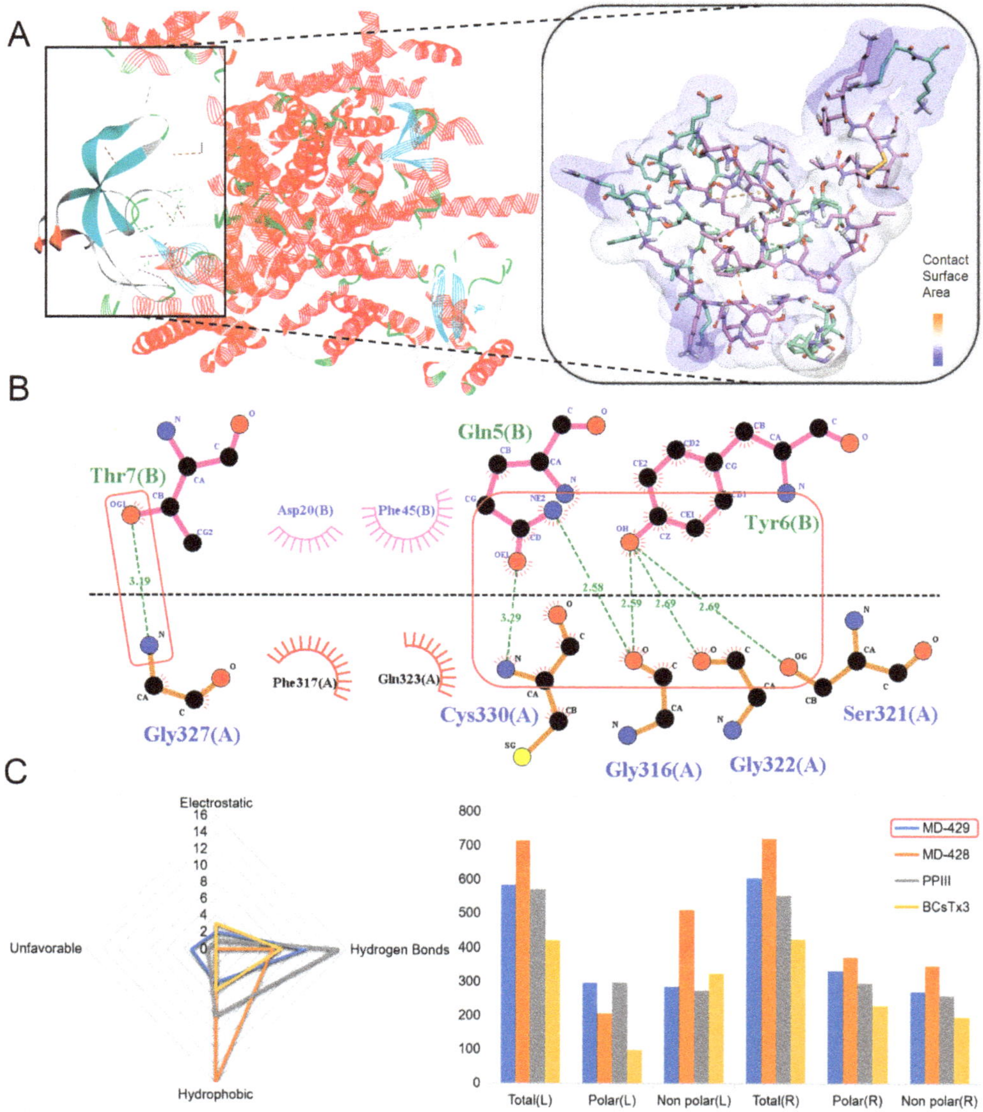

Figure 11. Molecular docking analyses revealed the binding interaction between ICK peptides and the Nav1.7 channel. (**A**) A docking pattern diagram illustrating MD-429 (flat representation) and the Nav1.7 channel (PDB: 7W9L; A chain) (line representation), along with a more comprehensive 3D diagram displaying the docking locations and contact surface area of MD-429 (colored purple) and the Nav1.7 channel (colored green). (**B**) An eyelash diagram depicting the key docking sites of MD-429 (colored blue) and the Nav1.7 channel (colored green). (**C**) A radar chart illustrating the number of non-bond interactions between ICK peptides and the Nav1.7 channel, along with a 3D surface plot depicting the contact surface area.

2.4.6. Kazal-like

In the peptides of sea anemones, serine protease inhibitors predominantly fall under Kazal-like and Kunitz-type categories. We mainly introduce Kazal-like peptides here. This type of peptide is a single-chain peptide rich in cysteine residues in sea anemones, generally composed of about 50 amino acid residues. It has three pairs of disulfide bonds formed by six cysteine residues that are cross-linked with each other. The connection positions of the disulfide bonds within the molecule are Cys1-Cys5, Cys2-Cys4, and Cys3-Cys6, forming the cysteine pattern CX3C-C-C-C-C-C [56–58].

This study selected three representative peptides from known Kazal-like peptides for comparative analysis (AEI, TATI, and H3G689). Amino acid sequence analysis has revealed two putative peptides, MD-350 and MD-351, which exhibit close alignment with AEI, TATI, and H3G689, thereby highlighting the conserved structural motifs characteristic of this family of serine protease inhibitors (Figure S4A). Based on the amino acid sequence of each putative peptide, AlphaFold2 was used for modeling to predict their respective 3D structures. All prediction models possess the basic structural framework characteristics of Kazal-like peptides, as well as a conservative β-hairpin motif and a hydrophobic core α-β-β motif (Figure S4B). Since Kazal-like peptides are serine protease inhibitors, a molecular docking of a total of four Kazal-like peptides with the serine proteinase receptor was conducted to demonstrate the potential interactions and biological potential of these peptides with the serine proteinase receptor. Among the Kazal-like peptides identified in our transcriptome screening, MD-350 was found to form the most stable peptide–protein complex. Predictions indicate the formation of 13 favorable non-bonding interactions, comprising 10 hydrogen bonds, akin to the outcomes observed for AEI and TATI. Consequently, MD-350 potentially possesses a certain inhibitory effect on serine protease (Figure S5A).

2.4.7. EGF

Most EGF peptides are composed of around 50 amino acids, predominantly found within animal proteins. They contain three pairs of conserved disulfide bonds, typically formed between cysteine residues at the Cys1-Cys3, Cys2-Cys4, and Cys5-Cys6 positions. The defining architectural element of this family is a central double-stranded β-sheet, bordered by the Cys1-Cys3 and Cys2-Cys4 regions, which gives rise to a β-hairpin structure [59,60].

This study selected seven representative peptides from known EGF peptides for comparative analysis (Mgla, Gigt I, Gigt 4, BCs III 15.09, SHTX-5, human EGF, and mouse EGF). Through amino acid sequence analysis, nine putative peptides, designated MD-316 through MD-324, were identified. The analysis results indicate that Gigt I, Gigt 4, BCs III 15.09, and SHTX-5 exhibit the highest sequence similarity and identity (Figure S6A). Based on the amino acid sequence of each putative peptide, AlphaFold2 was used for modeling to predict their respective 3D structures. All prediction models encompassed β-sheets and an α-helix, emphasizing a consistently preserved structural core featuring α-β-β motifs, characteristic of EGF peptides. Since EGF peptides are epidermal growth factor receptor (EGFR) inhibitors, a molecular docking of all the above EGF peptides with EGFR was conducted to demonstrate the potential interactions and biological potential of these peptides with EGFR. Among the EGF peptides identified in our transcriptome screen, MD-323 was found to form the most stable peptide–protein complex, and MD-323 was discovered to constitute the most robust peptide–protein complex, with the prediction of the formation of 11 favorable non-covalent bonds, among which 10 are hydrogen bonds, resembling the outcomes observed for human EGF. Consequently, MD-323 potentially exhibits a certain inhibitory effect on EGFR (Figure S7A).

2.4.8. Waprin

The waprin peptides represent a class of cationic peptides that are rich in disulfide bonds. They are distinguished by eight cysteine residues forming four intramolecular disulfide bridges, which give rise to the so-called four-disulfide core. These conserved cysteine residues are crucial for the formation of the disulfide bonds, and they typically follow the arrangement Cys1-Cys6, Cys2-Cys7, Cys3-Cys5, and Cys4-Cys8, conforming to the cysteine pattern C-C-C-C-CCX3CX3C [61,62].

This study selected four representative peptides from known waprin peptides for comparative analysis (omwaprin-a, supwaprin-a-like, supwaprin-a, and nawaprin). The homology analysis of the primary structure confirmed that the MD-436 and MD-437 putative peptides belong to the waprin family. MD-437 exhibits the highest sequence similarity and identity to the supwaprin-a-like peptide (Figure S8A). Based on the amino acid sequence of each putative peptide, AlphaFold2 was used for modeling to predict their respective 3D structures. All predictive models exhibited a fundamental structural framework characteristic of waprin peptides, which includes a sequence of loop, α-helix, β-turn, β-sheet, and β-sheet from the N-terminus to the C-terminus (Figure S8B). Furthermore, all of these models consistently featured a conserved β-hairpin motif, as well as a hydrophobic core α-β-β motif.

3. Discussion

In this study, HTS was used to conduct a comprehensive transcriptomic analysis of the tentacles and column of *M. doreensis* to explore the distribution characteristics and diversity of proteins and peptides, thereby establishing a functional database of putative proteins and peptides. Transcriptomics and bioinformatics were deeply combined to predict the family, 3D structure, and potential activity of putative peptides based on the primary (amino acid sequence analysis), secondary and tertiary (AlphaFold2 modeling), and quaternary structures (molecular docking) of the protein in order to screen putative peptides with potentially high activity.

With the aid of the UniProt database and a focus on primary structure homology, we achieved a comprehensive alignment of amino acid sequences. This database contains a total of 4239 sequences of putative proteins and peptides, which were divided into 216 known families. Most putative protein components matched those of immunoglobulins and a spectrum of protein kinases. The putative peptide components were related to the ShK domain, EGF, Kazal-like, β-defensins, Kunitz-type, and insulin-like peptides, echoing the characteristic venom components commonly identified in a multitude of known sea anemones [63]. Our data shows that within the tentacles, the prevalence of the primary constituents of sea anemone venom, such as the ShK domain, EGF, Kazal-like, and Kunitz-type peptides, is markedly higher than that noted in the column. Conversely, the immunoglobulin content in this column is more than twice that of the tentacles. This distribution pattern suggests a functional difference, in which the tentacles are usually used as hunting tools for sea anemones, injecting peptides with various biological activities such as paralysis and lethality into the prey through stinging cells, mainly related to predation behaviors [21–29]. The column is more focused on protective mechanisms, where the high level distribution of immunoglobulin plays a crucial role in resisting invading pathogens or foreign antigens [64].

The four common sea anemones in the South China Sea are *M. doreensis*, *H. crispa*, *E. diaphana*, and *H. magnifica*. Transcriptomic analysis shows that their peptide proportions are 18.7%, 9%, 27.02%, and 13.46%, respectively. They contain 12, 11, 11, and 13 ShK domain peptides; 7, 7, 10, and 12 Kunitz-type peptides; and 2, 11, 5, and 23 β-defensins peptides in the typical peptide families [17–19]. Through the above comparative analysis, it was discovered that teh ShK domain and Kunitz-type peptides are highly distributed in the sea anemones mentioned above, indicating that these two may be important peptides for the survival of sea anemone species. The high distribution of β-defensins peptides only noted in *H. magnifica* suggests that different sea anemones also exhibit diversity in their selection of peptides during their own evolution. The ShK domain peptides exhibit various biological activities, such as analgesia and lethality, and are more focused on competition and hunting [21–24], while the β-defensins peptides play a crucial role in immune responses and are more focused on defense and self-protection [38–41], and the

Kunitz-type peptides have dual functions [32–35]. Furthermore, a phylogenetic analysis was conducted by integrating the ShK domain peptides, Kunitz-type peptides, and β-defensins peptides of these four sea anemones (Figure S9). In regards to the ShK domain peptides, *M. doreensis* is more closely related to *E. diaphana*.

Extensive identification of putative peptides across different regions revealed highly diverse peptides, particularly those forming multiple disulfide bonds. Based on amino acid sequence homology, this study identified 40 putative peptides across eight families, including the ShK domain, Kunitz-type, and Kazal-like peptides, among others. Our initial screening for these putative peptides was based on the homology to known peptide sequences within the same family. The 3D structure characterized intrinsic structural motifs, providing a clearer understanding of the molecular frameworks of these putative peptides. Subsequently, the binding strength of the quaternary structure was predicted using molecular docking technology and was compared with that of known analogues to evaluate the activity of the putative peptide. This visualization approach demonstrates the application of computational biology and structural bioinformatics in investigating molecular interactions and predicting conformations within biological functions.

The molecular underpinnings unveil an intricate map of peptide sequences spanning multiple families, each with unique biological characteristics and potential pharmacological applications. We elaborate on the relevant findings within eight major peptide families, including the ShK domain, Kunitz-type, β-defensins, and ICK. First, cysteine residues within the amino acid sequence were utilized as the primary criterion for categorizing peptides due to the fundamental role that the primary structure plays in predicting putative peptide structure. Highly conserved among various peptide families, cysteine residues are crucial for the formation of disulfide bonds that confer stability to the peptide's tertiary structure, maintaining specific conformations and spatial arrangements which are critical for determining peptide stability and functional specificity. The distribution characteristics and resultant effects of other amino acid residues within the peptide sequences of distinct families have also been investigated. For example, the occurrence of hydrophobic amino acids such as leucine, glycine, valine, and isoleucine (L/G/V/I), which are more prevalent in the β-defensins family than in others, correlated with the formation and maintenance of their characteristic defensin motif and the consequent broad-spectrum and salt-stable antimicrobial activity [65]. Similarly, an elevated distribution of lysine (Lys) has been observed in peptides belonging to the Kunitz-type family, which has been associated with their capacity to bind with the Kv1.1 channel [66–68].

The 3D structure complements these findings by providing a framework for understanding how motifs within these putative peptides might fold into biologically active conformations. For example, the 3D structural model of Kunitz-type peptides is relatively stable, highlighting the conserved cysteine pattern and providing support for their ability to inhibit serine proteases and the Kv1.2 channel [69,70]. Similarly, the β-hairpin structure appeared as a frequent feature, hinting at the evolutionary conservation of these molecules, both within and outside the β-defensins [71,72]. The 3D structure of the putative peptides belonging to the eight major families demonstrates a certain level of conservation. Specifically, the ShK domain is characterized by a helix-loop-helix motif. In contrast, both β-defensins and BBH exhibit a conserved core structure, primarily composed of β-folds. The EGF peptides stand out due to their distinctive double-stranded β-sheet architecture, where β-sheets serve as the fundamental structural backbone. Furthermore, the α-β-β motif consistently appears within the 3D architecture of the putative peptides from the Kunitz-type, Kazal-like, ICK, and waprin families, creating a stable hydrophobic core typically structured by an α-helix positioned between two β-strands that may be either parallel or antiparallel. Upon aligning the models of all the putative peptides, we found that the fragments of proteins or peptides with the most similar 3D structures to those of the putative peptides predominantly originate from various types of sea anemones, with a smaller proportion derived from marine organisms such as sea cucumbers and sponges. Notably, there are some exceptional cases, such as the optimal putative peptide MD-312 from the β-defensins family, whose corresponding peptide is derived from terrestrial chickens, highlighting the diversity of the distribution of β-defensins in terrestrial and marine organisms [37–39]. In addition, all putative peptides and their corresponding peptides or protein fragments exhibit consistent cysteine patterns, indicating that cysteine indeed plays a very important role in maintaining the 3D structure of proteins [21,31,37,46,51,56–62]. The β-strands, through hydrogen bonding, create a fixed stratified structure, while interactions between side chains of the α-helix and β-sheet (involving hydrogen bonds, electrostatic forces, and hydrophobic interactions) contribute to enhanced stability in the peptide architecture and also play a role in its functionality. For instance, the α-helix may be involved in substrate recognition or the formation of catalytic sites, and this particular structure is found in numerous enzymes with catalytic activity [73]. The Kunitz-type and Kazal-like peptides, which act as serine protease inhibitors, share a highly conserved 3D structure (an α-helix and a multi-stranded anti-parallel β-sheet, with a stable central hydrophobic core) [18,19].

Through detailed molecular docking simulations, we assess the binding strength of the quaternary structure and analyze putative peptides that surpass known peptides in predicting activity. The ShK domain peptides primarily inhibit the Kv channels, with ShK and BgK serving as representative peptides of their family, both exhibiting inhibitory effects on the Kv1.3 channel [22,25]. Molecular docking simulations have differentiated between peptides within the ShK domain. MD-384 and MD-374 demonstrate a high affinity for the Kv1.3 channel and can form relatively stable peptide–protein complexes. Notably, MD-374 is predicted to form 27 favorable non-covalent bonds, including 18 hydrogen bonds, which are indicators of its potent inhibitory capability on the Kv 1.3 channel. Pharmaceuticals used to treat autoimmune disorders often cause systemic immunosuppressive effects, increasing the risk of opportunistic infections and neoplastic progression. By targeting the Kv1.3 channel, ShK and its derivatives specifically suppress CCR7-TEM lymphocytes, providing a strategy to mitigate widespread immunosuppression [23,74]. Molecular docking simulations indicate that MD-374 exhibits a potentially superior binding affinity for the Kv1.3 channel compared to that of ShK and BgK. Consequently, MD-374 may exhibit stronger inhibitory effects on the Kv1.3 channel, positioning it as a more promising lead compound for the treatment of autoimmune diseases than either ShK or BgK. Kunitz-type peptides, originating from the alkaline protease inhibitor family, have been identified across various taxonomic landscapes, indicating a ubiquitous biological role [75]. The bifunctional nature of Kunitz-type peptides, which involves serine protease inhibition and voltage-gated ion channel antagonism, represents a multifaceted mechanism potentially employed for predation and defense [76,77]. The complex formed by MD-364 and serine protease is the most stable, predicted to establish 16 favorable non-covalent bonds, including 10 hydrogen bonds. This finding is consistent with the known peptides of the Kunitz-type family evaluated in this study (Figure 7). Furthermore, MD-364 is anticipated to exhibit some binding affinity for the Kv1.2 channel, which is associated with pain [34]. Consequently, MD-364 emerges as a promising lead compound for the development of multifaceted drugs, focusing on serine protease inhibition and pain alleviation. β-defensins peptides play a crucial role in immune responses, exhibiting broad-spectrum antimicrobial and immunomodulatory activities [78,79]. MD-312 has been identified as a peptide belonging to the β-defensins family. It is speculated that MD-312 can form relatively stable protein–peptide complexes with the Kv1.2 or Nav1.7 channels, which may exert an inhibitory effect on both channels (Figure 9). Both the Kv1.2 and Nav1.7 channels are related to the process of pain production; consequently, MD-312 may have a strong analgesic effect. ICK peptides, characterized by four pairs of disulfide bonds, play a key role in modulating the Nav and Cav channels [51–53]. MD-428 and MD-429 have been identified as peptides belonging to this family. Based on the predicted docking results, these putative peptides can form stable complexes with the Nav1.7 channel (Figure 11), potentially exerting a strong inhibitory effect on the Nav1.7 channel. This could provide valuable insights for the development of analgesics, as the Nav1.7 channel is associated with the pain-production process. The advantage of these biologically derived peptide substances is that they can potentially avoid many of the side effects commonly seen with clinically used analgesics by inhibiting the ion channels [80]. Molecular docking simulation serves as a key bridge between structural prediction and practical biological significance [81], with molecular docking technology used to compare putative peptides within the family and highlight top candidate peptides from different families.

4. Materials and Methods

4.1. Specimen Collection

These sea anemones were collected from the Paracel Islands (7.8 km^2), located at [15°46′ N, 111°11′ E] in the South China Sea. A total of three sea anemones were collected and identified as *M. doreensis*. The sea anemone *M. doreensis* has been identified based on its external morphological features. It closely resembles the sea anemone *Cribinopsis crassa*, but the tentacles of *Cribinopsis crassa* are more numerous and stiffer. Its mouthparts are usually visible, and the adult diameter can usually reach up to 50 cm. Its tentacles are spaced farther apart than are those of similar sea anemones. They present in various colors, but the base is usually orange to red. The specimens were cultivated in an indoor water tank for at least three weeks, and then tweezers and a surgical knife were used to cut the tentacles and columnar tissues of three individuals, which were then mixed separately before being stored in liquid nitrogen.

4.2. Transcriptome Construction and Quality Checking

Total RNA was extracted using the RNeasy Mini Kit (Qiagen, Duesseldorf, North Rhine Westphalia, Germany). The Illumina sequencing library was constructed using the NEBNext® Ultra™ RNA Library Prep Kit for Illumina® (NEB, Beverly, MA, USA), and the qualified libraries were sequenced using the Illumina Novaseq6000 (Illumina, San Diego, CA, USA) high-throughput sequencing platform. The sequencing strategy was PE150 (Pair End 150), and the sequencing data volume for each sample was not less than 6 Gb [82,83].

4.3. Gene Annotation

The prediction of the coding region of the unigene uses three types of forward and reverse reading frames, which may generate a total of six types of encoded protein sequences to be predicted. After obtaining the encoded protein sequences, they were compared with the Nr (non-redundant protein sequences database) and the UniProt protein database. The encoding method of the gene with the best comparison (maximum comparison score) is selected. To identify potential toxin transcripts, the translated transcriptome data are cross-aligned with the assumed animal toxin sequences, retrieving information from the five major public databases (Nr, UniProt, KEGG, GO, and KOG/COG) [84–86].

4.4. Classification of Proteins and Peptides of *M. doreensis*

To identify the protein family (domain) in *M. doreensis*, we first retrieved the *M. doreensis* genome sequences, including 287,880 protein-coding genes from the Pfam database (https://pfam-legacy.xfam.org/, accessed on 8 October 2024) using local BLASTP (with an e-value of 1×10^{-10}, -max_target_seqs 2). The related protein-coding genes from *M. doreensis* were divided by SignalP 5.0 (https://services.healthtech.dtu.dk/services/SignalP-5.0/, accessed on 1 September 2024) as signal and mature regions. The putative proteins and peptides must be sequences containing predicted signal regions and mature regions. A total of 4239 protein-coding genes were collected and then aligned using MUSCLE [87] to build a hidden den markov model (HMM) profile using hmmbuild in HMMER v3.0 (http://hmmer.janelia.org/static/binaries/hmmer3.0_windows.zip, accessed on 4 October 2024). The complete proteins family (domain) in *M. doreensis* was identified using hmmsearch program (with an e-value of 1×10^{-3}) by searching the full protein-coding gene sequences of the *M. doreensis* genome, based on the HMM profile. The candidate proteins were further confirmed using Pfam [88] programs, combining them with their corresponding cysteine frameworks to predict the family of sea anemone proteins and peptides. The redundant proteins were manually removed (Table S2). In our current research, peptides with less than 120 amino acids are defined as peptides.

4.5. Phylogenetic Analyses

The mature regions of 40 putative peptides were aligned using MEGA 7.0.14. A phylogenetic tree was established using the Maximum Likelihood (Bootstrap Method 2000).

4.6. Alignment

Multiple sequence alignment was performed using Snapgene Viewer (aligned using MUSCLE [87], with the results marked with color highlighting based on the physicochemical properties and the consensus, with a threshold of >50%), and the sequence was exported in the FASTA format [89]. The sequence homology (similarity and identity) was analyzed using the manual calculation method.

4.7. AlphaFold2 Modeling and Model Alignment

We logged in to the AlphaFold Colab website and entered the sequence of the estimated protein to obtain its model [90] (https://colab.research.google.com/github/sokrypton/ColabFold/blob/main/AlphaFold2.ipynb, accessed on 6 October 2024). Using the Foldseek tool for model alignment, we accessed the Foldseek website [91] (https://search.foldseek.com/search, accessed on 8 October 2024), uploaded the model to be aligned (in PDB or CIF format), opened the Databases & Search settings, selected eight databases (AlphaFold/Proteome, AlphaFold/Swiss-Prot, AlphaFold/UniProt50, BFMD, CATH50, GMGCL, MGnify-ESM30, PDB100), selected the 3Di/AA mode, and clicked "search" to get the results.

4.8. Molecular Docking and Eyelash Diagram

The interaction between the key targets and the main active components was further investigated using molecular docking. Energy minimization of the 3D structures was performed using ChemBio 3D. The crystal structures of the receptors were downloaded from the RCSB Protein Data Bank (http://www.pdb.org/, accessed on 8 October 2024) and subsequently modified using Discovery Studio 2019. The modifications included hydrogen addition, ligand and water removal, amino acid optimization, and charge computation. Discovery Studio 2019 was employed for the docking and visualization of the docking results. The required functional modules include Dock Proteins with ZDOCK (rigid protein docking algorithm) and Refinement with RDOCK (evaluating the binding configuration of the complex), and the data files are the receptor proteins and peptides. The ZDOCK parameter is set to Angular Step Size 15, with a sample number of 3600 binding configurations (when the sampling angle is selected as 6 degrees, there are 54,000 samples, resulting in more accurate prediction results, but requiring more time). In addition, for such a small binding configuration, for improved results, the RMSD Cutoff should be set to 6.0, the Interface Cutoff to 9.0, and the Maximum Number of Clusters to 60. The Zrank parameter should be set to False, while the other parameters will remain at their default values. Select the pose with the highest score from the 10 largest clusters in the ZDOCK results required by the RDOCK (default parameters). The smaller the RDOCK score, the closer it is to the true docking conformation. The results were analyzed using LigPlot+ to evaluate the accurate binding interactions between the ligands and the receptors. All the targets and their PDB IDs used in the modeling are included in Table S3.

5. Conclusions

This study reports, for the first time, a comprehensive transcriptomic analysis of different tissues of *Macrodactyla doreensis*. A total of 4239 transcripts were identified, and a database of putative proteins (81.53%) and peptides (18.47%) was preliminarily constructed. Immunoglobulin, mainly distributed in the column, comprised the greatest content of the putative protein, along with putative peptides containing the ShK domain, which were mainly distributed in the tentacle. Bioinformatics was further integrated to screen for putative peptides with potentially high activity. The optimal peptides selected from this database are MD-374, MD-364, MD-312, and MD-429, which belong to the ShK domain, Kunitz-type, β-defensins, and ICK families, respectively. It is through such research that we can bridge the gap between the sequence, structure, and target, providing direction for further in-depth research on the peptides of *Macrodactyla doreensis*.

Supplementary Materials: The following supporting information can be downloaded at: https://www.mdpi.com/article/10.3390/md22100470/s1, Table S1. Sequencing statistics and assembly summary of *M. doreensis* transcriptome; Table S2. Classification of proteins and peptides of *M. doreensis*; Table S3. The targets and their PDB IDs used in docking; Table S4. The protein or peptide exhibiting the most structurally similar fragment to each of the putative peptides; Figure S1. The number of gene annotations between different databases (GO; KEGG; Nr; KOG) for the *M. doreensis* transcriptome; Figure S2. Deduced BBH transcripts from *M. doreensis*; Figure S3. Molecular docking analyses revealed the binding interaction between BBH peptides and the Kv1.1 channel; Figure S4. Deduced Kazal-like transcripts from *M. doreensis*; Figure S5. Molecular docking analyses revealed the binding interaction between Kazal-like peptides and the serine proteinase receptor; Figure S6. Deduced EGF transcripts from *M. doreensis*; Figure S7. Molecular docking analyses revealed the binding interaction between EGF peptides and EGFR; Figure S8. Deduced waprin transcripts from *M. doreensis*; Figure S9. Phylogenetic analysis of the putative ShK domain, Kunitz-type, and β-defensins peptides derived from four common sea anemones in the South China Sea.

Author Contributions: Conceptualization, B.G. and M.H.; methodology, Z.H. and Y.L.; data curation, Q.X., X.L. and L.L.; writing—original draft preparation, Z.H. and B.G.; writing—review and editing, J.F. and M.H.; funding acquisition, B.G. All authors have read and agreed to the published version of the manuscript.

Funding: This research was supported by the National Natural Science Foundation of China (no. 82060686), the Hainan Provincial Key Point Research and Invention Program (no. ZDYF2022SHFZ309), the Hainan Provincial Natural Science Foundation of China (no. 424QN258), and by a special scientific research project of the Hainan Academician Innovation Platform (no. YSPTZX202132).

Institutional Review Board Statement: Not applicable.

Informed Consent Statement: Not applicable.

Data Availability Statement: The transcriptome data of *M. doreensis* have been uploaded to the National Center for Biotechnology Information (NCBI) repository (BioProject: PRJNA1134430; SRA accession: SRR29774726 and SRR29774727).

Conflicts of Interest: The authors declare no conflicts of interest.

References

1. Prentis, P.J.; Pavasovic, A.; Norton, R.S. Sea Anemones: Quiet Achievers in the Field of Peptide Toxins. *Toxins* **2018**, *10*, 36. [CrossRef] [PubMed]
2. Jouiaei, M.; Yanagihara, A.A.; Madio, B.; Nevalainen, T.J.; Alewood, P.F.; Fry, B.G. Ancient Venom Systems: A Review on Cnidaria Toxins. *Toxins* **2015**, *7*, 2251–2271. [CrossRef] [PubMed]
3. Garese, A.; Goes Correa, F.; Acuña, F.H.; Stampar, S.N. Cnidom in Ceriantharia (Cnidaria, Anthozoa): New findings in the composition and micrometric variations of cnidocysts. *PeerJ* **2023**, *11*, e15549. [CrossRef] [PubMed]
4. La Corte, C.; Catania, V.; Dara, M.; Parrinello, D.; Staropoli, M.; Trapani, M.R.; Cammarata, M.; Parisi, M.G. Equinins as Novel Broad-Spectrum Antimicrobial Peptides Isolated from the Cnidarian *Actinia equina* (Linnaeus, 1758). *Mar. Drugs* **2024**, *22*, 172. [CrossRef] [PubMed]
5. Domínguez-Pérez, D.; Campos, A.; Alexei Rodríguez, A.; Turkina, M.V.; Ribeiro, T.; Osorio, H.; Vasconcelos, V.; Antunes, A. Proteomic Analyses of the Unexplored Sea Anemone Bunodactis verrucosa. *Mar. Drugs* **2018**, *16*, 42. [CrossRef]
6. Brinzeu, A.; Berthiller, J.; Perreton, N.; Subtil, F.; Gervaise, C.; Luaute, J.; Mertens, P. SPIDOL study protocol for the assessment of intrathecal ziconotide antalgic efficacy for severe refractory neuropathic pain due to spinal cord lesions. *Trials* **2024**, *25*, 595. [CrossRef]

7. Tarcha, E.J.; Olsen, C.M.; Probst, P.; Peckham, D.; Muñoz-Elías, E.J.; Kruger, J.G.; Iadonato, S.P. Safety and pharmacodynamics c dalazatide, a Kv1.3 channel inhibitor, in the treatment of plaque psoriasis: A randomized phase 1b trial. *PLoS ONE* **2017**, *1* e0180762. [CrossRef]
8. Mitchell, M.L.; Hossain, M.A.; Lin, F.; Pinheiro-Junior, E.L.; Peigneur, S.; Wai, D.C.C.; Delaine, C.; Blyth, A.J.; Forbes, B.E.; Tytga J.; et al. Identification, Synthesis, Conformation and Activity of an Insulin-like Peptide from a Sea Anemone. *Biomolecules* **2021**, *1* 1785. [CrossRef]
9. Elnahriry, K.A.; Wai, D.C.C.; Ashwood, L.M.; Naseem, M.U.; Szanto, T.G.; Guo, S.; Panyi, G.; Prentis, P.J.; Norton, R.S. Structura and functional characterisation of Tst2, a novel TRPV1 inhibitory peptide from the Australian sea anemone *Telmatactis stephenson* *Biochim. Biophys. Acta Proteins Proteom.* **2024**, *1872*, 140952. [CrossRef]
10. Gladkikh, I.N.; Klimovich, A.A.; Kalina, R.S.; Kozhevnikova, Y.V.; Khasanov, T.A.; Osmakov, D.I.; Koshelev, S.G.; Monastyrnay M.M.; Andreev, Y.A.; Leychenko, E.V.; et al. Anxiolytic, Analgesic and Anti-Inflammatory Effects of Peptides Hmg 1b-2 and Hm 1b-4 from the Sea Anemone *Heteractis magnifica*. *Toxins* **2023**, *15*, 341. [CrossRef]
11. Mackieh, R.; Abou-Nader, R.; Wehbe, R.; Mattei, C.; Legros, C.; Fajloun, Z.; Sabatier, J.M. Voltage-Gated Sodium Channels: Prominent Target of Marine Toxins. *Mar. Drugs* **2021**, *19*, 562. [CrossRef] [PubMed]
12. Zhao, R.; Qasim, A.; Sophanpanichkul, P.; Dai, H.; Nayak, M.; Sher, I.; Chill, J.; Goldstein, S.A.N. Selective block of human Kv1. channels and an epilepsy-associated gain-of-function mutation by AETX-K peptide. *FASEB J. Off. Publ. Fed. Am. Soc. Exp. Bio* **2024**, *38*, e23381. [CrossRef]
13. Ferreira Junior, W.A.; Zaharenko, A.J.; Kazuma, K.; Picolo, G.; Gutierrez, V.P.; de Freitas, J.C.; Konno, K.; Cury, Y. Periphera 5-HT3 Receptors Are Involved in the Antinociceptive Effect of Bunodosine 391. *Toxins* **2017**, *10*, 12. [CrossRef] [PubMed]
14. Kalina, R.S.; Gladkikh, I.N.; Klimovich, A.A.; Kozhevnikova, Y.V.; Kvetkina, A.N.; Rogozhin, E.A.; Koshelev, S.G.; Kozlov S.A.; Leychenko, E.V. First Anti-Inflammatory Peptide AnmTX Sco 9a-1 from the Swimming Sea Anemone Stomphia coccinea *Biomolecules* **2022**, *12*, 1705. [CrossRef]
15. García-Montoya, C.; Heras-Márquez, D.; Amigot-Sánchez, R.; García-Linares, S.; Martínez-Del-Pozo, Á.; Palacios-Ortega, Sticholysin recognition of ceramide-phosphoethanolamine. *Arch. Biochem. Biophys.* **2023**, *742*, 109623. [CrossRef]
16. Beeton, C.; Pennington, M.W.; Wulff, H.; Singh, S.; Nugent, D.; Crossley, G.; Khaytin, I.; Calabresi, P.A.; Chen, C.Y.; Gutman, G.A et al. Targeting effector memory T cells with a selective peptide inhibitor of Kv1.3 channels for therapy of autoimmune diseases *Mol. Pharmacol.* **2005**, *67*, 1369–1381. [CrossRef]
17. Fu, J.; He, Y.; Peng, C.; Tang, T.; Jin, A.; Liao, Y.; Shi, Q.; Gao, B. Transcriptome Sequencing of the Pale Anemones (*Exaiptasia diaphana*) Revealed Functional Peptide Gene Resources of Sea Anemone. *Front. Mar. Sci.* **2022**, *9*, 856501. [CrossRef]
18. Guo, Q.; Fu, J.; Yuan, L.; Liao, Y.; Li, M.; Li, X.; Yi, B.; Zhang, J.; Gao, B. Diversity analysis of sea anemone peptide toxins ir different tissues of *Heteractis crispa* based on transcriptomics. *Sci. Rep.* **2024**, *14*, 7684. [CrossRef]
19. Li, M.; Mao, K.; Huang, M.; Liao, Y.; Fu, J.; Pan, K.; Shi, Q.; Gao, B. Venomics Reveals the Venom Complexity of Sea Anemone *Heteractis magnifica*. *Mar. Drugs* **2024**, *22*, 71. [CrossRef]
20. Rojas, L.; Cabrera-Muñoz, A.; Espinosa, L.A.; Montané, S.; Alvarez-Lajonchere, L.; Mojarena, J.D.; Moya, G.; Lorenzo, J.; González L.J.; Betzel, C.; et al. CogiTx1: A novel subtilisin A inhibitor isolated from the sea anemone Condylactis gigantea belonging to the defensin 4 protein family. *Biochimie* **2023**, *213*, 41–53. [CrossRef]
21. Sanches, K.; Prypoten, V.; Chandy, K.G.; Chalmers, D.K.; Norton, R.S. Interaction of the Inhibitory Peptides ShK and HmK with the Voltage-Gated Potassium Channel K(V)1.3: Role of Conformational Dynamics. *J. Chem. Inf. Model.* **2023**, *63*, 3043–3053 [CrossRef] [PubMed]
22. Kim, C.H.; Lee, Y.J.; Go, H.J.; Oh, H.Y.; Lee, T.K.; Park, J.B.; Park, N.G. Defensin-neurotoxin dyad in a basally branching metazoan sea anemone. *FEBS J.* **2017**, *284*, 3320–3338. [CrossRef] [PubMed]
23. Beeton, C.; Pennington, M.W.; Norton, R.S. Analogs of the sea anemone potassium channel blocker ShK for the treatment of autoimmune diseases. *Inflamm. Allergy Drug Targets* **2011**, *10*, 313–321. [CrossRef]
24. Castañeda, O.; Sotolongo, V.; Amor, A.M.; Stöcklin, R.; Anderson, A.J.; Harvey, A.L.; Engström, A.; Wernstedt, C.; Karlsson, E Characterization of a potassium channel toxin from the Caribbean Sea anemone *Stichodactyla helianthus*. *Toxicon Off. J. Int. Soc. Toxinol.* **1995**, *33*, 603–613. [CrossRef]
25. Racapé, J.; Lecoq, A.; Romi-Lebrun, R.; Liu, J.; Kohler, M.; Garcia, M.L.; Ménez, A.; Gasparini, S. Characterization of a novel radiolabeled peptide selective for a subpopulation of voltage-gated potassium channels in mammalian brain. *J. Biol. Chem.* **2002**, *277*, 3886–3893. [CrossRef]
26. Schweitz, H.; Bruhn, T.; Guillemare, E.; Moinier, D.; Lancelin, J.M.; Béress, L.; Lazdunski, M. Kalicludines and kaliseptine. Two different classes of sea anemone toxins for voltage sensitive K+ channels. *J. Biol. Chem.* **1995**, *270*, 25121–25126. [CrossRef] [PubMed]
27. Gendeh, G.S.; Young, L.C.; de Medeiros, C.L.; Jeyaseelan, K.; Harvey, A.L.; Chung, M.C. A new potassium channel toxin from the sea anemone *Heteractis magnifica*: Isolation, cDNA cloning, and functional expression. *Biochemistry* **1997**, *36*, 11461–11471. [CrossRef]
28. Minagawa, S.; Ishida, M.; Nagashima, Y.; Shiomi, K. Primary structure of a potassium channel toxin from the sea anemone *Actinia equina*. *FEBS Lett.* **1998**, *427*, 149–151. [CrossRef]

29. Orts, D.J.; Peigneur, S.; Madio, B.; Cassoli, J.S.; Montandon, G.G.; Pimenta, A.M.; Bicudo, J.E.; Freitas, J.C.; Zaharenko, A.J.; Tytgat, J. Biochemical and electrophysiological characterization of two sea anemone type 1 potassium toxins from a geographically distant population of *Bunodosoma caissarum*. *Mar. Drugs* **2013**, *11*, 655–679. [CrossRef]
30. Navarro-Pérez, M.; Capera, J.; Benavente-Garcia, A.; Cassinelli, S.; Colomer-Molera, M.; Felipe, A. Kv1.3 in the spotlight for treating immune diseases. *Expert. Opin. Ther. Targets* **2024**, *28*, 67–82. [CrossRef]
31. Mishra, M. Evolutionary Aspects of the Structural Convergence and Functional Diversification of Kunitz-Domain Inhibitors. *J. Mol. Evol.* **2020**, *88*, 537–548. [CrossRef] [PubMed]
32. Minagawa, S.; Sugiyama, M.; Ishida, M.; Nagashima, Y.; Shiomi, K. Kunitz-type protease inhibitors from acrorhagi of three species of sea anemones. *Comp. Biochem. Physiol. Part B Biochem. Mol. Biol.* **2008**, *150*, 240–245. [CrossRef] [PubMed]
33. Mourão, C.B.; Schwartz, E.F. Protease inhibitors from marine venomous animals and their counterparts in terrestrial venomous animals. *Mar. Drugs* **2013**, *11*, 2069–2112. [CrossRef] [PubMed]
34. Imredy, J.P.; MacKinnon, R. Energetic and structural interactions between delta-dendrotoxin and a voltage-gated potassium channel. *J. Mol. Biol.* **2000**, *296*, 1283–1294. [CrossRef]
35. Honma, T.; Kawahata, S.; Ishida, M.; Nagai, H.; Nagashima, Y.; Shiomi, K. Novel peptide toxins from the sea anemone Stichodactyla haddoni. *Peptides* **2008**, *29*, 536–544. [CrossRef]
36. Xie, C.; Kessi, M.; Yin, F.; Peng, J. Roles of KCNA2 in Neurological Diseases: From Physiology to Pathology. *Mol. Neurobiol.* **2024**. [CrossRef]
37. Zaharenko, A.J.; Ferreira, W.A., Jr.; de Oliveira, J.S.; Konno, K.; Richardson, M.; Schiavon, E.; Wanke, E.; de Freitas, J.C. Revisiting cangitoxin, a sea anemone peptide: Purification and characterization of cangitoxins II and III from the venom of *Bunodosoma cangicum*. *Toxicon Off. J. Int. Soc. Toxinol.* **2008**, *51*, 1303–1307. [CrossRef]
38. Rodríguez, A.A.; Garateix, A.; Salceda, E.; Peigneur, S.; Zaharenko, A.J.; Pons, T.; Santos, Y.; Arreguín, R.; Ständker, L.; Forssmann, W.G.; et al. PhcrTx2, a New Crab-Paralyzing Peptide Toxin from the Sea Anemone *Phymanthus crucifer*. *Toxins* **2018**, *10*, 72. [CrossRef]
39. Sintsova, O.; Gladkikh, I.; Kalinovskii, A.; Zelepuga, E.; Monastyrnaya, M.; Kim, N.; Shevchenko, L.; Peigneur, S.; Tytgat, J.; Kozlovskaya, E.; et al. Magnificamide, a β-Defensin-Like Peptide from the Mucus of the Sea Anemone *Heteractis magnifica*, Is a Strong Inhibitor of Mammalian α-Amylases. *Mar. Drugs* **2019**, *17*, 542. [CrossRef]
40. Yeung, S.Y.; Thompson, D.; Wang, Z.; Fedida, D.; Robertson, B. Modulation of Kv3 subfamily potassium currents by the sea anemone toxin BDS: Significance for CNS and biophysical studies. *J. Neurosci. Off. J. Soc. Neurosci.* **2005**, *25*, 8735–8745. [CrossRef]
41. Oliveira, J.S.; Zaharenko, A.J.; Ferreira, W.A., Jr.; Konno, K.; Shida, C.S.; Richardson, M.; Lúcio, A.D.; Beirão, P.S.; de Freitas, J.C. BcIV, a new paralyzing peptide obtained from the venom of the sea anemone *Bunodosoma caissarum*. A comparison with the Na+ channel toxin BcIII. *Biochim. Biophys. Acta* **2006**, *1764*, 1592–1600. [CrossRef] [PubMed]
42. Wulff, H.; Castle, N.A.; Pardo, L.A. Voltage-gated potassium channels as therapeutic targets. *Nat. Rev. Drug Discov.* **2009**, *8*, 982–1001. [CrossRef] [PubMed]
43. Matsumura, K.; Shimomura, T.; Kubo, Y.; Oka, T.; Kobayashi, N.; Imai, S.; Yanase, N.; Akimoto, M.; Fukuda, M.; Yokogawa, M.; et al. Mechanism of hERG inhibition by gating-modifier toxin, APETx1, deduced by functional characterization. *BMC Mol. Cell Biol.* **2021**, *22*, 3. [CrossRef] [PubMed]
44. Castañeda, O.; Harvey, A.L. Discovery and characterization of cnidarian peptide toxins that affect neuronal potassium ion channels. *Toxicon Off. J. Int. Soc. Toxinol.* **2009**, *54*, 1119–1124. [CrossRef]
45. Kamei, T.; Kudo, T.; Yamane, H.; Ishibashi, F.; Takada, Y.; Honda, S.; Maezawa, Y.; Ikeda, K.; Oyamada, Y. Unique electrophysiological property of a novel Nav1.7, Nav1.8, and Nav1.9 sodium channel blocker, ANP-230. *Biochem. Biophys. Res. Commun.* **2024**, *721*, 150126. [CrossRef]
46. Osmakov, D.I.; Kozlov, S.A.; Andreev, Y.A.; Koshelev, S.G.; Sanamyan, N.P.; Sanamyan, K.E.; Dyachenko, I.A.; Bondarenko, D.A.; Murashev, A.N.; Mineev, K.S.; et al. Sea anemone peptide with uncommon β-hairpin structure inhibits acid-sensing ion channel 3 (ASIC3) and reveals analgesic activity. *J. Biol. Chem.* **2013**, *288*, 23116–23127. [CrossRef]
47. Honma, T.; Hasegawa, Y.; Ishida, M.; Nagai, H.; Nagashima, Y.; Shiomi, K. Isolation and molecular cloning of novel peptide toxins from the sea anemone *Antheopsis maculata*. *Toxicon Off. J. Int. Soc. Toxinol.* **2005**, *45*, 33–41. [CrossRef]
48. Zaharenko, A.J.; Ferreira, W.A., Jr.; Oliveira, J.S.; Richardson, M.; Pimenta, D.C.; Konno, K.; Portaro, F.C.; de Freitas, J.C. Proteomics of the neurotoxic fraction from the sea anemone *Bunodosoma cangicum* venom: Novel peptides belonging to new classes of toxins. *Comp. Biochem. Physiology. Part D Genom. Proteom.* **2008**, *3*, 219–225. [CrossRef]
49. Logashina, Y.A.; Mosharova, I.V.; Korolkova, Y.V.; Shelukhina, I.V.; Dyachenko, I.A.; Palikov, V.A.; Palikova, Y.A.; Murashev, A.N.; Kozlov, S.A.; Stensvåg, K.; et al. Peptide from Sea Anemone *Metridium senile* Affects Transient Receptor Potential Ankyrin-repeat 1 (TRPA1) Function and Produces Analgesic Effect. *J. Biol. Chem.* **2017**, *292*, 2992–3004. [CrossRef]
50. Sun, W.B.; Fu, J.X.; Chen, Y.L.; Li, H.F.; Wu, Z.Y.; Chen, D.F. Both gain- and loss-of-function variants of KCNA1 are associated with paroxysmal kinesigenic dyskinesia. *J. Genet. Genom. Yi Chuan Xue Bao* **2024**, *51*, 801–810. [CrossRef]
51. Kasheverov, I.E.; Logashina, Y.A.; Kornilov, F.D.; Lushpa, V.A.; Maleeva, E.E.; Korolkova, Y.V.; Yu, J.; Zhu, X.; Zhangsun, D.; Luo, S.; et al. Peptides from the Sea Anemone *Metridium senile* with Modified Inhibitor Cystine Knot (ICK) Fold Inhibit Nicotinic Acetylcholine Receptors. *Toxins* **2022**, *15*, 28. [CrossRef] [PubMed]

52. Meissner, G.O.; de Resende Lara, P.T.; Scott, L.P.; Braz, A.S.; Chaves-Moreira, D.; Matsubara, F.H.; Soares, E.M.; Trevisan-Silva, D.; Gremski, L.H.; Veiga, S.S.; et al. Molecular cloning and in silico characterization of knottin peptide, U2-SCRTX-Lit2, from brown spider (*Loxosceles intermedia*) venom glands. *J. Mol. Model.* **2016**, *22*, 196. [CrossRef] [PubMed]
53. Bende, N.S.; Kang, E.; Herzig, V.; Bosmans, F.; Nicholson, G.M.; Mobli, M.; King, G.F. The insecticidal neurotoxin Aps III is an atypical knottin peptide that potently blocks insect voltage-gated sodium channels. *Biochem. Pharmacol.* **2013**, *85*, 1542–1554. [CrossRef]
54. Orts, D.J.; Moran, Y.; Cologna, C.T.; Peigneur, S.; Madio, B.; Praher, D.; Quinton, L.; De Pauw, E.; Bicudo, J.E.; Tytgat, J.; et al. BcsTx3 is a founder of a novel sea anemone toxin family of potassium channel blocker. *FEBS J.* **2013**, *280*, 4839–4852. [CrossRef] [PubMed]
55. Columbus-Shenkar, Y.Y.; Sachkova, M.Y.; Macrander, J.; Fridrich, A.; Modepalli, V.; Reitzel, A.M.; Sunagar, K.; Moran, Y. Dynamics of venom composition across a complex life cycle. *eLife* **2018**, *7*, e35014. [CrossRef]
56. Hemmi, H.; Kumazaki, T.; Yoshizawa-Kumagaye, K.; Nishiuchi, Y.; Yoshida, T.; Ohkubo, T.; Kobayashi, Y. Structural and functional study of an Anemonia elastase inhibitor, a "nonclassical" Kazal-type inhibitor from *Anemonia sulcata*. *Biochemistry* **2005**, *44*, 9626–9636. [CrossRef]
57. Xue, G.; Zou, X.; Zhou, J.Y.; Sun, W.; Wu, J.; Xu, J.L.; Wang, R.P. Raddeanin A induces human gastric cancer cells apoptosis and inhibits their invasion in vitro. *Biochem. Biophys. Res. Commun.* **2013**, *439*, 196–202. [CrossRef]
58. Huang, Y.; Zhou, W.; Liu, S.; Zeng, D.; Zhou, W. Association between polymorphisms and atopic dermatitis susceptibility: A systematic review and meta-analysis. *Gene* **2024**, *913*, 148397. [CrossRef]
59. Cuypers, E.; Peigneur, S.; Debaveye, S.; Shiomi, K.; Tytgat, J. TRPV1 Channel as New Target for Marine Toxins: Example of Gigantoxin I, a Sea Anemone Toxin Acting Via Modulation of the PLA2 Pathway. *Acta Chim. Slov.* **2011**, *58*, 735–741.
60. Sampoli Benitez, B.A.; Komives, E.A. Disulfide bond plasticity in epidermal growth factor. *Proteins* **2000**, *40*, 168–174. [CrossRef]
61. Saldarriaga-Córdoba, M.; Clavero-León, C.; Rey-Suarez, P.; Nuñez-Rangel, V.; Avendaño-Herrera, R.; Solano-González, S.; Alzate, J.F. Unveiling Novel Kunitz- and Waprin-Type Toxins in the *Micrurus mipartitus* Coral Snake Venom Gland: An In Silico Transcriptome Analysis. *Toxins* **2024**, *16*, 224. [CrossRef] [PubMed]
62. Nair, D.G.; Fry, B.G.; Alewood, P.; Kumar, P.P.; Kini, R.M. Antimicrobial activity of omwaprin, a new member of the waprin family of snake venom proteins. *Biochem. J.* **2007**, *402*, 93–104. [CrossRef] [PubMed]
63. Norton, R.S. Structures of sea anemone toxins. *Toxicon Off. J. Int. Soc. Toxinol.* **2009**, *54*, 1075–1088. [CrossRef] [PubMed]
64. Atkinson, T.P. Immunoglobulins, Structure, and Function. In *Paraproteinemia and Related Disorders*; Ragab, G., Quartuccio, L., Goubran, H., Eds.; Springer International Publishing: Cham, Germany, 2022; pp. 27–36. [CrossRef]
65. Gerdol, M.; Schmitt, P.; Venier, P.; Rocha, G.; Rosa, R.D.; Destoumieux-Garzón, D. Functional Insights From the Evolutionary Diversification of Big Defensins. *Front. Immunol.* **2020**, *11*, 758. [CrossRef]
66. Harvey, A.L.; Robertson, B. Dendrotoxins: Structure-activity relationships and effects on potassium ion channels. *Curr. Med. Chem.* **2004**, *11*, 3065–3072. [CrossRef] [PubMed]
67. AlShammari, A.K.; Abd El-Aziz, T.M.; Al-Sabi, A. Snake Venom: A Promising Source of Neurotoxins Targeting Voltage-Gated Potassium Channels. *Toxins* **2023**, *16*, 12. [CrossRef]
68. Biji, C.A.; Balde, A.; Nazeer, R.A. Anti-inflammatory peptide therapeutics and the role of sulphur containing amino acids (cysteine and methionine) in inflammation suppression: A review. *Inflamm. Res.* **2024**, *73*, 1203–1221. [CrossRef]
69. Banijamali, S.E.; Amininasab, M.; Elmi, M.M. Characterization of a new member of kunitz-type protein family from the venom of Persian false-horned viper, Pseudocerastes persicus. *Arch. Biochem. Biophys.* **2019**, *662*, 1–6. [CrossRef]
70. Yee, K.T.; Rojnuckarin, P. Complementary DNA library of Myanmar Russell's viper (*Daboia russelii siamensis*) venom gland. Comparative biochemistry and physiology. *Toxicol. Pharmacol. CBP* **2020**, *227*, 108634. [CrossRef]
71. Gao, B.; Zhu, S. An insect defensin-derived β-hairpin peptide with enhanced antibacterial activity. *ACS Chem. Biol.* **2014**, *9*, 405–413. [CrossRef]
72. Wang, X.; Li, S.; Du, M.; Liu, N.; Shan, Q.; Zou, Y.; Wang, J.; Zhu, Y. A Novel β-Hairpin Peptide Z-d14CFR Enhances Multidrug-Resistant Bacterial Clearance in a Murine Model of Mastitis. *Int. J. Mol. Sci.* **2022**, *23*, 4617. [CrossRef] [PubMed]
73. Shao, X.; Wang, C.; Wang, C.; Bai, M.; Hou, T.; Wang, X.; Yan, C.; Guan, P.; Hu, X. Novel photocatalytic carbon dots: Efficiently inhibiting amyloid aggregation and quickly disaggregating amyloid aggregates. *Nanoscale* **2024**, *16*, 8074–8089. [CrossRef] [PubMed]
74. Selvakumar, P.; Fernández-Mariño, A.I.; Khanra, N.; He, C.; Paquette, A.J.; Wang, B.; Huang, R.; Smider, V.V.; Rice, W.J.; Swartz, K.J.; et al. Structures of the T cell potassium channel Kv1.3 with immunoglobulin modulators. *Nat. Commun.* **2022**, *13*, 3854. [CrossRef] [PubMed]
75. Gladkikh, I.; Peigneur, S.; Sintsova, O.; Lopes Pinheiro-Junior, E.; Klimovich, A.; Menshov, A.; Kalinovsky, A.; Isaeva, M.; Monastyrnaya, M.; Kozlovskaya, E.; et al. Kunitz-Type Peptides from the Sea Anemone *Heteractis crispa* Demonstrate Potassium Channel Blocking and Anti-Inflammatory Activities. *Biomedicines* **2020**, *8*, 473. [CrossRef]
76. Kvetkina, A.; Leychenko, E.; Chausova, V.; Zelepuga, E.; Chernysheva, N.; Guzev, K.; Pislyagin, E.; Yurchenko, E.; Menchinskaya, E.; Aminin, D.; et al. A new multigene HCIQ subfamily from the sea anemone *Heteractis crispa* encodes Kunitz-peptides exhibiting neuroprotective activity against 6-hydroxydopamine. *Sci. Rep.* **2020**, *10*, 4205. [CrossRef]

7. Sintsova, O.; Gladkikh, I.; Monastyrnaya, M.; Tabakmakher, V.; Yurchenko, E.; Menchinskaya, E.; Pislyagin, E.; Andreev, Y.; Kozlov, S.; Peigneur, S.; et al. Sea Anemone Kunitz-Type Peptides Demonstrate Neuroprotective Activity in the 6-Hydroxydopamine Induced Neurotoxicity Model. *Biomedicines* **2021**, *9*, 283. [CrossRef]
8. Biswas, S.; Sarojini, S.; Jayaram, S.; Philip, I.; Umesh, M.; Mascarenhas, R.; Pappuswamy, M.; Balasubramanian, B.; Arokiyaraj, S. Understanding the Role of Antimicrobial Peptides in Neutrophil Extracellular Traps Promoting Autoimmune Disorders. *Life* **2023**, *13*, 1307. [CrossRef]
9. Li, S.; Mu, R.; Guo, X. Defensins regulate cell cycle: Insights of defensins on cellular proliferation and division. *Life Sci.* **2024**, *349*, 122740. [CrossRef]
0. Yu, X.; Zhao, X.; Li, L.; Huang, Y.; Cui, C.; Hu, Q.; Xu, H.; Yin, B.; Chen, X.; Zhao, D.; et al. Recent advances in small molecule Nav 1.7 inhibitors for cancer pain management. *Bioorganic Chem.* **2024**, *150*, 107605. [CrossRef] [PubMed]
1. Vincenzi, M.; Mercurio, F.A.; Leone, M. Virtual Screening of Peptide Libraries: The Search for Peptide-Based Therapeutics Using Computational Tools. *Int. J. Mol. Sci.* **2024**, *25*, 1798. [CrossRef]
2. Macrander, J.; Brugler, M.R.; Daly, M. A RNA-seq approach to identify putative toxins from acrorhagi in aggressive and non-aggressive *Anthopleura elegantissima* polyps. *BMC Genom.* **2015**, *16*, 221. [CrossRef] [PubMed]
3. Li, B.; Dewey, C.N. RSEM: Accurate transcript quantification from RNA-Seq data with or without a reference genome. *BMC Bioinform.* **2011**, *12*, 323. [CrossRef] [PubMed]
4. Liao, Q.; Gong, G.; Poon, T.C.; Ang, I.L.; Lei, K.M.; Siu, S.W.I.; Wong, C.T.T.; Rádis-Baptista, G.; Lee, S.M.-Y. Combined transcriptomic and proteomic analysis reveals a diversity of venom-related and toxin-like peptides expressed in the mat anemone *Zoanthus natalensis* (Cnidaria Hexacorallia). *Arch. Toxicol.* **2019**, *93*, 1745–1767. [CrossRef]
5. Jungo, F.; Bougueleret, L.; Xenarios, I.; Poux, S. The UniProtKB/Swiss-Prot Tox-Prot program: A central hub of integrated venom protein data. *Toxicon* **2012**, *60*, 551–557. [CrossRef]
6. Kanehisa, M.; Furumichi, M.; Sato, Y.; Kawashima, M.; Ishiguro-Watanabe, M. KEGG for taxonomy-based analysis of pathways and genomes. *Nucleic Acids Res.* **2023**, *51*, D587–D592. [CrossRef]
7. Edgar, R.C. MUSCLE: Multiple sequence alignment with high accuracy and high throughput. *Nucleic Acids Res.* **2004**, *32*, 1792–1797. [CrossRef]
8. Finn, R.D.; Mistry, J.; Tate, J.; Coggill, P.; Heger, A.; Pollington, J.E.; Gavin, O.L.; Gunasekaran, P.; Ceric, G.; Forslund, K.; et al. The Pfam protein families database. *Nucleic Acids Res.* **2009**, *33*, 211–222. [CrossRef]
9. Kumar, S.; Stecher, G.; Tamura, K. MEGA7: Molecular Evolutionary Genetics Analysis Version 7.0 for Bigger Datasets. *Mol. Biol. Evol.* **2016**, *33*, 1870–1874. [CrossRef] [PubMed]
0. Jumper, J.; Evans, R.; Pritzel, A.; Green, T.; Figurnov, M.; Ronneberger, O.; Tunyasuvunakool, K.; Bates, R.; Žídek, A.; Potapenko, A.; et al. Highly accurate protein structure prediction with AlphaFold. *Nature* **2021**, *596*, 583–589. [CrossRef]
1. van Kempen, M.; Kim, S.S.; Tumescheit, C.; Mirdita, M.; Lee, J.; Gilchrist, C.L.M.; Söding, J.; Steinegger, M. Fast and accurate protein structure search with Foldseek. *Nat. Biotechnol.* **2024**, *42*, 243–246. [CrossRef]

marine drugs

MDPI

Article

Proteomic Analysis and Biochemical Characterization of the Nematocyst Extract of the Hydrozoan *Velella velella*

Eleonora Tassara [1], Ivan Mikšík [2], Petr Pompach [3], Gian Luigi Mariottini [1], Liang Xiao [4], Marco Giovine [1,*] and Marina Pozzolini [1,*]

[1] Department of Earth, Environment and Life Sciences (DISTAV), University of Genova, Via Pastore 3, 16132 Genova, Italy; eleonora.tassara@edu.unige.it (E.T.); gian.luigi.mariottini@unige.it (G.L.M.)
[2] Department of Analytical Chemistry, Faculty of Chemical Technology, University of Pardubice, Studentska 573, 532 10 Pardubice, Czech Republic; imiksik@seznam.cz
[3] Institute of Biotechnology, Czech Academy of Sciences, 252 50 Vestec, Czech Republic; petr.pompach@ibt.cas.cz
[4] Faculty of Naval Medicine, Naval Medical University, Shanghai 200433, China; hormat830713@hotmail.com
* Correspondence: mgiovine@unige.it (M.G.); marina.pozzolini@unige.it (M.P.)

Abstract: The venom contained within cnidarian nematocysts has a complex composition and holds significant potential for biotechnological applications. In this context, one of the most effective methods for studying nematocyst contents is the proteomic approach, which can detect even trace amounts of compounds while minimizing the need for large-scale animal collection, thus helping to preserve ecosystem integrity. This study aimed to provide a comprehensive proteomic and biochemical characterization of the crude nematocyst extract from the common hydrozoan *Velella velella*. Despite not being harmful to humans, the analysis of the crude venom extract from *V. velella* brought to the identification of 783 different proteins, categorized into structural components, enzymes, and potential toxins, revealing a qualitative composition of the venom similar to that of other more toxic cnidarians. Biochemical assays confirmed the presence of various active hydrolytic enzymes within the extract, including proteases, phospholipases, hyaluronidases, DNases, and chitinases. These findings pave the road for future studies involving the pharmacological applications of *Velella velella* venom components through recombinant production and functional testing.

Keywords: hydrozoan; venom; proteome; toxins

1. Introduction

Cnidaria is an ancient and worldwide spread phylum of mostly aquatic animals, present in both seawater and freshwater [1,2]. More than 12,000 species of cnidarians have been officially described to date [World Register of Marine Species, https://www.marinespecies.org/index.php, accessed on 19 April 2024], distributed in six classes: Scyphozoa (jellyfish), Anthozoa (anemones, sea pens), Hydrozoa (hydras), Cubozoa (box jellyfish), Staurozoa (stalked jellyfish), and Myxozoa (parasitic organisms with terrestrial hosts) [3,4]. The common feature among all members of this phylum is the presence of nematocysts, specialized venom-containing organelles used for prey capture and defense [1,5].

The composition of Cnidarian venom varies across different classes and species. This venomous arsenal contains a wide range of compounds, both proteinaceous and non-proteinaceous [6]. The first category includes enzymes that disrupt extracellular matrices and cellular membranes through lipolytic and proteolytic activities [7], along with proteinase inhibitors that enhance the venom's effect and spread [8]. Additionally, it contains non-enzymatic proteins and peptides with toxic activities, such as blocking voltage-gated channels [9] or forming pores in phospholipid bilayers [10]. The non-proteinaceous components can be represented by purines that act, for instance, as adenosine receptor antagonists [11], by biogenic amines (such as histamine and serotonin), platelet-activating factors, prostaglandins, leukotrienes [12] with inflammatory effects, by neurotoxic quaternary ammonium compounds [13], and by betaines [14,15]. In humans, exposure to cnidarian venoms can trigger a large spectrum of symptoms, spanning from localized acute skin inflammation to more serious outcomes, such as cardiac and respiratory failure, neurotoxic effects, and, in extreme cases, even death occurring within minutes post-sting [16–18]. This represents a potential risk to human activities, especially in case of sudden jellyfish outbreaks occurrence, which frequency has increased since the last decades [17]. On the other hand, cnidarian venoms have led to ever-increasing attention towards their biotechnological and pharmacological potential: in fact, both crude and purified nematocyst extracts from many species have shown, among others, anti-cancer activity, anti-arrhythmic activity, insecticidal activity, radical-scavenging activity, anti-microbial activity, and immunomodulatory activity [18–20].

The great molecular complexity of cnidarian venom, also accounting how challenging it is to obtain enough fresh specimens, makes the proteomic approach one of the most effective methods for its study. The growing advancement of Next-Generation Sequencing (NGS) techniques and omics sciences has made the study of cnidarian venom and its pharmacological potential more accessible through proteomic approaches; this method bypasses time-consuming extraction processes and helps preserve the integrity of ecosystems, as it avoids the withdrawal of many animals from their habitat. Proteomic approaches have been successfully utilized to purify and characterize the venom components of many other venomous creatures besides jellyfish, such as snakes, spiders, scorpions, and cone snails [21–24]. The initial application of high-throughput mass spectrometry to characterize jellyfish venom identified the putative toxins extracted from the nematocysts of *Olindias sambaquiensis* [25], which was followed by numerous other studies [26–28]. When characterizing venoms, relying on peptide mass fingerprinting with databases primarily composed of protein sequences from model organisms can be limiting, particularly when the goal is to capture protein diversity. Utilizing species-specific transcriptomes addresses the absence of unique peptide sequences in these databases. Thus, combining transcriptomics and proteomics, the accuracy of venom profiling is significantly enhanced compared to using either method alone [29,30]. In this paper, we employed this approach to study the venom of the cnidarian *Velella velella*.

Velella velella (Linnaeus, 1758) is a blue-hued cosmopolitan hydrozoan present in open-ocean fauna [31]. Adult specimens are polymorphic colonies housed within a flattened oval chitinous pneumatophore. Hanging centrally beneath the sail is a large gastrozooid, involved in feeding and digestion [32], surrounded by hundreds of smaller gastro-gonozooids responsible for both digestive processes and reproduction [31,32]; these specialized polyps are enclosed by an additional ring of dactylozooids, which are involved in prey capture and contain the highest number of nematocysts, being mainly of the eurytele and stenotele types [17]. Each specimen contains about 3×10^5 nematocysts; moreover, preliminary studies showed that crude venom extracted from these organelles displayed cytotoxic activity on L929 murine fibroblasts [17].

Building on previous findings that suggested a biotechnological potential for *V. velella*'s venom, this study aims to deliver a comprehensive proteomic characterization of the nematocyst extract from this species. By integrating high-throughput mass spectrometry with species-specific transcriptomic data, we intended to uncover the full range of protein diversity within the venom. Additionally, we performed enzymatic assays to experimentally validate the presence and activity of key enzymes identified in the proteomic analysis, further substantiating the functional components of this venom.

2. Results

2.1. Quantification of Total Proteins in the Crude Nematocyst Extract

The total protein content in the crude venom extract was quantified as described in Section 4 (see Scheme 1), comparing a sample where the nematocysts suspension was subjected to a cycle of sonication (S) with a non-sonicated (NS) one. In the NS sample, the protein concentration resulted being 0.10 ± 0.04 mg/mL, while in the S sample, the concentration grew to 1.30 ± 0.26 mg/mL (see Figure S1). As reported by [17], the chosen methodology of extraction caused the discharge of the nematocyst's thread and, consequently, the release of the nematocyst's content. Therefore, the difference in protein concentration is mainly ascribable to the soluble protein fraction coming from the nematocysts after the treatment, as the structural proteins that constitute the nematocyst capsule are insoluble and were removed from the sample through centrifugation.

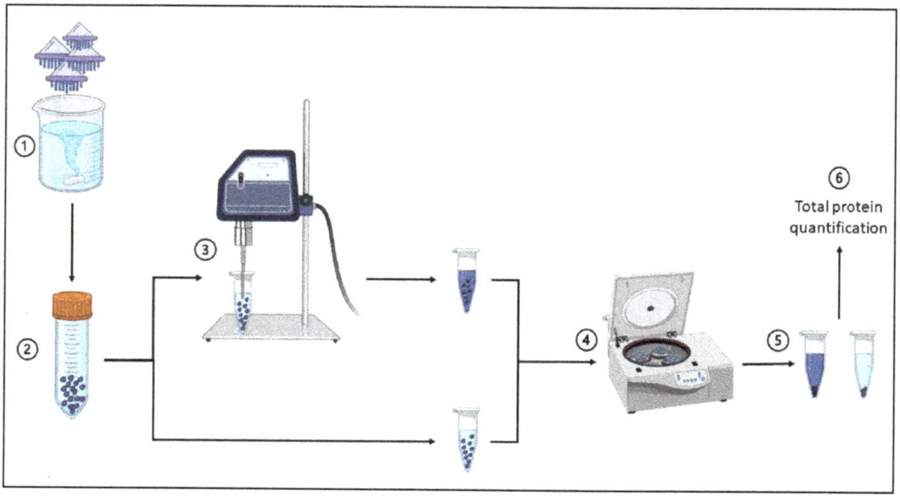

Scheme 1. Process of extraction of *V. velella* venom starting from whole animals. The steps are represented in chronological order. ①: tissue autolysis of specimens; ②: recovering of the nematocysts; ③: sonication of the nematocysts; ④: centrifugation; ⑤: recovery of the supernatant.

2.2. Transcriptome Assembly and Analysis

The results of the transcriptomic assembly performed with the Trinity-v2.14.0 software are presented in Figure S2, which reports the statistics of the assembled sequences' length, their composition, and the quality and completeness assessment of the transcriptome using BUSCO and gVolante analyses. The assembly produced a total of 234,123 sequences. According to the quality and completeness analyses, 78.62% of the core genes queried were complete, 11.43% were fragmented, and 9.96% were missing. This transcriptome was subsequently used as a species-specific database for protein identification in the proteomic analysis.

2.3. Sodium Dodecyl Sulphate–PolyAcrylamide Gel Electrophoresis (SDS–PAGE) of the Crude Venom Extract

The protein content of the crude nematocyst extract from *Velella velella* was separated using Sodium Dodecyl Sulphate–PolyAcrylamide Gel Electrophoresis (SDS–PAGE). Due to the highly concentrated starting material, the gel appeared smeared, making it difficult to distinguish discrete protein bands. To ensure no information was lost, we excised the entire lanes rather than targeting individual bands. Each lane was divided into 11 distinct sections, as shown in Figure 1, and each section was prepared for further mass spectrometry analysis, allowing for a comprehensive analysis of all proteins present in the extract (see Section 4 for further details). The results of the mass spectrometry analysis are discussed in subsequent sections.

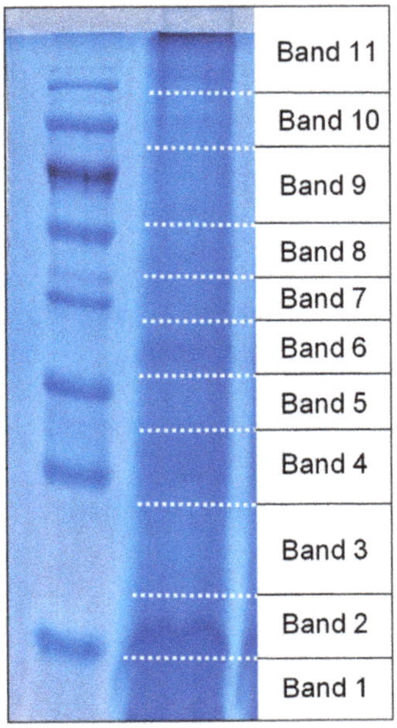

Figure 1. SDS–PAGE of the *V. velella* nematocyst extract. The white dotted lines represent the excision points, from which the 11 bands were obtained. Excision points were established in a way that allowed one to include the entire lane, following at the same time the protein bands pattern observed in the SDS–PAGE.

2.4. Characterization of Proteins Obtained from Proteomic Analysis

Mass spectrometry analysis identified 783 different proteins within the raw nematocyst extract from *Velella velella,* whose sequences are reported in Supplementary Figure S3. The proteins were first separated using electrophoresis, and specific protein bands were excised from the gel, as shown in Figure 1. These bands were then adequately treated and subjected to mass spectrometry, which enabled the identification of the proteins present in the extract. The proteomics results were mapped against the assembled transcriptome, which served as a species-specific reference database. This approach allowed for more accurate identification and annotation of the proteins. After mapping, the identified proteins were annotated using BLAST and categorized into functional groups based on Gene Ontology (GO) terms (see Section 4 for details).

In a total of 783 proteins, 116 did not match against the queried BLAST database; they were thus subjected to a more in-depth analysis using modeling software (see the Section 4) and subsequently categorized as well based on modeling homology results. Table 1 presents the categories used to characterize proteins, and Table 2 shows the characterization of the previously unknown proteins, which were grouped into the same categories where applicable for ease of understanding. The categories "Toxins", "Symbiont contaminants", and "Fluorescent proteins" were added in Table 2.

Table 1. Functional categories of proteins identified within the crude nematocyst extract of *V. velella.*

Category	GO Analysis	N° of Proteins
Uncharacterized	-	116
Structural Components and Extracellular Matrix	GO:0005198—Structural molecule activity GO:0005201—Extracellular matrix structural constituent	65

Table 1. *Cont.*

Category	GO Analysis	N° of Proteins
Membrane Transport and Channel Activities	GO:0016020—Membrane GO:0015297—Antiporter activity GO:0022803—Passive transmembrane transporter activity GO:0005216—Ion channel activity	76
Cytoskeletal Structure and Dynamics	GO:0005856—Cytoskeleton GO:0003779—Actin binding	43
Regulation and Modification of Cellular Processes	GO:0006464—Cellular protein modification process GO:0003677—DNA binding transcription factor activity GO:0004842—Ubiquitin-protein transferase activity GO:0050789—Regulation of biological process GO:0065007—Biological regulation GO:0019222—Regulation of metabolic process	170
Cell-Cell Communication and Adhesion	GO:0007155—Cell adhesion GO:0005921—Gap junction	19
Signal Transduction and Kinase Activity	GO:0007165—Signal transduction GO:0004672—Protein kinase activity	44
Lectins and Sugar-Binding proteins	GO:0030246—Carbohydrate binding	11
Ribosomal proteins	GO:0005840—Ribosome GO:0003735—Structural constituent of ribosome	58
Immune System and Response	GO:0006955—Immune response GO:0002376—Immune system process	19
Hydrolytic Enzymes	GO:0016787—Hydrolase activity	47
Other Enzymes	GO:0003824—Catalytic activity	115

Table 2. Functional characterization of proteins that did not match against the queried BLAST database. Results are presented based on modeling homology from SMART, Swiss Model, and/or Phyre2 analysis.

Category	N° of Proteins
Structural Components and Extracellular Matrix	12
Membrane Transport and Channel Activities	6
Cytoskeletal Structure and Dynamics	3
Regulation and Modification of Cellular Processes	5
Cell-Cell Communication and Adhesion	3
Signal Transduction and Kinase Activity	7
Lectins and Sugar-Binding Proteins	6
Ribosomal Proteins	4
Immune System and Response	5
Hydrolytic Enzymes	16
Other Enzymes	10
Toxins	14
Symbiont Contaminants	24
Fluorescent Proteins	1

A qualitative analysis of protein categories in the total extract was performed after annotating the unknown proteins and is shown in Figure 2, while the percentage of protein categories corresponding to each band excised from the SDS–PAGE is provided in Table S1.

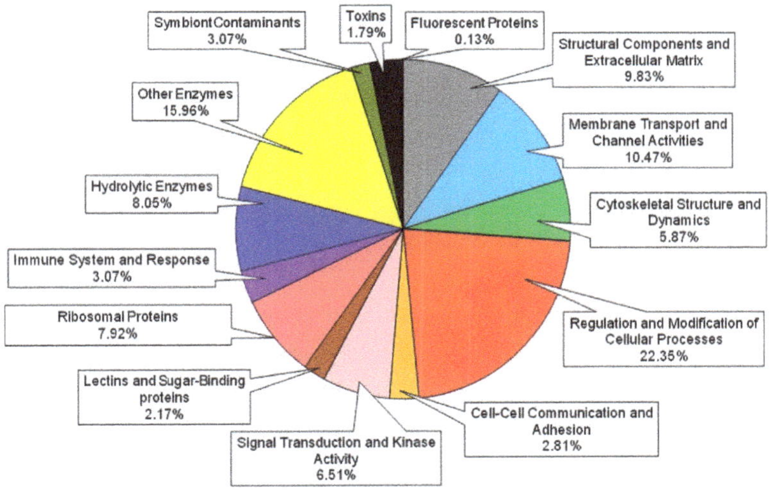

Figure 2. Qualitative analysis showing the percentage composition of various protein categories found in the crude nematocyst extract from *V. velella*.

The most represented protein categories enlightened by the proteomic analysis of the crude nematocyst extract regard regulation or metabolic processes or constitute structural elements (see Figure 2). To obtain a better insight into the possible actual venom components, the categories that according to the literature [15,33–35] were more likely to display toxic activity were extrapolated from the total and reported in Table 3. Figure 3 shows the percentage distribution of potentially toxic elements in the extract. Notably, it was possible to identify 15 different classes of potentially toxic compounds, of which 50% are constituted by hydrolytic enzymes.

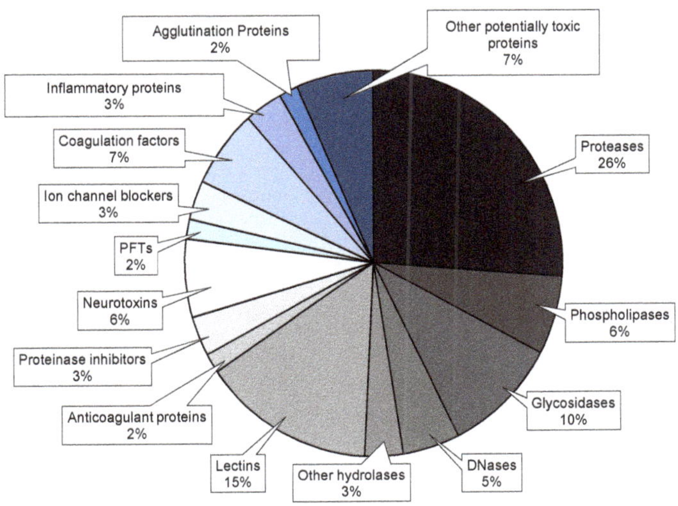

Figure 3. Percentage composition of the potentially venom-related proteins in the crude *V. velella* nematocyst extract.

Table 3. Diversity of potential venom-related components identified in the crude nematocyst extract of *V. velella* obtained with combined functional annotation.

Category of Potential Venom-Related Element	Proteins (Best Hit in Terms of Homology and Structure)
Proteases	A disintegrin and metalloproteinase Aminopeptidase N Pro-X carboxypeptidase Aminopeptidase 1-like *N*-acetylated-alpha-linked acidic dipeptidase 2-like Trypsin-1 isoform X3-like Carboxypeptidase D-like Carboxypeptidase B-like Neprilysin-4 isoform X2 Puromycin-sensitive aminopeptidase-like Zinc metalloproteinase nas-13-like CAAX prenyl protease 1 Furin-like protease 1 C5a peptidase Gelatinase
Phospholipases	Phospholipase D1-like PI-Phospholipase-C X domain-containing protein 3-like Group XIIA secretory phospholipase A2-like *N*-acyl-phosphatidylethanolamine-hydrolyzing phospholipase D-like
Glycosidases	Alpha-mannosidase-like Chitinase Chitooligosaccharidolytic beta-*N*-acetylglucosaminidase Beta-hexosaminidase Non-lysosomal glucosylceramidase-like Chitotriosidase-1
DNases	TatD hydrolase Endonuclease/exonuclease/phosphatase family protein Diphosphonucleotide phosphatase-1
Other hydrolases	Ectonucleoside triphosphate diphosphohydrolase-2 Angiotensin-converting enzyme
Lectins	Maltose-binding periplasmic protein Lectin-2 FBP32 f-lectin Cation-independent mannose-6-phosphate receptor Macrophage mannose receptor 1 SUEL-type lectin D-galactoside-specific lectin-like Hemolectin, putative L-rhamnose-binding lectin CSL3-like
Anticoagulant proteins	Urokinase plasminogen activator
Proteinase inhibitors	Kunitz-type protease inhibitor-1 Pappalysin1
Neurotoxins	Ly-6/neurotoxin-like protein-1 Alpha-elapitoxin-Dpp2a Long neurotoxin OH-55 Antigen 5 like allergen Cul-n-1
Pore-forming toxins (PFTs)	XaxB pore forming toxin
Ion channel blockers	Kappa-stichotoxin-She3a ShKT peptide ShKT-Ts1

Table 3. *Cont.*

Category of Potential Venom-Related Element	Proteins (Best Hit in Terms of Homology and Structure)
Coagulation factors	Coagulation factor FVIII-Fc-XTEN Coagulation factor V Coagulation factor VIII Properdin
Inflammatory response-related proteins	Complement C1q and TNF-related protein 9B-like Neutrophil cytosol factor-2-like
Agglutination and Haemostasis Proteins	Hemagglutinin/amebocyte aggregation factor-like
Other potentially toxic proteins	Merozoite surface protein-1 25 kDa ookinete surface antigen Cry1K-like protein chimera Pesticidal crystal protein Cry11Ba

2.5. Evaluation of the Main Venom Hydrolytic Enzymes Activities in the Crude Nematocyst Extract

The activity of the main hydrolytic enzymes present in the crude venom extract of *Velella velella* was evaluated using in vitro assays. The results are presented in two sections: an evaluation of protease activity (Section 2.5.1) and an analysis of other hydrolytic enzymes (Section 2.5.2). Graphs and gel images are provided in Figure S1, while the data are summarized in Table 4 (for general protease activity with and without specific inhibitors) and Table 5 (for other enzymatic activities).

Table 4. Azocasein protease activity assay. Protease units were calculated by comparing the OD values with those of a positive control (trypsin), as the OD values are directly proportional to the amount of substrate released by protease activity in the sample. The data are expressed as the mean of three independent experiments.

Sample	Corresponding Protease Units (U)/mg
Crude nematocyst extract	306.66 ± 11.00 (SD)
Crude nematocyst extract treated with PMSF	178.90 ± 10.90 (SD)
Crude nematocyst extract treated with EDTA	189.93 ± 7.03 (SD)

Table 5. Enzymatic activities (excluding serine proteases and metalloproteases, which are presented in Table 4) were evaluated in the crude venom extract of *Velella velella*. Chitinase and phospholipase A2 activities are expressed in U/mg, while hyaluronidase activity is expressed in µg/mg.

Enzymatic Test	Enzymatic Activity
Collagenase	undetected
Hyaluronidase	2.00 ± 0.03 (SD) µg/mg
Phospholipase A2	12.11 ± 2.20 (SD) U/mg
DNase	detectable
Chitinase	0.070 ± 0.001 (SD) U/mg

2.5.1. Proteases Activity Evaluation

Generic Protease Activity Evaluation

The generic protease activity was quantified both on the crude venom extract alone and in the presence of specific protease inhibitors, as described in Section 4.7.1. The protease activity was quantified with respect to the absorbance of the positive control, containing 562 units of trypsin; considering that this amount of enzyme could digest an amount of substrate able to give a mean absorbance of 1.07 ± 0.01 (Standard Deviation, SD) at 440 nm with this assay, it is possible to say that the amount of crude nematocyst extract used for the assay contains a protease activity comparable to 306.66 ± 11.00 (SD) units of trypsin, equivalent to 1202.60 ± 47.00 (SD) protease units/mg of crude extract. When treated with specific inhibitors, in the assay this value goes down to 178.90 ± 10.90 (SD) protease units after phenylmethylsulfonyl fluoride (PMSF) treatment, equivalent to 701.60 ± 43.00 (SD) protease units/mg of crude extract, and to 189.93 ± 7.03 (SD) protease units after ethylenediamine tetraacetic acid (EDTA) treatment, equivalent to 744.80 ± 27.60 (SD) protease units/mg of crude extract, respectively (see Table 4). These results highlight the fact that the *V. velella* crude venom extract contains both serine-protease and metalloprotease activity in similar amounts.

Collagenase Activity Evaluation

To evaluate the presence of collagenase activity in the *Velella velella* sample, an in vitro assay was performed using the synthetic peptide Fa-LGPA as a substrate. The results are graphically presented in Figure S1 and reported in Table 5. The positive control showed a consistent decrease in absorbance throughout the assay, indicating active collagenase-mediated cleavage of Fa-LGPA. In contrast, the absorbance values of the crude venom extract remained relatively stable, suggesting that no detectable collagenase activity was present in the nematocyst extract of *V. velella* under the tested conditions.

Zymographic Analysis of Active Proteases

To qualitatively identify the active proteases in the *Velella velella* nematocyst extract, a zymography analysis was performed using two different substrates: casein, to detect serine proteases, and gelatin, to detect metalloproteases (see Section 4). The results are presented in Figure S1. In both gels, proteolytic activity is visualized by clear bands against the blue background, indicating where proteases have digested the substrate. These results confirm, as previously demonstrated by the azocasein assay with specific inhibitors, that the crude venom extract contains both active serine proteases and metalloproteases. Specifically, proteases with a molecular weight of approximately 60 kDa were observed digesting casein, while proteases with a molecular weight of around 50 kDa were responsible for gelatin digestion.

2.5.2. Evaluation of Other Enzymatic Activities

Hyaluronidase Activity Evaluation

As explained in the Section 4, a turbidimetric assay was conducted to quantify the potential hyaluronidase activity within the *V. velella* crude venom extract. The absorbance differences between the negative control and the *V. velella* venom extract are presented in Figure S1. The absorbance of the nematocyst extract sample is significantly lower than that of the negative control ($p < 0.05$), indicating that a portion of the hyaluronic acid in the solution was digested by the crude venom extract. By comparing the absorbance values to a standard curve of hyaluronic acid, it was determined that the crude venom extract contains approximately 2.00 ± 0.03 (SD) µg of hyaluronidase per mg of crude venom extract, as reported in Table 5.

Phospholipase A2 Activity Evaluation

The PLA2 in vitro turbidimetric assay was conducted using a principle similar to that of the collagenase and hyaluronidase assays, where enzymatic activity is indicated by a reduction in absorbance over time. Specifically, the absorbance value of the nematocyst extract sample at 740 nm decreased by an average of 71.53 ± 21.70% (SD) after incubation. This reduction corresponds to a mean activity of 12.11 ± 2.20 (SD) enzyme units per mg of nematocyst extract. The quantification of PLA2 activity in the crude venom extract is presented in Table 5.

DNase Activity Evaluation

The presence of DNase activity was assessed by treating *E. coli* genomic DNA with a known amount of nematocyst extract. Under our experimental conditions, 50 μg of *Velella velella* nematocyst extract contained sufficient DNase activity to completely degrade 250 ng of *E. coli* genomic DNA. The gel image showing the results is provided in Figure S1.

Chitinase Activity Evaluation

Chitinase activity in the crude venom extract was evaluated as described in Section 4.7.7. The activity was quantified by comparing the absorbance to a positive control containing 0.02 units of commercial chitinase. In this assay, 0.02 units of chitinase produced a mean absorbance of 1.37 ± 0.04 (SD) at 562 nm. Based on this, the samples containing crude nematocyst extract showed chitinase activity equivalent to 0.009 ± 0.001 (SD) units of commercial chitinase. This corresponds to a chitinase activity of 0.070 ± 0.001 (SD) units per mg of crude extract (see Table 5).

3. Discussion

This study presents the first comprehensive proteomic and biochemical characterization of the crude nematocyst extract from the hydrozoan *Velella velella*. Public SRA data were assembled to create a transcriptome database that is nearly 80% complete, making it a suitable species-specific resource for subsequent proteomic analysis. This high level of completeness enabled more accurate interpretation and identification of proteomics results compared to using generalized databases. Through mapping the proteomics data to the assembled transcriptome, a total of 783 unique proteins were identified. The most represented protein categories found in the extract include those involved in the regulation and modification of cellular processes, structural components, and extracellular matrix elements (see Figure 2). The predominance of these proteins is probably due to the methodology used for crude venom extraction: the sonication process, used for inducing nematocyst lysis, not only permits the release of the venom but also tends to disrupt the stinging organelle's structural and functional elements. Consequently, many of the proteins identified are not directly associated with venom function but are more likely derived from the structure of the capsule and the thread. Indeed, among others, the typical short cysteine-rich fibrillary collagens constituting the capsule [36,37] were detected, as well as ion channels, transporters, and pumps which are thought to be essential in maintaining the internal osmotic pressure of the cyst [38–40]. However, a significant portion of the identified proteins actually turned out to be potential venom-related molecules according to the literature [33–35].

Considering the potentially toxic elements, hydrolases are the most represented group, comprising approximately 50% of the identified potential venom-related elements (see Figure 3). These hydrolases, half of which are proteases, were further evaluated through in vitro biochemical analyses to obtain better insight into their effective enzymatic activity within the crude venom extract.

Proteases are known to be an important fraction in the venom of many animals, such as snakes, spiders, and scorpions, as they play crucial roles in predation, digestion processes, and defense mechanisms [41]. The proteomic analysis of the *V. velella* crude venom extract (see Table 3) reveals the presence especially of metalloproteases (which, in venoms, are related to inflammatory and necrotic effects [42,43]) and, in lesser amounts, serine proteases (which have been linked to various physiological functions like platelet aggregation, fibrinolytic activity, enhancement of venom spreading in the target tissues, and post-translational modifications of other toxins [44]); the prevalence of these two categories of proteases has been observed in the venom of other cnidarians [45,46]. Notably, metalloproteases have been extensively studied in snake venoms and are often produced recombinantly to study their role in cancer progression and tissue remodeling, making them valuable for potential therapeutic use [42]. Serine proteases, known for their involvement in blood coagulation and fibrinolysis, are also studied in many venomous animals, including spiders and snakes, and have been recombinantly produced for use in anti-coagulant therapies [47]. In the crude venom extract of *Velella velella*, the azocasein assay revealed an overall protease activity almost comparable to that of purified trypsin (see Table 4); to better assess the contribution of each category to the total protease activity, the same assay was repeated in the presence of metalloprotease and serine protease-specific inhibitors (EDTA and PMSF, respectively). Notably, despite the transcriptomic analysis indicating a higher number of metalloproteases compared to serine proteases, the enzymatic activities of both categories were similar. This could be explained by different catalytic efficiencies or inactive metalloproteases within the extracted venom. Moreover, post-translational modifications or the presence of endogenous inhibitors [48] could also play a role in balancing the overall enzymatic activities of the two categories. Curiously, no collagenase activity was detected among the metalloproteases using our methods, despite collagenases being identified in other hydrozoans [49]. However, zymographic analysis revealed enzymes capable of digesting gelatine, a denatured form of collagen. This suggests the presence of other enzymes that, under physiological conditions, break down collagen in prey to facilitate gelatinase access to the already denatured molecule. The absence of detectable collagenase activity could be due to a low amount of collagenase in the crude venom, making the assay insufficiently sensitive. Alternatively, collagenase activity could be inhibited by quenching effects, which involve the reduction of enzyme activity due to interaction with other substances or intrinsic inhibitors present in the venom.

The other half of the hydrolases identified by proteomic analysis consists of a diverse assortment of enzymes confirmed to be active in the extract. These enzymes likely play crucial roles in the predation, feeding processes, and/or defense mechanisms of *V. velella*. For example, hyaluronidases, while not toxic themselves, act as spreading factors by enhancing the diffusion of toxins through the prey's tissues [44]. Our methods detected hyaluronidase at a concentration of 2 μg/mL, corresponding to 1.2 μg of enzyme for each mg of total protein in the extract. This enzymatic activity aligns with the roles of phospholipases, which break down phospholipid bilayers, potentially enhancing the spreading of venom-related elements, as observed in many cnidarian extracts [50,51]. In particular, phospholipase D and Group XIIA secretory phospholipase A2, both found in many venoms, including that of *V. velella*, have been produced recombinantly and are valuable in lipid metabolism research. Group XIIA PLA2, in particular, has significant potential in anti-inflammatory drug development, as it can be used to study its mechanism of action and develop targeted inhibitors [52]. The presence of DNase activity, which is believed to assist in comprehensive cellular disruption, was also detected in venoms [15]. In the case of *V. velella*, an endonuclease activity capable of digesting prokaryotic genomic DNA was detected; however, our analysis was purely qualitative, and further studies will be necessary to quantify this enzyme. Nonetheless, the presence of this kind of hydrolase aligns with previous findings in *Physalia physalis* [53] and *Chrysaora quinquecirrha* [54], where endonucleases were purified and characterized. Finally, the presence of glycosidases—particularly chitinases—seems to be relevant considering the ecological niche of this hydrozoan, as a significant part of its diet consists of planktonic crustacean prey [55], whose exoskeletons are made of chitin. Chitinases have been recombinantly produced from bacterial and fungal sources, with applications in biopesticides and antifungal drug development [56]. These findings suggest that *V. velella* chitinases are worthy of further investigation to explore their potential for similar biotechnological applications.

The remaining 50% of potentially venom-related proteins in the *V. velella* nematocyst extract are non-enzymatic and were identified through proteomic analysis either as toxins or molecules capable of causing toxic effects in their targets. Of these, 30% are lectins. Lectins are carbohydrate-binding proteins, whose role in the venoms of many animals is to specifically target glycoproteins and glycolipids on the surface of cells, contributing to various physiological effects; more specifically, these molecules can facilitate the delivery of other harmful compounds by causing hemagglutination, thus impairing oxygen transport and causing tissue damage [57–59]. These molecules play roles in cell targeting, immune modulation, and hemagglutination, making them promising candidates for drug delivery systems and immunomodulatory therapies [60]. For instance, the Macrophage mannose receptor 1-like, found in *V. velella*, could be studied recombinantly produced to explore its structural or functional similarities to human receptors, potentially offering insights into novel mechanisms of immune modulation or therapeutic targeting in infections and cancers. Likewise interesting is the identification of homologs of toxins in the *V. velella* nematocyst extract, which have counterparts in many other venomous animals. These findings are significant because venoms exemplify evolutionary convergence, having evolved independently over 100 times across major animal groups, such as snakes, insects, arachnids, and jellyfish. These venoms have developed similar biochemical arsenals—especially neurotoxins, protease inhibitors, and pore-forming toxins—to immobilize prey and deter predators [61]. Notably, the *V. velella* nematocyst extract contains homologs of the Ly-6 neurotoxin, alpha-elapitoxin-Dpp2a, XaxB pore-forming toxin, kappa-stichotoxin-She3a, ShKT peptide ShKT-Ts1, and the Kunitz-type protease inhibitor-1 (see Table 3). The XaxB pore-forming toxin is a type of cytolysin that was first isolated from the bacterium *Xenorhabdus nematophila* [62]; however, sea anemones [63], spiders [64], and even some vertebrates [65] can also produce pore-forming toxins (PFTs), which share functional similarities with the bacterial ones. Regarding toxins distinctive of cnidarians, kappa-stichotoxin-She3a and ShK peptides are channel blockers found in the sea anemone *Stichodactyla helianthus*, whose toxic action can disrupt normal nervous and muscular function, leading to prey paralysis [66]; while these compounds are typical of sea anemones (Anthozoa), many homologues or similar domains in bigger proteins have been found in other cnidarians, such as *Hydra magnapapillata* and *Aurelia aurita* [51]. From a biotechnological point of view, neurotoxins such as Ly-6 and alpha-elapitoxin-Dpp2a target ion channels, leading to paralysis and neurological effects, making them interesting candidates for neurological drug development, particularly in targeting ion channels for pain management and neurological disorders [67]. Kunitz-type protease inhibitors, found in spiders, scorpions, and cnidarians, have been recombinantly produced and are highly relevant for developing cancer, thrombosis, and inflammation treatments [47].

Compared to other cnidarians, the venom of *Velella velella* displays both similarities and distinct features. Like *Physalia physalis* and *Chrysaora quinquecirrha*, *V. velella* venom contains a diverse array of enzymes, including proteases, phospholipases, and glycosidases [53,54]. Notably, *V. velella* venom also includes phospholipase D and Group XIIA secretory phospholipase A2, both of which are found in other cnidarians [50,51]. Additionally, homologous of neurotoxins and ion channel blockers in *V. velella* recall similarities with venom components found in sea anemones like *Stichodactyla helianthus* [66] and *Hydra magnipapillata* [51], which also target ion channels and disrupt nervous system function. However, the venom of *V. velella* seems to have a milder effect on humans compared to more potent cnidarian venoms, such as that of *Physalia physalis*, which is known to cause severe envenomation symptoms [53]. This suggests that while *V. velella* shares many venom components with other cnidarians, its lower toxicity may be due to differences in concentration or potency of the toxic elements; for instance, the mild effect of *V. velella*'s venom on humans may be due to the fact that, although it contains homologues of dangerous toxins, these compounds would need to be present in much higher concentrations to harm larger organisms. However, for *V. velella*'s purposes, the venom is likely potent enough to be lethal to its prey or deter attackers. For example, proteins containing ShK-like domains can block K+ channels, but while they may not affect humans at lower concentrations, they are likely sufficient to have a strong impact on smaller organisms [51,68]. Nonetheless, based on the current information, the precise physiological role of these substances in the organism remains unclear.

In conclusion, the presence of such compounds in an animal that does not provoke medically relevant effects on humans suggests a potential for biotechnological applications: molecules with specific targets but low toxicity are particularly interesting for drug discovery and development, as they are less likely to cause harmful side effects, crucial for developing safe therapeutic agents with a wider therapeutic window, especially for chronic use [69]. Further research should focus on selecting high-potential molecules from the nematocyst arsenal of *V. velella* for recombinant production and subsequent functional testing, which could reveal novel insights into their roles and contribute to the development of innovative therapeutic strategies. In particular, metalloproteases, phospholipase A2, chitinases, Kunitz-type protease inhibitors, and lectins stand out as the most promising compounds, given their established biotechnological applications and therapeutic potential in areas such as cancer, inflammation, thrombosis, and drug delivery systems.

4. Materials and Methods

4.1. Chemicals

Unless otherwise stated, all reagents were acquired from SIGMA-ALDRICH (Milan, Italy).

4.2. *Velella velella* Sampling

A small floating colony of *V. velella* was recovered in shallow coastal waters in the Gulf of Genoa area (Italy) in April 2023. The specimens were preserved in fresh seawater during their transfer to the laboratory, where they were promptly stored at −20 °C until future processing.

4.3. Crude Venom Extraction

The isolation of the nematocysts from the tissue was conducted as described in [17]. A group of about 20–25 specimens were soaked in a beaker containing artificial seawater and left to stir overnight at 4 °C. This process promotes tissue autolysis, leading to the subsequent release of intact nematocysts into the surrounding medium. Afterward, the sails were removed, and the whole content of the beaker was filtered through a cloth filter to remove chitin and tissue residues. The obtained suspension was then washed thrice by centrifuging for 10 min at 4000 rpm and resuspending the nematocyst pellet in clean seawater. After the last wash, the pellet was resuspended in 10 mL of a saline buffer (10 mM NaH_2PO_4 and 0.9% NaCl, pH 7.4). The nematocyst suspension was sonicated using a UP50H ultrasonic processor (Hielscher Ultrasonic, Teltow, Germany), subjecting the sample to 60 cycles of sonication of 10 s each and with a frequency of 30 kHz, always keeping it in ice. Finally, the suspension was centrifuged at 4 °C for 20 min at 18,000 rpm; the supernatant of the sonicated sample (S) was recovered and transferred into a clean tube. A non-sonicated sample (NS) was also centrifuged, and its supernatant was recovered to use it as a comparison and to confirm the breaking of the nematocysts and the release of the venom (see Scheme 1). The crude extract thus obtained was then preserved at −20 °C until used for further experiments.

4.4. Quantification of Total Proteins in the Crude Nematocyst Extract

As cnidarian venoms are mainly constituted of proteinaceous compounds, such as enzymes, non-catalytic proteins, and peptides [15], to verify the release of the venom from the *V. velella* nematocysts, the total protein content in the soluble supernatant of both the non-sonicated (NS) and the sonicated (S) samples, obtained as described in Section 4.3, was assayed with the Bicinchoninic Acid Protein Assay kit, following the manufacturer's instructions, in comparison to a BSA (Bovine Serum Albumin) standard curve. The absorbance of each sample was read at 562 nm using a Beckman spectrophotometer (DU 640, Beckman Coulter SpA, Milan, Italy). The procedure was carried out in triplicate.

4.5. Transcriptome Assembly and Putative Protein Database Construction

Sequence Read Archive (SRA) data available on NCBI (https://www.ncbi.nlm.nih.gov/, accessed on 6 June 2024; accession number: SRX4941821) were used to obtain a de novo assembled transcriptome for *V. velella*. A quality assessment of the raw reads was made with FastQC (v0.12.0) (https://www.bioinformatics.babraham.ac.uk/projects/

fastqc/, accessed on 6 June 2024). Considering that a *Phred* score equal to 30 corresponds to a 99.9% base accuracy and a probability of 10^{-3} of incorrect base calling, the terminal bases resulted from the FastQC analysis with a Phred quality score < 30 were rejected using Trimmomatic (http://www.usadellab.org/cms/index.php?page=trimmomatic, accessed on 6 June 2024). Trimmed reads were re-screened using FastQC to check for residual adapters or primer sequences, which, if present, were discarded with Trimmomatic during a second round of processing. De novo assembling with clean data sets was performed with the Trinity software (https://github.com/trinityrnaseq/trinityrnaseq, accessed on 6 June 2024). The quality of the assembly was checked with Bowtie2 (https://bowtie-bio.sourceforge.net/bowtie2/index.shtml, accessed on 6 June 2024) and with the BUSCO tool (https://busco.ezlab.org/, accessed on 6 June 2024), together with a gVolante analysis (https://gvolante.riken.jp/, accessed on 6 June 2024). Finally, the transcript database was translated into proteins using the TransDecoder tool (https://github.com/TransDecoder/TransDecoder/wiki, last accessed 6 June 2024), where for each transcript, only the best open reading frame was retained; additionally, transcripts were filtered by length, with 20 amino acids set as the minimum, as some toxins can be very short. The translated transcript database was used as a reference database to search for protein identifications generated by the mass spectrometry (MS) analysis.

4.6. Proteomic Analysis

To obtain a complete protein profile of the nematocyst extract of *Velella velella*, a mass spectrometry-based proteomic analysis was performed.

4.6.1. Sample Preparation

The sonicated extract (see Section 4.3) was further concentrated following the procedure described in [70]. In short, cold acetone was added to an aliquot of the extract in a 4:1 ratio; proteins were thus precipitated overnight at −20 °C. Samples were centrifuged for 10 min at 13,000 rpm and the supernatant was discarded, letting the protein pellet dry in uncapped tubes for 30 min and then resuspending it in an adequate volume of 10 mM NaH_2PO_4 and 0.9% NaCl, pH 7.4. The concentrated extract was subsequently used to perform two SDS–PAGE with a Mini-Protean 3 (Bio-Rad Laboratories, Hercules, CA, USA), following the method previously described in [71]. A 5 mg/mL solution of crude extract (value referred to the total protein content after acetone concentration) was mixed at a 3:1 (*v*/*v*) ratio with a 4× gel loading buffer (1 M Tris–HCl buffer, pH 6.8, 10% 2-mercaptoethanol, 40% glycerol, 0.2% bromophenol blue and 20% sodium dodecyl sulfate solution) and heated at 90 °C for 10 min. Consequently, 40 µL of the resulting solution (thus containing a total of 150 µg of proteins) were loaded into 14 wells of the two 12% polyacrylamide gels and run at 60 mA with constant amperage, using 10 µL of PageRuler™ Prestained Protein Ladder as standard. At the end of the run, gels were briefly rinsed with distilled water and fixed for 1 h in a fixing solution containing 10% (*v/v*) acetic acid and 40% (*v/v*) ethanol, then washed twice for 10 min with distilled water and stained overnight with a staining solution containing a 1:4 (*v*/*v*) *ratio* of methanol and colloidal Coomassie solution (0.1% (*w/v*) Coomassie Brilliant Blue G250, 2% (*w/v*) orthophosphoric acid, 10% (*w/v*) ammonium sulfate). Each one of these steps was conducted at room temperature. Finally, gels were destained with a 5% acetic acid solution. As the crude venom extract contains a mixture of proteins, to obtain an adequate concentration of each protein suitable for further mass spectrometry analysis, the 14 resulting lanes were precisely excised at identical points using a sterile scalpel. This process yielded 11 bands, each representing the combined slices from the same points across all 14 lanes. Figure 1 reports a single SDS–PAGE lane and the corresponding bands into which it was divided. Slices were cut into cubes, transferred into 1.5 mL tubes, and prepared for mass spectrometry analysis following the method reported in [72]; after reduction, alkylation, and destaining (see [72]) proteins within the gel pieces were digested overnight at 37 °C with a 13 ng/µL trypsin solution dissolved into 10 mM ammonium bicarbonate containing 10% (*v/v*) acetonitrile. Then, peptide digestion products were recovered by adding into each tube 100 μL of extraction buffer (1:2 5% *(v/v)* formic acid/acetonitrile) and incubating for 15 min at 37 °C in a shaker, further recovering the supernatant in a PCR tube. This step was repeated twice. Finally, samples were dried down in a vacuum centrifuge before mass spectrometry analysis.

4.6.2. Mass Spectrometry Analysis

Samples digested by trypsin were analyzed using a liquid chromatography system Vanquish (Thermo Scientific, Waltham, MA, USA) connected to the timsToF SCP mass spectrometer equipped with Captive spray (Bruker Daltonics, Billerica, MA, USA). The mass spectrometer was operated in a positive data-dependent mode. One microliters of peptide mixture were injected by autosampler on the C18 trap column (Pepmap Neo C18 5 μm, 0.3 × 5 mm, Thermo Scientific). After trapping, peptides were eluted from the trap column and separated on a C18 column (Pepsep C18 150 × 0.15 mm, 1.5 μm, Bruker Daltonics) by a linear 35 min water-acetonitrile gradient from 5% (*v/v*) to 35% (*v/v*) acetonitrile at a flow rate of 1.5 μL/min. The trap and analytical columns were both heated to 50 °C. Parameters from the standard proteomics PASEF method were used to set timsTOF SCP. The target intensity per individual PASEF precursor was set to 20,000, and the intensity threshold was set to 1500. The scan range was set between 0.6 and 1.6 V s/cm^2 with a ramp time of 100 ms. The number of PASEF MS/MS scans was 10. Precursor ions in the m/z range between 100 and 1700 with charge states ≥2+ and ≤6+ were selected for fragmentation. The active exclusion was enabled for 0.4 min. The raw data were processed by PeaksStudio 11.0 software (Bioinformatics Solutions, Waterloo, ON, Canada). The search parameters were set as follows: enzyme—trypsin (specific), carbamidomethylation as a fixed modification, oxidation of methionine, and acetylation of protein *N*-terminus as variable modifications. The MS outcome was searched against the *V. velella* putative protein database obtained by translating the assembled transcriptome (as described in Section 4.5).

4.6.3. Protein Characterization

The data set resulting from mass spectrometry analysis was investigated using the Basic Local Alignment Search Tool (BLAST, https://blast.ncbi.nlm.nih.gov/Blast.cgi, accessed on 13 June 2024) against a UniProtKB/Swiss Invertebrate proteins database (https://www.uniprot.org/help/downloads, accessed on 13 June 2024), and the UniProtKB/Swiss-Prot Tox-Prot database [73] using default settings, considering an e-value $\leq 1 \times 10^{-5}$. Then, proteins were functionally classified using Gene Ontology categories (Gene Ontology Resource, accessed on 13 June 2024). The proteins that remained uncharacterized after the BLAST analysis were manually identified using SMART (http://smart.embl-heidelberg.de/, accessed on 13 June 2024), Swiss Model (https://swissmodel.expasy.org/, accessed on 13 June 2024), and/or Phyre2 (http://www.sbg.bio.ic.ac.uk/~phyre2/html/page.cgi?id=index, accessed on 13 June 2024), by singularly subjecting the uncharacterized sequences to the above-mentioned prediction tools.

4.7. *In Vitro Evaluation of the Main Crude Venom Hydrolytic Enzymes Activities*

The presence of some classes of hydrolytic enzymes commonly found in the venom of cnidarians [15] was experimentally evaluated by some in vitro trials on the crude extract obtained from *V. velella*. More specifically, the activity of generic proteases (with and without specific protease inhibitors), collagenases, hyaluronidases, phospholipases, chitinases, and DNases was investigated.

4.7.1. Generic Protease Activity

The generic protease activity was evaluated and quantified following the method described by [74], using azocasein as an enzymatic substrate. As an enzymatic source, 150 μL of 1.7 mg/mL crude venom was added to 600 μL of a solution, previously heated up to 37 °C, composed of 225 μL of reaction buffer (100 mM Tris-HCl, pH 8; 20 mM $CaCl_2$) and 375 μL of 2% azocasein solution, pH 8. Positive and negative controls were also prepared in the same way using, respectively, 0.5 mg/mL trypsin and a thermally inactivated aliquot of the crude venom extract itself. For the blank, the enzymatic source was replaced by the same volume of reaction buffer. The samples were incubated at 37 °C for 30 min. After the incubation, 500 µL of each sample was transferred into clean tubes, and an equal volume of 110 mM trichloroacetic acid was added; then, they were centrifuged at 16,000 rpm for 10 min. The supernatant was recovered and added to 500 µL of 500 mM NaOH. Finally, samples were read against blank at a 440 nm wavelength using a Beckman spectrophotometer (DU 640, Beckman Coulter Spa, Milan, Italy). The procedure was carried out in triplicate.

To better determine what kind of proteases were contained in the sample, the same experiment was repeated after treating the crude venom extract with specific protease inhibitors, following the example of [75]. Ethylenediaminetetraacetic Acid (EDTA), which inhibits metalloproteases, or phenylmethylsulfonyl fluoride (PMSF), as a serin-proteases inhibitor, was added to the *V. velella* extract to reach a final concentration of 5 mM. In both cases, the extract was incubated with the specific inhibitor for 1 h at 4 °C before proceeding with the assay. The procedure was carried out in triplicate. Results were expressed in Units of enzymatic activity/mg of crude extract with respect to the activity of positive control.

4.7.2. Collagenase Activity

The collagenase activity was measured with a turbidimetric assay as explained in [76], using the synthetic peptide 2-furanacryloyl-l-leucylglycyl-l-prolyl-l-alanine (Fa-LGPA) as an enzymatic substrate. A volume of 20 µL of 1.7 mg/mL crude venom extract was added to 40 µL of a 5 mM Fa-LGPA solution, prepared by dissolving the peptide in a reaction buffer (50 mM tricine, 10 mM $CaCl_2$, 400 mM NaCl, pH 7.5). As a positive control, 20 µL of a 2 U/mL aqueous solution of commercial collagenase from *Clostridium histolyticum* was used. A negative control was also prepared by thermally inactivating an aliquot of the crude venom extract, specifically by treating it at 95 °C for 15 min. The samples were read at a wavelength of 345 nm every 15 s for 15 min, using a Beckman spectrophotometer (DU 640, Beckman Coulter SpA, Milan, Italy). The units of enzyme in the crude venom extract were calculated as follows:

$$Units/mL\ enzyme = \frac{(\Delta A345/\text{min Sample} - \Delta A345/\text{min Blank}\ (0.06)}{(0.53)\ (0.02)}$$

where

$\Delta A345$ = delta of absorbance between t_1 and t_0
0.06 = Total volume (mL) of the reaction mixture
0.53 = Millimolar extinction coefficient of Fa-LGPA at 345 nm
0.02 = Volume (mL) of enzyme solution/sample used

4.7.3. Zymographic Analysis of Active Proteases

A zymography assay was conducted to detect the proteolytic activity of *V. velella* crude venom using casein and gelatin as substrates. This method, outlined in prior studies [27,77], was slightly modified. Gelatin (1.2 mg/mL) or casein (1.2 mg/mL) were copolymerized with 12% SDS–polyacrylamide gels. Samples were prepared under non-reducing conditions by dissolving them in a loading buffer with final concentrations of 0.8% *(w/v)* SDS, 4% *(v/v)* glycerol, 0.002% *(w/v)* bromophenol blue, and 25 mM Tris-HCl at pH 6.8. Each well was loaded with 30 µg of proteins, including both the positive control (protease from *Streptomyces griseus*) and the crude venom extract. Electrophoresis was carried out at 4 °C with a constant voltage of 200 V. At the end of the running, gels were rinsed with deionized water and washed twice with 2.5% *(v/v)* Triton-X for 30 min and rinsed again; then, it was reconditioned for 5 min with an incubation buffer (0.2 M NaCl, 50 mM Tris-Base, 5 mM $CaCl_2$, 5 µM $ZnCl_2$, pH 7.6). The development of the gel was carried out overnight in the same incubation buffer, at 37 °C. The gel was then rinsed and stained with 0.8% *(w/v)* Coomassie Blue—G250 in 45 mM HCl, putting it in a microwave oven for about 10 s at a power of 500 W. Before staining, the lanes containing the pre-stained molecular weight standards were separated from the rest of the gel to prevent their reduced visibility due to overall staining caused by the embedded protein substrates. After staining, the gel portions containing the standards (whose background thus remained clear) were reassembled with the rest of the gel for visualization.

Finally, the gel was destained with deionized water until clear zones of hydrolysis in the gel against the blue background showed gelatin proteolytic activity.

4.7.4. Hyaluronidase Activity

The presence of hyaluronidases was evaluated using the method described in [78], with some modifications. A total of 100 μL of 1.7 mg/mL crude venom extract was mixed with 400 μL of a 0.5 mg/mL hyaluronic acid solution, prepared by dissolving hyaluronic acid in acetate buffer (0.2 M $C_2H_3NaO_2$ in acetic acid, 0.15 M NaCl, pH 6.0). An aliquot of the crude venom extract was thermally inactivated by keeping it at 95 °C for 15 min and used as a negative control. The samples were confronted with a standard curve of commercial hyaluronidase from *Streptomyces hyalurolyticus*. The hyaluronic acid solution is substituted for the blank with the same volume of acetate buffer. All samples were incubated at 37 °C for 15 min; consequently, the reaction was stopped with 2 mL of cetyltrimethylammonium bromide (CTAB) solution (2.5% CTAB in 2% NaOH, pH 12). After 10 min at room temperature, samples were read at a 400 nm wavelength against blank with a Beckman spectrophotometer (DU 640, Beckman Coulter SpA, Milan, Italy). As the CTAB solution forms an insoluble product reacting with hyaluronic acid, the lower the absorbance value, the higher is the amount of hyaluronic acid digested. The experiment was performed in triplicate.

4.7.5. Phospholipase A2 Activity

The presence of PLA2 in the crude venom extract from *V. velella* was investigated following the method of [79]. A solution of egg yolk substrate was created by dissolving the egg yolk in a 0.9% NaCl solution and adjusting the absorbance to 1 optical density (OD) at 740 nm, using a 20 mM Tris-HCl buffer (pH 7.4). A volume of 60 μL of 1.2 mg/mL crude venom extract was added to 250 μL of the egg yolk solution, and the absorbance was read at 740 nm (t_0) with a Beckman spectrophotometer (DU 640, Beckman Coulter SpA, Milan, Italy). Then, the sample was incubated at 37 °C for 60 min and the absorbance was read again (t_1). The procedure was carried out in triplicate, using a commercial phospholipase from *Apis mellifera* as a positive control and a thermally inactivated *V. velella* crude venom sample as a negative control. The unit activity of PLA2 activity is described as the amount of crude venom PLA2 which decreases the turbidity of the solution by 0.01 absorbance unit per minute at 740 nm. The unit activity of PLA2 was then calculated as the quantity of crude venom PLA2 that reduces the turbidity of the solution by 0.01 absorbance unit per minute at 740 nm.

4.7.6. DNase Activity

The DNase activity in the nematocyst extract of *V. velella* was researched as described in [79]. A quantity of 50 μg of nematocyst extract (quantity referred to the total protein content) was incubated at 37 °C for 1 h together with 250 ng of genomic DNA from *Escherichia coli*. An untreated sample of DNA was used as a negative control. The assay mixtures were subjected to 0.8% agarose horizontal gel electrophoresis and its image was acquired with a ChemiDoc Imaging System (Bio-Rad, Milan, Italy).

4.7.7. Chitinase Activity

The presence of chitinase activity was investigated using a combination of methods described by [80,81], with some modifications. A colloidal chitin solution was prepared by dissolving commercial crab chitin in ten volumes of 37% (*v*/*v*) HCl for 1 h at room temperature. Afterward, 1000 volumes of cold ethanol were slowly added, and the suspension was centrifuged at 8000× *g* for 20 min. The resulting pellet was collected and washed with distilled water until reaching a neutral pH. The assay reaction mixture consisted of equal volumes (50 μL each) of colloidal chitin solution, 50 mM sodium acetate buffer, pH 5.2, and *V. velella* nematocyst extract at a concentration of 2.4 mg/mL. For the positive control, a 1.2 mg/mL chitinase solution from *Streptomyces griseus* was used, while the negative control contained the same volume of sodium acetate buffer. Samples were then incubated at 37 °C for 1 h. Following incubation, 100 μL of 5% trichloroacetic acid were added to each sample to precipitate all proteins. The samples were centrifuged at 12,000× *g* for 10 min, the supernatant was recovered into clean tubes and the pH was adjusted to neutral. Finally, the reducing sugars produced from the chitin digestion were evaluated using a modified bicinchoninic acid assay [81] to measure the activity of the enzyme. The absorbances were read at 562 nm using an AMR-100 microplate reader. The procedure was carried out in triplicate and results were expressed in units of enzyme/mg of crude extract with respect to the activity of positive control.

5. Conclusions

This study presents the first comprehensive evaluation of the nematocyst extract of the hydrozoan *Velella velella*, utilizing a combination of transcriptomic, proteomic, and biochemical approaches. The analysis revealed a complex mixture of components with a qualitative composition similar to that of more toxic cnidarians, including both spreading factors and homologues of well-known toxins. However, unlike more dangerous species, contact with *V. velella* does not cause medically significant effects in humans. While many effective drugs are derived from highly toxic compounds, their extreme toxicity often complicates clinical trials and requires modifications to reduce harmful effects. In contrast, low- or non-toxic molecules, such as those found in *V. velella*, may require fewer adjustments, making them more promising candidates for drug development with minimal side effects. Additionally, the species-specific database developed in this study allows for the identification of novel compounds through in silico methods; these compounds can be produced with recombinant methods and functionally tested without the need for laborious purification processes, offering potential as bioactive compounds in pharmaceutical applications. Future work could further enhance the transcriptome assembly by incorporating alternative assembly software to explore greater diversity among transcript families [82], aiding in building on the foundation laid in this study and potentially uncovering additional novel bioactive compounds.

Supplementary Materials: The following supporting information can be downloaded at: https://www.mdpi.com/article/10.3390/md22100468/s1, Figure S1: enzymatic activities; Figure S2: completeness assessment, length statistics and completeness of the assembly; Figure S3: protein sequences (fasta format); Table S1: percentage of different proteins found in each band.

Author Contributions: Conceptualization, M.P., M.G. and E.T.; methodology, E.T., I.M. and P.P.; 653 software, E.T. and P.P.; writing—original draft preparation, E.T.; writing—review and editing, M.P., M.G., G.L.M. and L.X.; supervision, M.P.; project administration, M.P.; funding acquisition, M.P. All authors have read and agreed to the published version of the manuscript.

Funding: This research received no external funding.

Data Availability Statement: The data presented in this study are available on request from the corresponding author.

Acknowledgments: Structural mass spectrometry core facility of CIISB, Instruct-CZ Centre, supported by MEYS CR (LM2023042) and European Regional Development Fund-Project "UP CIISB" (No. CZ.02.1.01/0.0/0.0/18_046/0015974). Marco Capello from Department of Earth, Environment and Life Sciences (DISTAV), University of Genova has provided us with lived specimens of *V. velella* from the open sea.

Conflicts of Interest: The authors declare no conflicts of interest.

References

1. Slobodkin, L.B.; Bossert, P.E. Chapter 5—Cnidaria. In *Ecology and Classification of North American Freshwater Invertebrates*, 3rd ed Thorp, J.H., Covich, A.P., Eds.; Academic Press: San Diego, CA, USA, 2010; pp. 125–142, ISBN 978-0-12-374855-3.
2. Technau, U.; Genikhovich, G.; Kraus, J.E.M. Cnidaria. In *Evolutionary Developmental Biology of Invertebrates 1: Introductior Non-Bilateria, Acoelomorpha, Xenoturbellida, Chaetognatha*; Wanninger, A., Ed.; Springer: Vienna, Austria, 2015; pp. 115–163 ISBN 978-3-7091-1862-7.
3. Atkinson, S.D.; Bartholomew, J.L.; Lotan, T. Myxozoans: Ancient Metazoan Parasites Find a Home in Phylum Cnidaria. *Zoology* **2018**, *129*, 66–68. [CrossRef] [PubMed]
4. Berzins, I.K.; Yanong, R.P.E.; LaDouceur, E.E.B.; Peters, E.C. Cnidaria. In *Invertebrate Histology*; LaDouceur, E.E.B., Ed.; Wiley Hoboken, NJ, USA, 2021; pp. 55–86, ISBN 978-1-119-50765-9.
5. Beckmann, A.; Özbek, S. The Nematocyst: A Molecular Map of the Cnidarian Stinging Organelle. *Int. J. Dev. Biol.* **2012**, *56* 577–582. [CrossRef] [PubMed]
6. Santhanam, R. Venomology of Marine Cnidarians. In *Biology and Ecology of Venomous Marine Cnidarians*; Santhanam, R., Ed. Springer: Singapore, 2020; pp. 287–320, ISBN 9789811516030.
7. Jouiaei, M.; Yanagihara, A.A.; Madio, B.; Nevalainen, T.J.; Alewood, P.F.; Fry, B.G. Ancient Venom Systems: A Review on Cnidaria Toxins. *Toxins* **2015**, *7*, 2251–2271. [CrossRef] [PubMed]
8. Trim, C.M.; Byrne, L.J.; Trim, S.A. Chapter One—Utilisation of Compounds from Venoms in Drug Discovery. In *Progress in Medicinal Chemistry*; Witty, D.R., Cox, B., Eds.; Elsevier: Amsterdam, The Netherlands, 2021; Volume 60, pp. 1–66.
9. Frazão, B.; Vasconcelos, V.; Antunes, A. Sea Anemone (Cnidaria, Anthozoa, Actiniaria) Toxins: An Overview. *Mar. Drugs* **2012**, *10* 1812–1851. [CrossRef]
10. Podobnik, M.; Anderluh, G. Pore-Forming Toxins in Cnidaria. *Semin. Cell Dev. Biol.* **2017**, *72*, 133–141. [CrossRef]
11. Cooper, R.A.; De Freitas, J.; Porreca, F.; Eisenhour, C.M.; Lukas, R.; Huxtable, R.J. The Sea Anemone Purine, Caissarone: Adenosine Receptor Antagonism. *Toxicon* **1995**, *33*, 1025–1031. [CrossRef]
12. Tibballs, J.; Yanagihara, A.A.; Turner, H.C.; Winkel, K. Immunological and Toxinological Responses to Jellyfish Stings. *Inflamm Allergy-Drug Targets (Former. Curr. Drug Targets-Inflamm. Allergy)* **2011**, *10*, 438–446. [CrossRef]
13. Lee, H.; Jung, E.; Kang, C.; Yoon, W.D.; Kim, J.-S.; Kim, E. Scyphozoan Jellyfish Venom Metalloproteinases and Their Role in the Cytotoxicity. *Toxicon* **2011**, *58*, 277–284. [CrossRef]
14. Mariottini, G.L.; Grice, I.D. Natural Compounds and Drug Discovery: Can Cnidarian Venom Play a Role? *Cent. Nerv. Syst. Agents Med. Chem.-Cent. Nerv. Syst. Agents)* **2019**, *19*, 114–118. [CrossRef]
15. D'Ambra, I.; Lauritano, C. A Review of Toxins from Cnidaria. *Mar. Drugs* **2020**, *18*, 507. [CrossRef]
16. Cegolon, L.; Heymann, W.C.; Lange, J.H.; Mastrangelo, G. Jellyfish Stings and Their Management: A Review. *Mar. Drugs* **2013**, *11*, 523–550. [CrossRef] [PubMed]
17. Killi, N.; Bonello, G.; Mariottini, G.L.; Pardini, P.; Pozzolini, M.; Cengiz, S. Nematocyst Types and Venom Effects of *Aurelia aurita* and *Velella velella* from the Mediterranean Sea. *Toxicon* **2020**, *175*, 57–63. [CrossRef] [PubMed]
18. Amreen Nisa, S.; Vinu, D.; Krupakar, P.; Govindaraju, K.; Sharma, D.; Vivek, R. Jellyfish Venom Proteins and Their Pharmacological Potentials: A Review. *Int. J. Biol. Macromol.* **2021**, *176*, 424–436. [CrossRef] [PubMed]
19. Maduraiveeran, H.; Raja, K.; Chinnasamy, A. Antiproliferative and Antioxidant Properties of Nematocysts Crude Venom from Jellyfish *Acromitus flagellatus* against Human Cancer Cell Lines. *Saudi J. Biol. Sci.* **2021**, *28*, 1954–1961. [CrossRef]
20. D'Ambra, I.; Merquiol, L. Jellyfish from Fisheries By-Catches as a Sustainable Source of High-Value Compounds with Biotechnological Applications. *Mar. Drugs* **2022**, *20*, 266. [CrossRef]
21. de Oliveira, S.A.M.; Magalhães, M.R.; de Oliveira, L.P.; da Cunha, L.C. Identification of Antinociceptive Fraction of Snake Venom from *Crotalus durissus collilineatus* Crotamine-Negative and Its Acute Toxicity Evaluation. *Toxicon* **2016**, *122*, 145–151. [CrossRef]
22. Cardoso, F.C.; Dekan, Z.; Rosengren, K.J.; Erickson, A.; Vetter, I.; Deuis, J.R.; Herzig, V.; Alewood, P.F.; King, G.F.; Lewis, R.J. Identification and Characterization of ProTx-III [μ-TRTX-Tp1a], a New Voltage-Gated Sodium Channel Inhibitor from Venom of the Tarantula *Thrixopelma pruriens*. *Mol. Pharmacol.* **2015**, *88*, 291–303. [CrossRef]
23. Lima, P.C.; Bordon, K.C.F.; Pucca, M.B.; Cerni, F.A.; Zoccal, K.F.; Faccioli, L.H.; Arantes, E.C. Partial Purification and Functional Characterization of Ts19 Frag-I, a Novel Toxin from *Tityus serrulatus* Scorpion Venom. *J. Venom. Anim. Toxins Incl. Trop. Dis.* **2015**, *21*, 49. [CrossRef]
24. Nguyen, B.; Le Caer, J.-P.; Aráoz, R.; Thai, R.; Lamthanh, H.; Benoit, E.; Molgó, J. Isolation, Purification and Functional Characterization of Alpha-BnIA from *Conus bandanus* Venom. *Toxicon* **2014**, *91*, 155–163. [CrossRef]
25. Weston, A.J.; Chung, R.; Dunlap, W.C.; Morandini, A.C.; Marques, A.C.; Moura-da-Silva, A.M.; Ward, M.; Padilla, G.; da Silva, L.F.; Andreakis, N.; et al. Proteomic Characterisation of Toxins Isolated from Nematocysts of the South Atlantic Jellyfish *Olindias sambaquiensis*. *Toxicon* **2013**, *71*, 11–17. [CrossRef]
26. Brinkman, D.L.; Jia, X.; Potriquet, J.; Kumar, D.; Dash, D.; Kvaskoff, D.; Mulvenna, J. Transcriptome and Venom Proteome of the Box Jellyfish *Chironex fleckeri*. *BMC Genom.* **2015**, *16*, 407. [CrossRef] [PubMed]
27. Choudhary, I.; Hwang, D.H.; Lee, H.; Yoon, W.D.; Chae, J.; Han, C.H.; Yum, S.; Kang, C.; Kim, E. Proteomic Analysis of Novel Components of *Nemopilema nomurai* Jellyfish Venom: Deciphering the Mode of Action. *Toxins* **2019**, *11*, 153. [CrossRef] [PubMed]

28. Leung, T.C.N.; Qu, Z.; Nong, W.; Hui, J.H.L.; Ngai, S.M. Proteomic Analysis of the Venom of Jellyfishes *Rhopilema esculentum* and *Sanderia malayensis*. *Mar. Drugs* **2020**, *18*, 655. [CrossRef] [PubMed]
29. Modahl, C.M.; Durban, J.; Mackessy, S.P. Exploring Toxin Evolution: Venom Protein Transcript Sequencing and Transcriptome-Guided High-Throughput Proteomics. In *Snake and Spider Toxins: Methods and Protocols*; Priel, A., Ed.; Springer: New York, NY, USA, 2020; pp. 97–127, ISBN 978-1-4939-9845-6.
30. Modahl, C.M.; Saviola, A.J.; Mackessy, S.P. Integration of Transcriptomic and Proteomic Approaches for Snake Venom Profiling. *Expert Rev. Proteom.* **2021**, *18*, 827–834. [CrossRef]
31. Betti, F.; Bo, M.; Enrichetti, F.; Manuele, M.; Cattaneo-Vietti, R.; Bavestrello, G. Massive Strandings of *Velella velella* (Hydrozoa: Anthoathecata: Porpitidae) in the Ligurian Sea (North-Western Mediterranean Sea). *Eur. Zool. J.* **2019**, *86*, 343–353. [CrossRef]
32. Kirkpatrick, P.A.; Pugh, P.R. *Siphonophores and Velellids: Keys and Notes for the Identification of the Species*; Brill Archive: Boston, MA, USA, 1984; ISBN 978-90-04-07470-5.
33. Jaimes-Becerra, A.; Chung, R.; Morandini, A.C.; Weston, A.J.; Padilla, G.; Gacesa, R.; Ward, M.; Long, P.F.; Marques, A.C. Comparative Proteomics Reveals Recruitment Patterns of Some Protein Families in the Venoms of Cnidaria. *Toxicon* **2017**, *137*, 19–26. [CrossRef]
34. Doonan, L.B.; Lynham, S.; Quinlan, C.; Ibiji, S.C.; Winter, C.E.; Padilla, G.; Jaimes-Becerra, A.; Morandini, A.C.; Marques, A.C.; Long, P.F. Venom Composition Does Not Vary Greatly Between Different Nematocyst Types Isolated from the Primary Tentacles of *Olindias sambaquiensis* (Cnidaria: Hydrozoa). *Biol. Bull.* **2019**, *237*, 26–35. [CrossRef]
35. Jaimes-Becerra, A.; Gacesa, R.; Doonan, L.B.; Hartigan, A.; Marques, A.C.; Okamura, B.; Long, P.F. "Beyond Primary Sequence"—Proteomic Data Reveal Complex Toxins in Cnidarian Venoms. *Integr. Comp. Biol.* **2019**, *59*, 777–785. [CrossRef]
36. Holstein, T.W.; Benoit, M.; Herder, G.V.; David, C.N.; Wanner, G.; Gaub, H.E. Fibrous Mini-Collagens in Hydra Nematocysts. *Science* **1994**, *265*, 402–404. [CrossRef]
37. David, C.N.; Özbek, S.; Adamczyk, P.; Meier, S.; Pauly, B.; Chapman, J.; Hwang, J.S.; Gojobori, T.; Holstein, T.W. Evolution of Complex Structures: Minicollagens Shape the Cnidarian Nematocyst. *Trends Genet.* **2008**, *24*, 431–438. [CrossRef]
38. Watson, G.M.; Mire-Thibodeaux, P. The Cell Biology of Nematocysts. In *International Review of Cytology*; Jeon, K.W., Jarvik, J., Eds.; Academic Press: Cambridge, MA, USA, 1994; Volume 156, pp. 275–300.
39. Kass-Simon, G.; Scappaticci, A.A., Jr. The Behavioral and Developmental Physiology of Nematocysts. *Can. J. Zool.* **2002**, *80*, 1772–1794. [CrossRef]
40. Berking, S.; Herrmann, K. Formation and Discharge of Nematocysts Is Controlled by a Proton Gradient across the Cyst Membrane. *Helgol. Mar. Res.* **2006**, *60*, 180–188. [CrossRef]
41. Sabatier, J.-M.; Kovacic, H.; Shahbazzadeh, D.; Wang, J.-L.; Cao, Z.; McNutt, P.M.; Wulff, H.; Utkin, Y.N. *Venoms, Animal and Microbial Toxins, Volume II*; Frontiers Media SA: Lausanne, Switzerland, 2022; ISBN 978-2-8325-0114-6.
42. Fox, J.W.; Serrano, S.M.T. Structural Considerations of the Snake Venom Metalloproteinases, Key Members of the M12 Reprolysin Family of Metalloproteinases. *Toxicon* **2005**, *45*, 969–985. [CrossRef]
43. Li, A.; Yu, H.; Li, R.; Yue, Y.; Yu, C.; Geng, H.; Liu, S.; Xing, R.; Li, P. Jellyfish *Nemopilema nomurai* Causes Myotoxicity through the Metalloprotease Component of Venom. *Biomed. Pharmacother.* **2022**, *151*, 113192. [CrossRef]
44. García-Arredondo, A.; Rojas-Molina, A.; Ibarra-Alvarado, C.; Lazcano-Pérez, F.; Arreguín-Espinosa, R.; Sánchez-Rodríguez, J. Composition and Biological Activities of the Aqueous Extracts of Three Scleractinian Corals from the Mexican Caribbean: *Pseudodiploria strigosa*, *Porites astreoides* and *Siderastrea siderea*. *J. Venom. Anim. Toxins Incl. Trop. Dis.* **2016**, *22*, 32. [CrossRef]
45. Liu, G.; Zhou, Y.; Liu, D.; Wang, Q.; Ruan, Z.; He, Q.; Zhang, L. Global Transcriptome Analysis of the Tentacle of the Jellyfish *Cyanea capillata* Using Deep Sequencing and Expressed Sequence Tags: Insight into the Toxin- and Degenerative Disease-Related Transcripts. *PLoS ONE* **2015**, *10*, e0142680. [CrossRef]
46. Ponce, D.; Brinkman, D.L.; Potriquet, J.; Mulvenna, J. Tentacle Transcriptome and Venom Proteome of the Pacific Sea Nettle, *Chrysaora fuscescens* (Cnidaria: Scyphozoa). *Toxins* **2016**, *8*, 102. [CrossRef]
47. Koh, C.Y.; Kini, R.M. From Snake Venom Toxins to Therapeutics—Cardiovascular Examples. *Toxicon* **2012**, *59*, 497–506. [CrossRef]
48. Ra, H.-J.; Parks, W.C. Control of Matrix Metalloproteinase Catalytic Activity. *Matrix Biol.* **2007**, *26*, 587–596. [CrossRef]
49. Lal, D.M.; Calton, G.J.; Neeman, I.; Burnett, J.W. Characterization of *Physalia physalis* (Portugese Man-o'war) Nematocyst Venom Collagenase. *Comp. Biochem. Physiol. Part B Comp. Biochem.* **1981**, *70*, 635–638. [CrossRef]
50. Nevalainen, T.J.; Peuravuori, H.J.; Quinn, R.J.; Llewellyn, L.E.; Benzie, J.A.H.; Fenner, P.J.; Winkel, K.D. Phospholipase A2 in Cnidaria. *Comp. Biochem. Physiol. Part B Biochem. Mol. Biol.* **2004**, *139*, 731–735. [CrossRef] [PubMed]
51. Rachamim, T.; Morgenstern, D.; Aharonovich, D.; Brekhman, V.; Lotan, T.; Sher, D. The Dynamically Evolving Nematocyst Content of an Anthozoan, a Scyphozoan, and a Hydrozoan. *Mol. Biol. Evol.* **2015**, *32*, 740–753. [CrossRef] [PubMed]
52. Murakami, T.; Kamikado, N.; Fujimoto, R.; Hamaguchi, K.; Nakamura, H.; Chijiwa, T.; Ohno, M.; Oda-ueda, N. A [Lys49]Phospholipase A2 from *Protobothrops flavoviridis* Venom Induces Caspase-Independent Apoptotic Cell Death Accompanied by Rapid Plasma-Membrane Rupture in Human Leukemia Cells. *Biosci. Biotechnol. Biochem.* **2011**, *75*, 864–870. [CrossRef] [PubMed]
53. Neeman, I.; Calton, G.J.; Burnett, J.W. Purification and Characterization of the Endonuclease Present in *Physalia physalis* Venom. *Comp. Biochem. Physiol. Part B Comp. Biochem.* **1980**, *67*, 155–158. [CrossRef]

54. Neeman, I.; Calton, G.J.; Burnett, J.W. Purification of an Endonuclease Present in *Chrysaora quinquecirrha* Venom. *Proc. Soc. Exp Biol. Med.* **1981**, *166*, 374–382. [CrossRef]
55. Purcell, J.E.; Clarkin, E.; Doyle, T.K. Foods of *Velella velella* (Cnidaria: Hydrozoa) in Algal Rafts and Its Distribution in Irish Sea *Hydrobiologia* **2012**, *690*, 47–55. [CrossRef]
56. Patil, R.S.; Ghormade, V.; Deshpande, M.V. Chitinolytic Enzymes: An Exploration. *Enzym. Microb. Technol.* **2000**, *26*, 473–48 [CrossRef]
57. Matsui, T.; Fujimura, Y.; Titani, K. Snake Venom Proteases Affecting Hemostasis and Thrombosis. *Biochim. Biophys. Act (BBA)-Protein Struct. Mol. Enzymol.* **2000**, *1477*, 146–156. [CrossRef]
58. Sulca, M.A.; Remuzgo, C.; Cárdenas, J.; Kiyota, S.; Cheng, E.; Bemquerer, M.P.; Machini, M.T. Venom of the Peruvian Snak *Bothriopsis oligolepis*: Detection of Antibacterial Activity and Involvement of Proteolytic Enzymes and C-Type Lectins in Growt Inhibition of *Staphylococcus aureus*. *Toxicon* **2017**, *134*, 30–40. [CrossRef]
59. Fiorotti, H.B.; Soares, T.G.; Borges, M.H.; Matavel, A.; Campos, F.V.; Figueiredo, S.G.D. Preliminary Report on the Hemagglutina ing Activity of the *Scorpaena plumieri* Fish Venom. *An. Acad. Bras. Ciênc.* **2022**, *94*, e20200976. [CrossRef]
60. East, L.; Isacke, C.M. The Mannose Receptor Family. *Biochim. Biophys. Acta (BBA)-Gen. Subj.* **2002**, *1572*, 364–386. [CrossRef]
61. von Reumont, B.M.; Anderluh, G.; Antunes, A.; Ayvazyan, N.; Beis, D.; Caliskan, F.; Crnković, A.; Damm, M.; Dutertre, S Ellgaard, L.; et al. Modern Venomics—Current Insights, Novel Methods, and Future Perspectives in Biological and Appliec Animal Venom Research. *GigaScience* **2022**, *11*, giac048. [PubMed]
62. Ribeiro, C.; Vignes, M.; Brehélin, M. *Xenorhabdus nematophila* (Enterobacteriacea) Secretes a Cation-Selective Calcium-Independen Porin Which Causes Vacuolation of the Rough Endoplasmic Reticulum and Cell Lysis. *J. Biol. Chem.* **2003**, *278*, 3030–3039 [CrossRef] [PubMed]
63. Anderluh, G.; Maček, P. Cytolytic Peptide and Protein Toxins from Sea Anemones (Anthozoa: Actiniaria). *Toxicon* **2002**, *40* 111–124. [CrossRef] [PubMed]
64. Kuhn-Nentwig, L.; Langenegger, N.; Heller, M.; Koua, D.; Nentwig, W. The Dual Prey-Inactivation Strategy of Spiders—In-Dept Venomic Analysis of *Cupiennius salei*. *Toxins* **2019**, *11*, 167. [CrossRef]
65. Campos, F.V.; Fiorotti, H.B.; Coitinho, J.B.; Figueiredo, S.G. Fish Cytolysins in All Their Complexity. *Toxins* **2021**, *13*, 877 [CrossRef]
66. Castañeda, O.; Sotolongo, V.; Amor, A.M.; Stöcklin, R.; Anderson, A.J.; Harvey, A.L.; Engström, Å.; Wernstedt, C.; Karlsson E. Characterization of a Potassium Channel Toxin from the Caribbean Sea Anemone *Stichodactyla helianthus*. *Toxicon* **1995**, *33* 603–613. [CrossRef]
67. Escoubas, P.; Rash, L. Tarantulas: Eight-Legged Pharmacists and Combinatorial Chemists. *Toxicon* **2004**, *43*, 555–574. [CrossRef]
68. Rangaraju, S.; Khoo, K.; Feng, Z.; Crossley, G.; Nugent, D.; Khaytin, I.; Pennington, M.; Norton, R.; Chandy, K.G. Potassium Channel Modulation by A Toxin Domain in Matrix Metalloprotease 23. *Biophys. J.* **2010**, *98*, 212a. [CrossRef]
69. Terstappen, G.C.; Schlüpen, C.; Raggiaschi, R.; Gaviraghi, G. Target Deconvolution Strategies in Drug Discovery. *Nat. Rev. Drug Discov.* **2007**, *6*, 891–903. [CrossRef]
70. Botelho, D.; Wall, M.J.; Vieira, D.B.; Fitzsimmons, S.; Liu, F.; Doucette, A. Top-Down and Bottom-Up Proteomics of SDS-Containing Solutions Following Mass-Based Separation. *J. Proteome Res.* **2010**, *9*, 2863–2870. [CrossRef] [PubMed]
71. Laemmli, U.K. Cleavage of Structural Proteins during the Assembly of the Head of Bacteriophage T4. *Nature* **1970**, *227*, 680–685 [CrossRef] [PubMed]
72. Shevchenko, A.; Tomas, H.; Havli, J.; Olsen, J.V.; Mann, M. In-Gel Digestion for Mass Spectrometric Characterization of Proteins and Proteomes. *Nat. Protoc.* **2006**, *1*, 2856–2860. [CrossRef] [PubMed]
73. Jungo, F.; Bougueleret, L.; Xenarios, I.; Poux, S. The UniProtKB/Swiss-Prot Tox-Prot Program: A Central Hub of Integrated Venom Protein Data. *Toxicon* **2012**, *60*, 551–557. [CrossRef]
74. Junior, N.; Barros, R.D.A.; Vital, C.E.; Barbosa, S.L.; de Oliveira, J.V.A.; da Rocha, G.C.; Oliveira, C.N.; Assis, J.V.M.G. *Enzymatic Assay of Protease Using Azocasein as Substrate*; Springer Nature: Berlin, Germany, 1970.
75. Fu, X.; Xue, C.; Miao, B.; Li, Z.; Gao, X.; Yang, W. Characterization of Proteases from the Digestive Tract of Sea Cucumber (*Stichopus japonicus*): High Alkaline Protease Activity. *Aquaculture* **2005**, *246*, 321–329. [CrossRef]
76. Van Wart, H.E.; Steinbrink, D.R. A Continuous Spectrophotometric Assay for *Clostridium histolyticum* Collagenase. *Anal. Biochem.* **1981**, *113*, 356–365. [CrossRef]
77. Wilkesman, J.; Kurz, L. Advances in Zymography Techniques and Patents Regarding Protease Analysis. *Recent Pat. Biotechnol.* **2012**, *6*, 106–114. [CrossRef]
78. Pukrittayakamee, S.; Warrell, D.A.; Desakorn, V.; McMichael, A.J.; White, N.J.; Bunnag, D. The Hyaluronidase Activities of Some Southeast Asian Snake Venoms. *Toxicon* **1988**, *26*, 629–637. [CrossRef]
79. Laxme, R.R.S.; Khochare, S.; de Souza, H.F.; Ahuja, B.; Suranse, V.; Martin, G.; Whitaker, R.; Sunagar, K. Beyond the 'Big Four': Venom Profiling of the Medically Important yet Neglected Indian Snakes Reveals Disturbing Antivenom Deficiencies. *PLoS Negl. Trop. Dis.* **2019**, *13*, e0007899. [CrossRef]
80. Shen, C.-R.; Chen, Y.-S.; Yang, C.-J.; Chen, J.-K.; Liu, C.-L. Colloid Chitin Azure Is a Dispersible, Low-Cost Substrate for Chitinase Measurements in a Sensitive, Fast, Reproducible Assay. *J. Biomol. Screen.* **2010**, *15*, 213–217. [CrossRef]

1. Hussain, M.H.M.; Ngaini, Z.; Chong, N.F.M. Modified Bicinchoninic Acid Assay for Accurate Determination of Variable Length Reducing Sugars in Carbohydrates. *Int. Food Res. J.* **2018**, *25*, 2614.
2. Holding, M.L.; Margres, M.J.; Mason, A.J.; Parkinson, C.L.; Rokyta, D.R. Evaluating the Performance of De Novo Assembly Methods for Venom-Gland Transcriptomics. *Toxins* **2018**, *10*, 249. [CrossRef]

marine drugs

Article

Exploring the Potential of *Crassostrea nippona* Hydrolysates as Dietary Supplements for Mitigating Dexamethasone-Induced Muscle Atrophy in C2C12 Cells

M. J. M. S. Kurera [1,2,†], D. P. Nagahawatta [1,†], N. M. Liyanage [1], H. H. A. C. K. Jayawardhana [1], D. S. Dissanayake [1], Hyo-Geun Lee [1], Young-Sang Kim [1,3], Sang In Kang [4,*] and You-Jin Jeon [1,3,*]

1 Department of Marine Life Sciences, Jeju National University, Jeju 63243, Republic of Korea; shehan.kurera@stu.jejunu.ac.kr (M.J.M.S.K.); pramuditha1992@jejunu.ac.kr (D.P.N.); liyanagenm@jejunu.ac.kr (N.M.L.); chathuri.k.j@stu.jejunu.ac.kr (H.H.A.C.K.J.); dinudissanayake95@stu.jejunu.ac.kr (D.S.D.); hyogeunlee92@jejunu.ac.kr (H.-G.L.); yskim@jejunu.ac.kr (Y.-S.K.)
2 Department of Biotechnology, Faculty of Agriculture and Plantation Management, Wayamba University of Sri Lanka, Makandura, Gonawila 60170, NWP, Sri Lanka
3 Marine Science Institute, Jeju National University, Jeju 63333, Republic of Korea
4 Seafood Research Center, Silla University, Busan 49277, Republic of Korea
* Correspondence: ftrnd5@silla.ac.kr (S.I.K.); youjinj@jejunu.ac.kr (Y.-J.J.); Tel.: +82-053-717-4312 (S.I.K.); +82-064-754-3475 (Y.-J.J.)
† These authors contributed equally to this work.

Abstract: Muscle atrophy is a detrimental and injurious condition that leads to reduced skeletal muscle mass and disruption of protein metabolism. Oyster (*Crassostrea nippona*) is a famous and commonly consumed shellfish in East Asia and has become a popular dietary choice worldwide. The current investigation evaluated the efficacy of *C. nippona* against muscle atrophy, which has become a severe health issue. Mammalian skeletal muscles are primarily responsible for efficient metabolism, energy consumption, and body movements. The proteins that regulate muscle hypertrophy and atrophy are involved in muscle growth. *C. nippona* extracts were enzymatically hydrolyzed using alcalase (AOH), flavourzyme (FOH), and protamex (POH) to evaluate their efficacy in mitigating dexamethasone-induced muscle damage in C2C12 cells in vitro. AOH exhibited notable cell proliferative abilities, promoting dose-dependent myotube formation. These results were further solidified by protein expression analysis. Western blot and gene expression analysis via RT-qPCR demonstrated that AOH downregulated MuRF-1, Atrogin, Smad 2/3, and Foxo-3a, while upregulating myogenin, MyoD, myosin heavy chain expression, and mTOR, key components of the ubiquitin–proteasome and mTOR signaling pathways. Finally, this study suggests that AOH holds promise for alleviating dexamethasone-induced muscle atrophy in C2C12 cells in vitro, offering insights for developing functional foods targeting conditions akin to sarcopenia.

Keywords: *Crassostrea nippona*; enzyme hydrolysates; functional foods; marine shellfish; muscle atrophy; skeletal muscles

1. Introduction

Sarcopenia is a health condition that leads to diminished skeletal muscle mass, muscle fiber number, and muscle area with increasing age. It destructively impacts a healthy life and is a high-risk factor for the physical disability known as muscle atrophy. Muscle atrophy leads to an interruption in the balance between protein synthesis and degradation, lessening the protein content, reducing fiber diameter, and forming fatigue resistance in the body [1]. At ages above 60, the potential occurrence of muscle atrophy-related sarcopenia increases to approximately 5–13% [2]. Hence, these unfavorable conditions influence the regular lifestyle of humans and impair their overall well-being. Therefore, functional foods, pharmaceuticals, and therapeutic products must address this health condition. Muscle protein synthesis and breakdown are the two significant processes of skeletal muscle growth. When conditions exceed protein synthesis during protein breakdown, net protein synthesis occurs in the body, leading to muscle growth [3]. Exercise is a key factor in muscle protein metabolism. Exception for exercise, proper nutrition, and hormonal stimulation are highly impacted by muscle development [4]. Skeletal muscles consist of bundles of myocytes, each of which contains myofibrils that permit muscle contraction. When myofibrils heave, they increase muscle strength and density [5]. Skeletal muscles comprise approximately two-fifths of the total body mass and contribute to the convention of creating movements, maintaining body posture and position, regulating body temperature, accumulating nutrients, soothing joints, and maintaining overall metabolic homeostasis. Therefore, proper muscle conditions are necessary to maintain a healthy body [6].

Marine organisms are potential candidates for producing valuable secondary metabolites, especially bioactive compounds such as alkaloids, flavonoids, phenolic compounds, terpenes, and steroids. These compounds have antioxidant, anti-microbial, anti-diabetic, anti-hypertension, anti-obesity, anti-cancer, and anti-aging properties [7–13]. Further, due to their high potential, they contribute to drug discovery, cancer therapies, anti-microbial functions, and many other applications in the current world effectively [14–20]. Shellfish play a crucial role in supporting muscle hypertrophy due to their rich nutritional profile. These marine delicacies are abundant sources of high-quality proteins, essential amino acids, and bioactive compounds, all of which are fundamental for muscle growth and repair. Proteins in shellfish contribute to synthesizing muscle proteins, fostering the development and maintenance of lean muscle mass. Additionally, shellfish are notable for providing important micronutrients such as zinc, selenium, and vitamin B12, which play key roles in various metabolic processes, including those involved in muscle function and growth [21,22]. The omega-3 fatty acids found in certain shellfish varieties further contribute to muscle health by reducing inflammation and supporting overall muscle function. Including shellfish in the diet can be a valuable strategy for individuals aiming to enhance muscle hypertrophy, making them a nutritious and flavorful addition to a muscle-building nutrition plan.

Mollusks are the second largest category of farmed seafood, accounting for 21% of all global aquaculture production, and oysters represent a key component in the industry [23]. Few studies have shown the potential of marine shellfish for muscle growth [6,24–26]. *Crassostrea nippona* is a shellfish that belongs to the class Bivalves, the phylum Mollusks, the order Ostreoida, and the family Ostreidae, and is located in the marine environments of East Asian countries such as Republic of Korea, China, and Japan [27]. In 1913, Fujita T. introduced the taxonomic categorization of *Crassostrea nippona*. A comprehensive study on the taxonomy and phylogeny of Crassostrea species, including *C. nippona*, was conducted by Liu et al. (2017) [28]. Due to the high glycogen component in their bodies, accompanied by a delicious taste and unique flavor, *C. nippona* is preferred in summer over other oyster species, especially the Pacific oysters (*Crassostrea gigas*) [29]. These oyster species contain high-quality proteins along with essential amino acids, vitamins, minerals, and omega-3 fatty acids [30]. Oyster proteins can be degraded into several peptides via enzymatic hydrolysis and have prominent physiological activities. Previous studies have shown that oyster hydrolysates have significant anti-cancer, anti-inflammatory, and antioxidant properties [31]. Nonetheless, the muscle growth effects and anti-muscle atrophy aptitude-related molecular mechanisms of enzyme hydrolysates isolated from shellfish species are not well understood. Hence, further feasibility studies are required to be carried out to explore the mechanism in different species.

Dexamethasone (DEXA) is a synthetic glucocorticoid widely used to induce proteolytic muscle atrophy in both in vivo and in vitro models and used to treat medical conditions such as autoimmune disorders and inflammations [32,33]. It also decreases protein synthesis and promotes the breakdown of proteins related to skeletal muscle mass via the ubiquitin–proteasome systems [34]. DEXA also delivers the necessary environmental conditions to downregulate the mammalian target of the rapamycin mTOR pathway to promote the breakdown of protein synthesis inside the cell [2]. Therefore, it is important to investigate therapeutic approaches to treat this condition to prevent DEXA-induced muscle atrophy, introduce anti-muscle atrophy compounds to prevent important clinical implications, and investigate anti-muscle atrophy drugs or foods to treat these conditions effectively. Numerous studies have shown that enzyme-assisted extraction methods provide outstanding outcomes, including efficient extraction of bioactive compounds with greater extraction yields and protein content, which are highly recognized for superlative muscle growth [6,35–39].

Therefore, this study aimed to prepare oyster hydrolysates from *C. nippona* hydrolyzed with three different enzymes, alcalase, flavourzyme, and protamex, and evaluate their ability to prevent or delay muscle atrophy in DEXA-induced C2C12 cells in vitro. Moreover, a comparative study was also performed to determine which *C. nippona* hydrolysates obtained from alcalase, flavourzyme, and protamex treatments can be applied convincingly for product development in downregulating DEXA-induced muscle atrophy in skeletal muscles of C2C12 cells in vitro. In addition, given that the development of functional foods and the demonstration of the suitability of enzyme hydrolysates of *C. nippona* is also necessary in the modern world, this study determined the effectiveness of these samples in the development of functional foods to treat atrophic disorders, such as sarcopenia, in a promising way.

2. Results

2.1. General Composition of the Enzyme Hydrolysates of Crassostrea Nippona

The proximate analyses of AOH, FOH, POH, and DWO are summarized in Table 1. Polysaccharides, polyphenols, proteins, lipids, ash, and yield were determined on a dry-weight basis. The yields of AOH, FOH, POH, and DWO were obtained as 75.33 ± 1.08%, 68.33 ± 0.58%, 61.33 ± 0.58%, and 14.00 ± 1.00%, respectively. The yields of all hydrolysates were significantly higher than those of the DWO sample, with that of AOH being the highest. The highest protein content was acquired in AOH as 45.90 ± 2.71%, while FOH, POH, and DWO had protein recovery percentages of 36.02 ± 2.42%, 43.68 ± 1.57%, and 40.05 ± 0.43%, respectively. AOH delivered a higher polysaccharide yield than FOH, POH, and DWO. As expected, the polyphenol content of all four samples was very low, and no significant differences were observed in the lipid profiles of all samples analyzed in this study.

Table 1. Hydrolysate yield and proximate analysis of the protein hydrolysate of *Crassostrea nippona* on muscle growth.

Component	Composition			
	DWO	AOH	FOH	POH
Polysaccharide	2.72 ± 0.47%	23.83 ± 1.42% ****	6.86 ± 0.45% ****	11.83 ± 0.50% ****
Polyphenol	0.47 ± 0.01%	0.51 ± 0.00% ****	0.45 ± 0.01%	0.50 ± 0.01% **
Proteins	40.05 ± 0.43%	45.90 ± 2.71% ****	36.02 ± 2.42%	43.68 ± 1.57% ****
Lipids	6.54 ± 1.56%	8.94 ± 1.67% ****	8.14 ± 1.42% ****	8.50 ± 1.22% ****
Ash	14.00 ± 1.00%	75.33 ± 1.08% ****	68.33 ± 0.58% ****	61.33 ± 0.58% ****
Yield	0.47 ± 0.01%	0.51 ± 0.00% ****	0.45 ± 0.01%	0.50 ± 0.01% **

The experiments were triplicated and data are represented as mean ± SE. Significance is denoted compared to control. ** $p < 0.01$., **** $p < 0.0001$; DWO: distilled water oyster; AOH: alcalase oyster hydrolysate; FOH: flavourzyme oyster hydrolysate; POH: protamex oyster hydrolysate.

2.2. Cell Cytotoxicity and Proliferation Activity of the Enzyme Hydrolysates of Crassostrea Nippona

The muscle growth activity of the three different enzyme hydrolysates obtained from *C. nippona* was evaluated by assessing cell cytotoxicity, proliferation, and differentiation. The toxic effects and proliferative ability of *C. nippona* in C2C12 cells were determined using CCK8 and BrdU assays, respectively, in a dose-dependent manner. According to the consideration of cell toxicity degrees, results confirmed that all three hydrolysate samples (AOH, FOH, and POH) and DWO were nontoxic to C2C12 cells at 25, 50, and 100 μg/mL concentrations (Figure 1a–d).

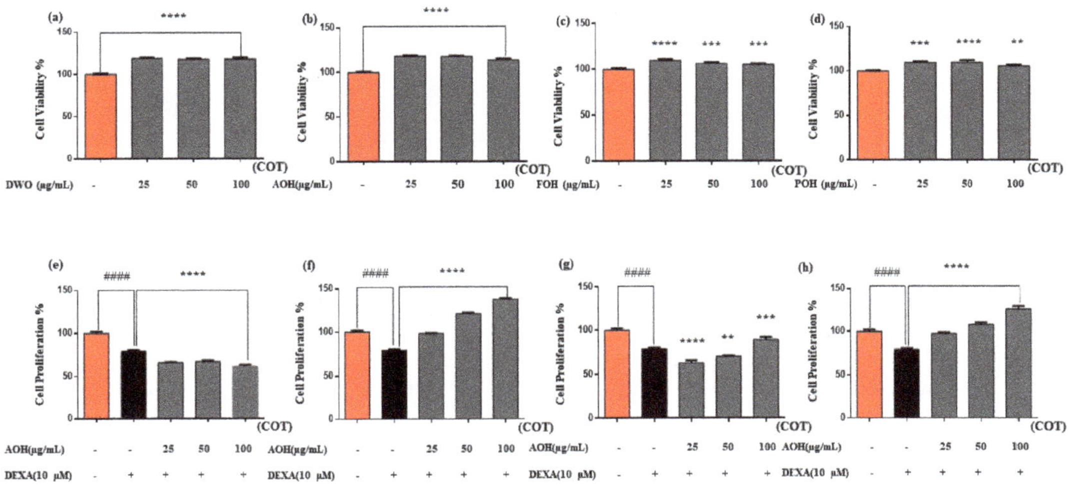

Figure 1. Cell viability % of enzyme hydrolysates of *C. nippona*. (**a**) DWO; (**b**) AOH; (**c**) FOH; (**d**) POH; cell proliferation effect of protein hydrolysate of *C. nippona* (**e**) DWO; (**f**) AOH; (**g**) FOH; (**h**) POH. Experiments were carried out in triplicate and the results are presented as means ± SD (n = 3). Significance is denoted compared to the control #### $p < 0.0001$ and DEXA groups; ** $p < 0.01$, *** $p < 0.001$, **** $p < 0.0001$. DWO: distilled water oyster; AOH: alcalase oyster hydrolysate; FOH: flavourzyme oyster hydrolysate; POH: protamex oyster hydrolysate; DEXA: dexamethasone; COT: concentration of treatments.

Cell proliferation results showed that both AOH and POH had a significant cell proliferation influence over the DEXA-treated group in C2C12 cells in a dose-dependent manner. FOH did not perform significantly on cell proliferation at the low-concentration ranges of 25 and 50 μg/mL. At the same time, the DWO group was unable to perform any cell-productive effect at any range over the DEXA-treated group. DEXA concentration was maintained constantly throughout the differentiation process at 10 μM [40]. The significance of each hydrolysate sample was assessed over the DEXA treatment group (Figure 1e–h).

2.3. Immunostaining and Myotube Diameters on MyHC Expression

Muscle growth was assessed by immunostaining with myosin heavy chain (MyHC) primary antibodies, followed by analysis of the myotube diameters of each stained caption relevant to specific hydrolysate groups and respective concentrations. MyHC observations, nuclear staining of 4′,6-diamidino-2-phenylindole (DAPI), and merged images of MyHC and DAPI for 25, 50, and 100 μg/mL concentrations for AOH and POH are shown in Figures 2a and 2b, respectively. Myotube quantification data for AOH and POH at 25, 50, and 100 μg/mL concentrations is shown in Figure 2c,d. The results indicate that MyHC expression and myotube diameters in differentiated C2C12 cells treated with AOH and POH were significantly upregulated by DEXA in a dose-dependent manner (Figure 2).

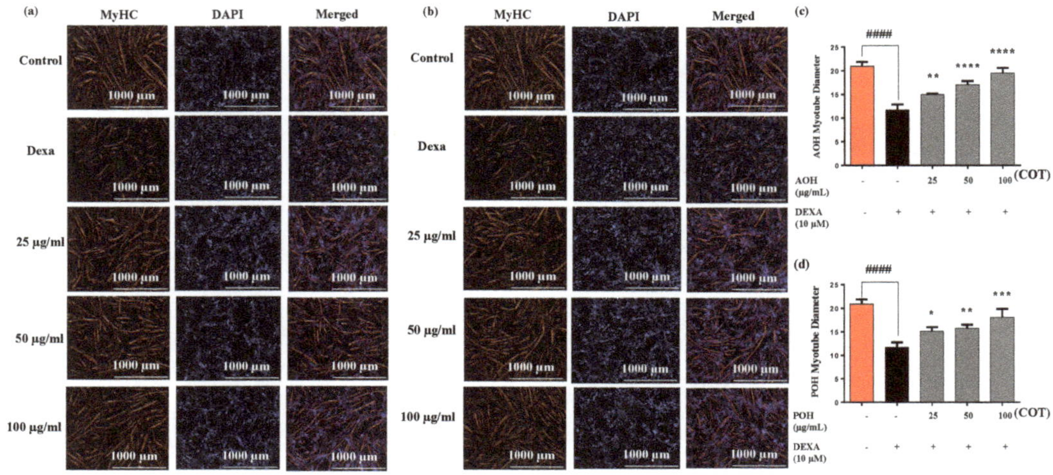

Figure 2. Immunofluorescence analysis of myosin heavy chains of enzyme hydrolysates of *Crassostrea nippona*. (**a**) AOH; (**b**) POH; (**c**) myotube diameter scoring analysis for AOH; (**d**) POH. Experiments were carried out in triplicate and the results are presented as means ± SD (n = 3). Significance is denoted compared to the control #### $p < 0.0001$ and DEXA group; ** $p < 0.01$, *** $p < 0.001$, **** $p < 0.0001$. DWO: distilled water oyster; AOH: alcalase oyster hydrolysate; POH: protamex oyster hydrolysate; DEXA: dexamethasone; COT: concentrations of treatments.

2.4. Molecular Weight Determination and Amino Acid Composition

The average molecular weight of the AOH-treated differentiated *C. nippona* was determined and is shown in Figure 3. The protein band was visualized as a smear distributed over the 15–75 kDa range and highly concentrated in the 63–75 kDa range. The amino acid compositions of the same sample (AOH) are presented in Table 2.

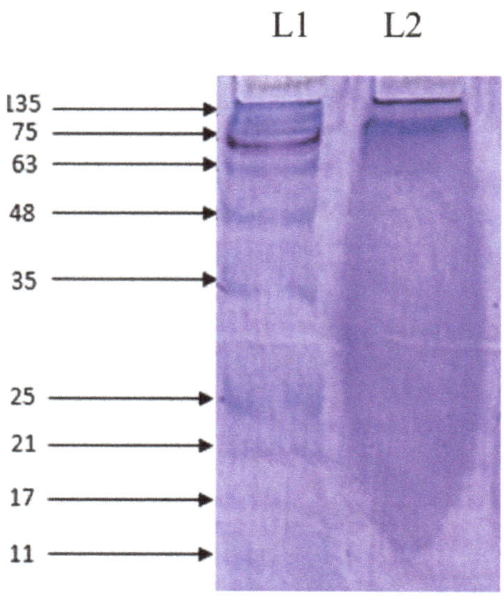

Figure 3. Determination of the molecular weight distribution of AOH by SDS-PAGE. Lane 1 (L1): protein marker (5–245 kDa) and Lane 2 (L2): alcalase oyster hydrolysate (AOH).

Table 2. Amino acid composition of AOH of *Crassostrea nippona*.

Amino Acid	Moles Percentage (%)	g/100 g
Cysteine and cystine	0.39%	0.223
Asparagine and asparatic acid	8.36%	3.429
Glutamine and glutamic acid	11.02%	4.997
Serine	4.62%	1.503
Glycine	10.21%	2.361
Histidine	1.50%	0.715
Arginine	15.43%	8.284
Threonine	4.66%	1.709
Alanine	9.70%	2.665
Proline	4.17%	1.478
Tyrosine	1.87%	1.046
Valine	5.46%	1.972
Methionine	2.19%	1.008
Isoleucine	4.65%	1.881
Leucine	6.95%	2.809
Phenylalanine	3.06%	1.558
Tryptophan	0.00%	0.000
Lysine	6.12%	2.758
Total	100.00%	40.173

2.5. Protein Expression after Treatment with AOH

Western blot analysis was conducted on the AOH samples to determine the expression of MyHC, myogenin, and MyoD, as shown in Figure 4. The protein expression of MuRF1, Atrogin, Smad 2/3, and Foxo-3a are illustrated in Figure 5. Relative expressions for each were analyzed through ImageJ. MyHC, myogenin, and MyoD were expressively upregulated over the DEXA-treated group, and MuRF-1, Atrogin, p-Smad 2/3, and p-foxo-3a were strongly downregulated in the AOH samples treated with DEXA (10 μM) in a dose-dependent manner. GAPDH was used as an internal reference.

2.6. Real-Time Quantitative Polymerase Chain Reaction (RT-qPCR)

Relative mRNA expression levels of myogenin, MyoD, mTOR, MuRF-1, smad 2/3, and foxo-3a were analyzed and are shown in Figure 6. GAPDH was used as an internal reference. In contrast to DEXA, AOH-treated samples showed significantly downregulated MuRF-1, Atrogin, and Foxo-3a mRNA expression and upregulated myogenin, MyoD, and mTOR-related mRNA expression.

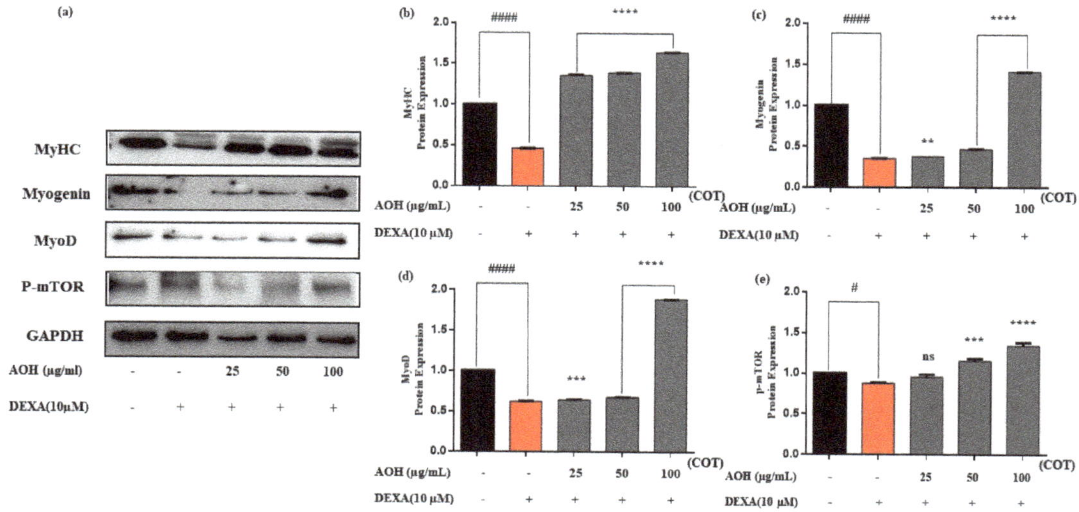

Figure 4. Evaluation of myogenic markers in differentiating myoblasts through myotubes in AOH of *C. nippona*. (**a**) Western blotting analysis, relative expression of (**b**) MyHC, (**c**) myogenin, (**d**) MyoD, (**e**) p-mTOR. Significance is denoted compared to control # $p < 0.05$, #### $p < 0.0001$ and compared to DEXA group ns $p > 0.05$, ** $p < 0.01$, *** $p < 0.001$, and **** $p < 0.0001$. AOH: alcalase oyster hydrolysate; MyHC: myosin heavy chain; mTOR: mammalian target of rapamycin; DEXA: dexamethasone; COT: concentration of treatments.

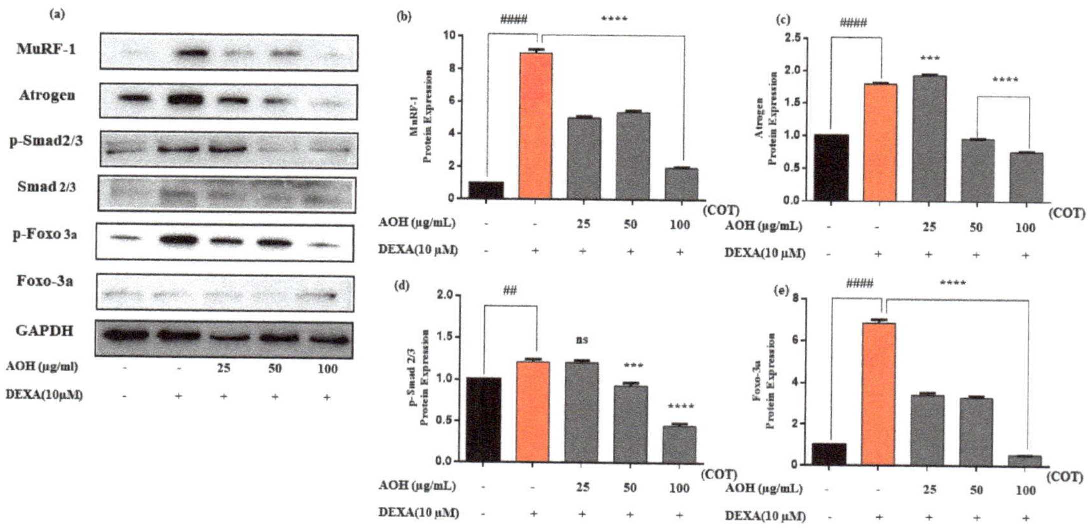

Figure 5. Evaluation of myogenic markers in differentiating myoblasts through myotubes in AOH of *C. nippona*. (**a**) Western blotting analysis, relative expression of (**b**) MuRF-1, (**c**) Atrogin, (**d**) Smad2/3, and (**e**) Foxo-3a. Significance is denoted compared to control ## $p < 0.01$, #### $p < 0.0001$ and compared to DEXA group ns $p > 0.05$, *** $p < 0.001$, and **** $p < 0.0001$. AOH: Alcalase oyster hydrolysate; MuRF-1: Muscle RING-finger protein-1; Foxo: Forkhead box protein; DEXA: dexamethasone; COT: concentration of treatments.

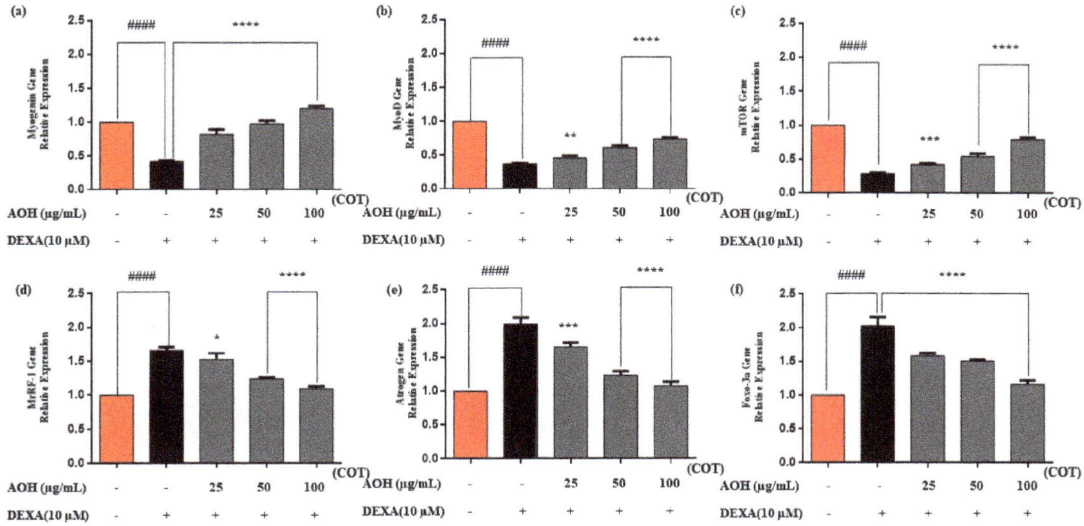

Figure 6. Gene expression of mRNA levels via RT qPCR analysis in AOH of *C. nippona*; (**a**) Myogenin, (**b**) MyoD, (**c**) mTOR, (**d**) MuRF-1), (**e**) Atrogin, (**f**) Foxo 3a. The data are represented as mean ± SE. Significance is denoted compared to control #### $p < 0.0001$ and compared to DEXA group * $p < 0.05$, ** $p < 0.01$, *** $p < 0.001$, and **** $p < 0.0001$. mTOR: mammalian target of rapamycin; MuRF-1: Muscle RING-finger protein-1; Foxo: Forkhead box protein; DEXA: dexamethasone; COT: concentration of treatments.

3. Discussion

Oyster (*C. nippona*) is a famous and commonly consumed shellfish in East Asian countries and has become a popular dietary choice worldwide [23]. Investigating the potential of marine shellfish in muscle growth has also become vital on the industrial scale to develop functional foods and pharmaceutical products [41]. Hence, the current study aimed to investigate the potential of *C. nippona* for human consumption to develop as an impactful functional food for sarcopenia-like muscle atrophic conditions effectively. Therefore, the present study examined the value of AOH on muscle atrophy. Specifically, the authors focused on its potential anti-atrophic effects and explored underlying mechanisms. The overall findings assuredly demonstrate that AOH declined muscle atrophy. Through comprehensive analysis of gene and protein expression levels, the potential of AOH for the regulation of key myokines involved in this process was revealed. Notably, both negative and positive myokines were affected, shedding light on the intricate regulatory network governing muscle preservation.

Proteins, lipids, and carbohydrates are critical determinants of skeletal muscle growth, with their composition playing a pivotal role in the development of functional foods and dietary supplements. Among these components, proteins are particularly essential for muscle growth and development. Conversely, polysaccharides serve as vital energy sources for cellular metabolism and play a crucial role in hormone modulation and regulation. Notably, the general composition of enzyme hydrolysates revealed significantly higher levels of proteins, lipids, and carbohydrates in AOH, encouraging further investigations into its anti-atrophic activity [41].

Considering both cytotoxicity and cell proliferation, this study shows that the AOH and POH are not toxic to differentiating cells and have significant effects on downregulating muscle atrophy. Based on these criteria, AOH and POH were selected for the immunostaining analysis. The present study reveals that myotube formation was repressed by DEXA (Figure 2). When AOH and POH were subjected to analysis of the formation of myotubes through immunofluorescence techniques, the results confirmed that both AOH and

POH were sufficient for their ability to downregulate muscle atrophy in a dose-dependent manner. MyHC was used as a marker to evaluate differentiation. The decline in MyHC expression suggests the incapacitation of myotube formation, which indicates declining muscle regeneration and strength. A similar trend was reported in recent studies of *Ishige okamurae* (IO) and diphloroethohydroxycarmalol (DPHC) against palmitic acid-impaired skeletal myogenesis in C2C12 cells [42,43]. A significant increase in myotube diameters in AOH and POH leads to enhanced myotube fusion of myoblasts to form active mature muscle fibers for treating muscle atrophy [43].

Comparative analysis revealed that protein expression was in an optimum state to abate muscle atrophy, detected only in AOH. Therefore, AOH was selected from the entire screening process to conduct this study. The enzyme hydrolysates of *C. nippona* contain a mixture of proteins with a wide range of molecular weights, and SDS-PAGE was used to analyze the molecular weight distribution of the AOH. This aids in our further studies regarding the peptide isolation from AOH depending on the significance of the biological activities. Amino acids also play a vital role in muscle development, and composition analysis of AOH revealed that arginine represented the highest MOL% (15.45%), which is significant for muscle growth, strength, and regeneration. Leucine, isoleucine, and valine are represented as branching amino acids in a considerable range compared to others. These amino acids are highly responsible for the downregulation of muscle breakdown and the assembly of proteins in muscles. Leucine activates mTOR, which is a necessary component of protein synthesis [28]. Isoleucine promotes myogenesis and intramyocellular lipid accumulation, which also facilitates significant protein turnover in the body. Moreover, valine affects the advancement of skeletal tissues, the synthesis of neurotransmitters, energy production, muscle synthesis, and maintenance of immunological parameters effectively [44–46]. Additionally, substantial amounts of alanine and glycine are also present in this sample, significantly contributing to muscle growth [47]. Hence, this amino acid profile obtained from AOH was highly beneficial for the development of muscles as well as showing their potential to suppress the atrophic conditions generated in the body due to various stress factors. Alcalase is a serine endopeptidase synthesized from *Bacillus lichenifomis* that has a superior potential for the effective breakdown of proteins from numerous sites, in the formation of a diverse range of peptide fragments, which leads to vital environmental conditions for treating muscle growth and abating muscle atrophy [48]. Therefore, in consideration of the broad view of our future studies regarding peptide isolation and purification, AOH shows high potential for the development of functional foods.

A Western blotting experiment was performed to evaluate the myokine expression and its regulation by AOH in the DEXA-induced C2C12 cells. Results obtained from Western blotting showed that AOH downregulated muscle atrophic conditions by synchronizing the major constituents contributing to the ubiquitin–proteasome and mTOR signaling pathways. MuRF-1 and Atrogin are E3 ubiquitin ligases involved in the ubiquitin–proteasome system, which leads to the degradation of proteins in the cells [49]. FoxO3a forms a complex with Smad2/3 and activates MuRF-1 transcription. An alternative study revealed that increasing the expression of phosphorylated forms of Smad2/3 and Foxo 3a leads to the upregulation of positive regulators of atrogein [50,51]. Therefore, overexpression of MuRF-1 and Atrogin activated the ubiquitin–proteasome pathway for protein degradation to promote muscle atrophy. The results obtained from Western blot analysis strongly expressed that, under DEXA-induced conditions, AOH has a significant effect on downregulating the ubiquitin ligases of MuRF-1 and Atrogin in a dose-dependent manner (Figure 4). One study reported similar effects by using a marine fish (*Lutjanus guttatus*), and another study detected quercetin-like flavonoids isolated from various plants for muscle atrophy potential [52,53]. MyoD synchronizes the myofibrillar gene expression to facilitate muscle contraction and regeneration. Furthermore, myogenic differentiation is increased by myogenin [54,55]. Due to the atrophy situation, Smad2 and Smad3 targeted MyoD and myogenin and inhibited myoblast and myofiber proliferation and differentiation [2].

The protein expression levels of MyoD and myogenin in the AOH-treated samples were significantly higher than in the DEXA group. Downregulation of the phosphorylated forms of Smad2/3 and Foxo-3a during this study signified an upsurge in the regulation of MyoD and myogenin to mitigate muscle atrophy in AOH-treated C2C12 cells. Hence, this study showed that AOH has a significant impact on abating muscle atrophic conditions and boosts muscle growth by facilitation through myoblast and myofiber proliferation, differentiation, muscle contraction, and regeneration effectively.

The major constituents required for protein synthesis in cells are mTOR [54] and the myosin-heavy chain which contributes to the contractile apparatus in muscle fibers [42]. Hence, protein expressions and mRNA gene expressions showed a significant upregulation of MyHC and mTOR even under DEXA-induced conditions, signifying that AOH can promote sufficient protein and fibers for muscles to avoid atrophic conditions effectively.

Therefore, the results indicated that AOH has an exceptional effect on attenuating DEXA-induced muscle atrophy in C2C12 cells in an in vitro model by impending negative regulators such as MuRF-1, Atrogin, Smad2/3, and FoxO 3a according to the ubiquitin–proteasome pathway and promoting the expression of positive regulators such as Myogenin, MyoD, mTOR, and MyHC, which correspond to the mTOR signaling pathway (Figures 4 and 5).

Analogous to the trend of protein expression, the AOH-treated differentiated *C. nippona* significantly downregulated the mRNA expression trend of MuRF-1, Atrogin, and FoxO 3a, and upregulated the expression of Myogenin, MyoD, and mTOR over in DEXA (Figure 5). Downregulation of Foxo-3a also signified that AOH effectively inhibited MuRF-1 and Atrogin in mRNA levels and amplified the gene levels of myogenin and MyoD.

Functional foods are the components that enclose higher nutritional value and maintain superior quality in terms of human health. Tasteful factors, along with the high capacity of safety levels and well-being impacts, are the profound beneficiaries of these products over others [56]. *C. nippona* is a potent oyster shellfish used as an edible portion for human consumption. The outcomes obtained from this study revealed that the AOH of *C. nippona* comprises high nutritional values and works as an outstanding component to fulfill nutrient requirements during the new product development stage.

Protein expressions at the Act-RIIB and PI3K pathways and their respective receptor level expression analyses were not extensively addressed in this study. Furthermore, previous studies by Gilson et al., 2007, revealed that myostatin contributes a vital role to the formation of stress conditions in the cells followed by the activation of phosphorylated Smad2/3 during muscle atrophic conditions. Hence, further feasibility studies are recommended and encouraged to distinguish the receptor-level mechanism analysis and execute experiments in vivo to confirm the suitability of these hydrolysates for abating muscle atrophy. The isolation of purified peptides will help categorize the molecular mechanism of the active agent of AOH and deliver effective applications in the functional food industry in a meaningful way.

4. Materials and Methods

4.1. Materials

The chemical constituents and mixtures necessary for all analyses were procured from Sigma-Aldrich (St. Louis, MO, USA). Dulbecco's Modified Eagle's medium (DMEM) (Cat. LM 001-05) was used as a growth media for cell culture obtained from WELGENE, Inc. (693, Namcheon-ro, Namcheon-myeon, Gyeongsan-si, Gyeongsangbuk-do 38695, Republic of Korea). C2C12 myoblast cell lines were obtained from the American Type Culture Collection (ATCC; VA, USA). Fetal Bovine Serum (FBS, WELGENE, Inc.). (Cat. S 101-01). and Horse Serum (Cat. 26170-043) (HS, Gibco, Invitrogen Inc.) were two types of serum-supplemented serum supplementing the DMEM. Differentiated C2C12 myotubes were treated with DEXA (Sigma Aldrich, D1756) (CAS-No 50-02-2). Penicillin and streptomycin were treated as antibiotics for the same medium from Invitrogen Inc. (Waltham, MA, USA). Goat anti-mouse IgG H&L (Alexa Fluor TM Plus 555; A32727) for immunofluorescence was obtained from Invitrogen (Cambridge, UK). Primary and secondary antibodies utilized for Western blotting were bought from Santa Cruz Biotechnology (Santa Cruz, CA, USA). Primers for RT-qPCR were purchased from Humanizing Genomics Macrogen, Seoul, Republic of Korea. *Crassostrea nippona* samples were acquired from a local market in Republic of Korea for experimental analysis. Alcalase, protamex, and flavourzyme were the digestive proteases used in this study and were purchased from Novo Co. (Novozyme Nordisk, Bagsvaerd, Denmark).

4.2. Preparation of Enzyme Hydrolysate of Crassostrea Nippona

Oysters (*C. nippona*) were purchased from the domestic seafood market in Tongyeong-si, Republic of Korea, from June to August 2022. The oysters' shells were carefully opened, and the shell was removed. The muscle tissue was then dissected from the remaining body, ensuring minimal inclusion of other tissues. To reduce salinity, the extracted muscle tissue underwent thorough rinsing with running water, followed by an additional rinse using distilled water. Subsequently, the oyster muscle tissue was ground into a fine paste, ensuring uniformity. Finally, the ground tissue was freeze-dried to remove moisture, resulting in powdered oyster samples utilized for further experiments. In this study, powdered oyster samples (20 g) were mixed with 200 mL of distilled water to achieve a 10% substrate concentration. The pH of the mixture was adjusted to 8. Next, three different enzymes—alcalase (7592870), flavourzyme (7612249), and protamex (PW2A1042)—were separately added to the sample, resulting in a final enzyme-substrate concentration of 5.00%. The prepared samples were placed in a shaking incubator (SFDSM06, Jeio Tech, Republic of Korea) and incubated at 55 °C for 24 h with a shaking speed of 130 rpm. Incubated enzyme-hydrolyzed samples were inactivated at 100 °C for 10 min followed by centrifugation at 10,000 rpm for 10 min at 4 °C. Supernatants from each enzymatic hydrolysis sample were collected through vacuum filtration by sourcing a Buchner funnel with Whatman number 4 double filter papers at a time (Cat:1004110, GE Healthcare, Amersham, UK) and triplicates of 1 mL samples were subjected to measure the yield and rest were freeze-dried for 72 h. Freeze-dried samples were collected and allocated to measure the general composition and muscle growth analysis and samples were kept at 4 °C until use for analysis.

4.3. General Composition of Crassostrea Nippona

Polysaccharides, polyphenols, proteins, lipids, ash, and yield content were analyzed in enzyme hydrolysates of oysters in accordance with the AOAC standards [6,57]. The ash content was determined by incinerating the samples in an electric muffle furnace at 555 °C for 5 h (JSMF-30T, JSR Research Inc., 69 Gumsang-gogaegil, Gongju-City, Chungchungnam-Do, Republic of Korea) and the Kjeldahl method (KjeltecTM 8100, FOSS Analytical Slangerupgade 69, Hilleroed, Denmark) was used to determine the protein content. The lipid content was evaluated gravimetrically by the Soxhlet method (FatExtractor E-500, BUCHI Labortechnik AG, Meierseggstrasse 40, Postfach, Flawil, Switzerland). Polysaccharides and polyphenols were measured by the phenol/sulfuric acid method and the Folin–Ciocalteu method, respectively [58,59].

4.4. Cell Culture and Cell Differentiation

DMEM supplemented with heat-inactivated 10% FBS, 1% penicillin, and streptomycin (Cat. LS 202-02) was used for cell culture experiments and DMEM enriched with 2% HS, (Ref. 26170-043) 1% streptomycin and penicillin was used to differentiate the cultured cells at 37 °C and in a 5% CO_2 controlled environment (60915672, Sanyo Electric Co., Ltd., Osaka, Japan) [1–4]. DEXA (CAS-No: 50-02-2) concentration was maintained at 10 μM throughout these experiments to determine whether selected enzyme hydrolysates of *C. nippona* had a significant effect against that in a dose-dependent way [25,60,61]. Initially, the C2C12 cells were seeded into the concentration of 1×10^5 cells/mL in the DMEM medium. Differentiation was initiated once the cells reached 80% confluence, by shifting the medium to DMEM containing 2% HS and 1% PS. The differentiation procedure was followed three times and differentiated cells were subjected to immunostaining, Western blot analysis, and RT-qPCR [62]. For immunostaining initially, cells were seeded into 48-well plates via Western blot and for RT-qPCR we used 5 mL cell culture dishes, respectively. Initially, the total volumes we used for the 48-well plates and 5 mL cell culture dishes were 250 μL and 4 mL, respectively.

4.5. Cytotoxicity Evaluation using Cell Counting Kit-8 (CCK-8 assay)

The cytotoxicity influence of oyster enzyme hydrolysates on the C2C12 cells was achieved using CCK8 assay [63]. C2C12 myoblasts were seeded in a 96-well plate and incubated for 24 h at 37 °C with 5% CO_2. Viable cell concentration was maintained at 1×10^5 cells/mL throughout the experiment period. Cells were counted by using a LUNAII automated cell counter machine. Here, we used Luna cell counting sides (Cat no: L12001). The concentrations of the enzyme extracts were maintained by considering the cytotoxicity results. The seeded cells were treated with three hydrolysates (AOH, FOH, POH) and DWO with three different concentrations of each extract, from 25 μg/mL, 50 μg/mL, and 100 μg/mL, and incubated for another 24 h at similar environmental conditions. Each well was treated with CCK8 solution and after 2 h the wells were evaluated by taking the absorbance at 450 nm by using a microplate reader (SynergyTM HT, Vermont, USA) [43]. The absence of sample-treated wells was used as a control and calculated as 100% cell viability, and all absorbance readings of sample-treated wells were stated as mean percentage values compared to the untreated wells [64].

4.6. Cell Proliferation Evaluation using 5-Bromo-2′-deoxyuridine (BrdU) Assay

To estimate the proliferation ability of AHO, FOH, POH, and DWO in C2C12 cells, the BrdU assay (Millipore, Billerica, MA, USA) was directed by handling Bromodeoxyuridine (BrdU) labeling and a detection kit (11647229001, RocheR Life Science Products, Penzberg, Upper Bavaria, Germany). Seeded C2C12 cells were treated with digested AHO, FOH, POH, and DWO during the cell differentiation period under a differentiation medium (DMEM + 2% HS + 1% PS). The proliferation capability of the cells treated with digested samples was compared with that of the DEXA-treated sample and DEXA was evaluated over control sample wells. The background effect of the media was eliminated by acquiring the absorbance by using the wells that contained only the media without any serum and cells. The absorbance of the samples and background were measured in an ELISA reader at 370 nm [62].

4.7. Immunostaining of Myosin Heavy Chain (MyHC)

Differentiated myotubes were washed three times with cold 1× PBS (20× PBS consisted of 2.7 M sodium chloride (NaCl), 54 mM potassium chloride (KCl), 86 mM sodium phosphate dibasic (Na_2HPO_4), and 28 mM potassium phosphate monobasic (KH_2PO_4), all dissolved in distilled water. The final pH was maintained at 7.2–7.5 and fixed with chilled methanol for 5 s. Fixed myotubes were washed three times with PBST (PBS + 1% Tween 20). Fixed and permeabilized cells were blocked with 1% BSA (Cat No. BSAS 0.1) containing 22.52 mg/mL glycine (CAS No. 56-40-6) and incubated with MHC-specific primary antibodies (Santa Cruz) (1:250 in 1% BSA) for 2 h at room temperature. After being washed three times with PBST, incubated cells were labeled with goat anti-mouse IgG H&L (Alexa Fluor TM Plus 555; A32727)-conjugated secondary antibodies (Santa Cruz) (1:1000 in 1% BSA) for 2 h at room temperature, followed by a similar washing step before nuclear staining. Cell nuclei were stained with 300 nm DAPI (40,6-diamidino-2-phenylindole; Sigma-Aldrich), and the final washing step was performed using PBST. Finally, fluorescent images of the stained myotubes were visualized using a confocal microscope (Carl Zeiss, Oberkochen, Germany) [44,65]. Myotube diameters were calculated for control, DEXA, and DEXA-treated samples of AOH and POH using ImageJ software 1.49v (National Institutes of Health, Bethesda, MD, USA) according to the confocal microscope captions referring to each well, and significance was calculated for each sample over the DEXA-treated control value [62].

4.8. Molecular Weight Determination and Amino Acid Composition

After confirmation of cytotoxicity, cell proliferation ability, and myotube diameter, AOH was selected for Western blotting and RT-qPCR analysis. Molecular weight determination was conducted using sodium dodecyl sulfate–polyacrylamide gel electrophoresis (SDS-PAGE) on 4% stacking and 12% separating polyacrylamide gels, and amino acid composition of the chosen sample was performed using Agilent 1260 series HPLC conditions [66,67]. An AOH sample (100 mg/mL) was subjected to SDS PAGE [45]. Waters Nova-Pak C18 4 μm (3.9 × 300 mm) was the column used during the HPLC analysis. The two mobile phases applied in this analysis were 140 mM Sodium Acetate trihydrate, 0.15% TEA, 0.03% EDTA, 6% CH3CN, pH 6.1, and 60% CH3CN, 0.015% EDTA. The flow rate, run time, and equilibration time were 1.0 mL/min, 30 min, and 10 min, respectively. The HP 1100 Series at 254 nm was used to detect the peaks. AOH (100 mg/mL) was used for the SDS-PAGE analysis [68].

4.9. Western Blot Analysis

The digested AOH-treated cells were harvested using PBS (Cat. PR2007-100-00)/trypsin (Cat LS 015-01) and lysed (RIPA Buffer (R4100-050). The total protein content (TPC) was calculated by BCA assay (23227, Thermo Fisher Scientific), and an equivalent amount of denatured protein (30 μg) was loaded into each well. The separated proteins were blotted onto a nitrocellulose membrane (Amersham™ ProtranTM 0.45 μm NC). A total of 5% skim milk in Tris-buffered saline (TBS) (0.05% Tween-20) was applied to block the protein-containing membrane for 2 h at room temperature. The blocked membrane was incubated overnight with primary antibodies (Myosin heavy chain, MyoD, Myogenin, phosphorylated Forkhead box protein (p-Foxo-3a), Forkhead box protein (Foxo-3a), Muscle RING-finger protein-1 (MuRF-1), and atrogin protein) which were diluted to 1:1000 in 5% skim milk in 1% TBS with 0.1% Tween 20 detergent (TBST) at 4 °C for 16 h followed by augmented fluorescent tagged respective mouse or rabbit secondary antibodies (1:3000) into the primary antibody-tagged proteins. The original concentration of TBS was 20× (20× TBS contains 160 g of NaCl, 4 g of KCl, and 60 g of Tris in 1000 mL of distilled water; the final pH was 7.6). Glyceraldehyde 3-phosphate (GAPDH) was used as a housekeeping protein during the analysis of samples using different antibodies. The membranes were imaged using a chemiluminescent substrate (Amersham, Arlington Heights, IL, USA) and quantified using ImageJ 1.49v software (National Institutes of Health, Bethesda, MD, USA) [63].

4.10. RT-qPCR

The selected AOH samples at 25, 50, and 100 μg/mL were differentiated as previously mentioned in Section 2.4. The cells were then harvested, and total Ribonucleic Acid (RNA) was extracted using TRIzol reagent (Invitrogen, CA, USA) following the manufacturer's instructions. RNA content was measured using a μDrop plate (Thermo Scientific, Waltham, MA, USA), and 2 μg of RNA was reverse-transcribed into cDNA using a first-strand cDNA synthesis kit (TaKaRa, Shiga, Japan). The primers used in the experiment are listed in Table 3. The PCR cycling conditions were as described by [65]. Here, initial denaturation was performed at 95 °C/30 s. Annealing and extension were conducted in 30 cycles (95 °C/5 s, 55 °C/10 s, and 72 °C/20 s. Final extension 95 °C/15 s, and 60 °C/30 s). Livak and Schmittgen (2001) implemented a scheme to quantify the relative gene expression [68]. GAPDH was used as an internal control to normalize the relative gene expression levels.

Table 3. Sequence of primers used in this study.

Gene	Primer	Sequence
GAPDH	Sense	5′-AAGGGTCATCATCTCTGCCC-3′
	Antisense	5′-CCACGATGGACGTAAGGGAG-3′
MyoD	Sense	5′-GCCGCCTGAGCAAAGTGAATG-3′
	Antisense	5′-CAGCGGTCCAGTGCGTAGAAG3′
Myogenin	Sense	5′-GTCCCAACCCAGGAGATCAT-3′
	Antisense	5′-CCACGATGGACGTAAGGGAG-3′
mTOR	Sense	5′-CACATCACTCCCTTCACCA-3′
	Antisense	5′-GCAACAACGGCTTTCCAC-3′
MuRF-1	Sense	5′-ATCTAGCCTGATTCCTGATGGA-3′
	Antisense	5′ACCACAGGCTTGGTAAACATCT3′
Foxo-3a	Sense	5ACCTTCGTCTCTGAACCTTG–3′
	Antisense	5′AGTGTGACACGGAAGAGAAGGT3′
Smad 2/3	Sense	5GTCCCAACCCAGGAGATCAT–3′
	Antisense	5′CCACGATGGACGTAAGGGAG3′

4.11. Statistical Analysis

To realize the final results, all the analyses were performed in triplicate. Data are expressed as means ± standard deviation. Statistical significance was evaluated using one-way analysis of variance (ANOVA) and Tukey's test. Two-tailed Student's *t*-tests were performed when two conditions were required for the analysis. The significance is denoted compared to the control; $p < 0.0001$ "####" and over DEXA level was defined; $p < 0.05$ "*", $p < 0.01$ "**", $p < 0.001$ "***", and $p < 0.0001$ "****" .

5. Conclusions

In conclusion, this study robustly supports the anti-muscle atrophic effect of AOH derived from *C. nippona*. Through a comprehensive analysis of gene and protein expression levels, AOH downregulates negative regulators (MuRF-1, Atrogin-1, Smad2/3, and FoxO 3a) associated with the ubiquitin–proteasome pathway. Simultaneously, AOH promotes the expression of positive regulators (Myogenin, MyoD, mTOR, and MHC) related to the mTOR signaling pathway. The study outcomes confirmed the muscle growth and anti-atrophy effects of AOH in a DEXA-induced muscle atrophy model using C2C12 cells. Amino acid composition and cell-protective aptitude analyses further highlight AOH's potential for treating muscle atrophy effectively compared to other hydrolysates. Receptor-level analysis and in vivo experiments are recommended to strengthen these findings and facilitate impactful industrial product development for muscle atrophy treatment.

Author Contributions: Conceptualization, Y.-J.J., S.I.K.; M.J.M.S.K. and D.P.N.; methodology, M.J.M.S.K., D.P.N., N.M.L., H.H.A.C.K.J., D.S.D. and H.-G.L.; software, M.J.M.S.K., D.P.N., N.M.L., H.H.A.C.K.J., D.S.D.; H.-G.L. and Y.-S.K.; validation, M.J.M.S.K. and D.P.N.; formal analysis, M.J.M.S.K. and D.P.N.; resources, Y.-J.J.; data curation, M.J.M.S.K., D.P.N., N.M.L., H.H.A.C.K.J., D.S.D., H.-G.L. and Y.-S.K.; writing—original draft preparation, M.J.M.S.K. and D.P.N.; writing—review and editing, Y.-J.J., M.J.M.S.K. and D.P.N.; visualization, M.J.M.S.K., D.P.N., N.M.L., H.H.A.C.K.J., D.S.D., H.-G.L. and Y.-S.K.; supervision, Y.-J.J. and S.I.K.; funding acquisition, Y.-J.J. All authors have read and agreed to the published version of the manuscript.

Funding: This research was supported by Development of technology for biomaterialization of marine fisheries by-products of Korea institute of Marine Scinece & Technology Promotion (KIMST) funded by the Ministry of Oceans and Fisheries (KIMST-20220128).

Institutional Review Board Statement: Not applicable.

Data Availability Statement: All data are contained within the manuscript.

Conflicts of Interest: The authors declare no conflicts of interest.

References

1. Jackman, R.W.; Kandarian, S.C. The molecular basis of skeletal muscle atrophy. *Am. J. Physiol. Cell Physiol.* **2004**, *287*, C834–C84 [CrossRef] [PubMed]
2. Morley, J.E.; Anker, S.D.; von Haehling, S. Prevalence, incidence, and clinical impact of sarcopenia: Facts, numbers, and epidemiology—Update 2014. *J. Cachexia Sarcopenia Muscle* **2014**, *5*, 253–259. [CrossRef]
3. Tipton, K.D.; Elliott, T.A.; Cree, M.G.; Aarsland, A.A.; Sanford, A.P.; Wolfe, R.R. Stimulation of net muscle protein synthesis by whey proteingestion before and after exercise. *Am. J. Physiol. Endocrinol. Metab.* **2007**, *292*, E71–E76. [CrossRef] [PubMed]
4. Schiaffino, S.; Dyar, K.A.; Ciciliot, S.; Blaauw, B.; Sandri, M. Mechanisms regulating skeletal muscle growth and atrophy. *FEBS J.* **2013**, *280*, 4294–4314. [CrossRef] [PubMed]
5. Sartori, R.; Milan, G.; Patron, M.; Mammucari, C.; Blaauw, B.; Abraham, R.; Sandri, M. Smad2 and 3 transcription factors control muscle mass in adulthood. *Am. J. Physiol. Cell Physiol.* **2009**, *296*, C1248–C1257. [CrossRef] [PubMed]
6. Nagahawatta, D.P.; Liyanage, N.M.; Jayawardena, T.U.; Jayawardhana, H.H.A.C.K.; Oh, J.-Y.; Sanjeewa, K.K.A.; Kang, S.I. Jeon, Y.-J. Skeletal muscle growth activity of Olive Flounder (*Paralichthys olivaceus*) meat digest. *Food Biosci.* **2023**, *53*, 102809 [CrossRef]
7. Jodral-Segado, A.M.; Navarro-Alarcón, M.; López-G de la Serrana, H.; López-Martínez, M.a.C. Magnesium and calcium contents in foods from SE Spain: Influencing factors and estimation of daily dietary intakes. *Sci. Total Environ.* **2003**, *312*, 47–58. [CrossRef]
8. Hao, L.; Wang, X.; Cao, Y.; Xu, J.; Xue, C. A comprehensive review of oyster peptides: Preparation, characterisation and bioactivities. *Rev. Aquac.* **2022**, *14*, 120–138. [CrossRef]
9. Rajapakse, N.; Mendis, E.; Jung, W.-K.; Je, J.-Y.; Kim, S.-K. Purification of a radical scavenging peptide from fermented mussel sauce and its antioxidant properties. *Food Res. Int.* **2005**, *38*, 175–182. [CrossRef]
10. Jung, W.-K.; Kim, S.-K. Isolation and characterisation of an anticoagulant oligopeptide from blue mussel, *Mytilus edulis*. *Food Chem.* **2009**, *117*, 687–692. [CrossRef]
11. Anderson, R.S.; Beaven, A.E. Antibacterial activities of oyster (*Crassostrea virginica*) and mussel (*Mytilus edulis* and *Geukensia demissa*) plasma. *Aquat. Living Resour.* **2001**, *14*, 343–349. [CrossRef]
12. Imsong, L.; Murali, M. DNA barcoding, determination of bioactive compounds, Antioxidant and anti-diabetic property in edible gastropod Brotia costula (Rafinesque, 1833) of dimapur district, Nagaland. *Int. J. Pharm. Sci. Res.* **2022**, *14*, 1795–1804.
13. Hu, X.; Tao, N.; Wang, X.; Xiao, J.; Wang, M. Marine-derived bioactive compounds with anti-obesity effect: A review. *J. Funct. Foods.* **2016**, *21*, 372–387. [CrossRef]
14. La Paglia, L.; Vazzana, M.; Mauro, M.; Urso, A.; Arizza, V.; Vizzini, A. Bioactive Molecules from the Innate Immunity of Ascidians and Innovative Methods of Drug Discovery: A Computational Approach Based on Artificial Intelligence. *Mar. Drugs* **2024**, *22*, 6. [CrossRef]
15. Librizzi, M.; Martino, C.; Mauro, M.; Abruscato, G.; Arizza, V.; Vazzana, M.; Luparello, C. Natural Anticancer Peptides from Marine Animal Species: Evidence from In Vitro Cell Model Systems. *Cancers* **2024**, *16*, 36. [CrossRef] [PubMed]
16. Punginelli, D.; Catania, V.; Abruscato, G.; Luparello, C.; Vazzana, M.; Mauro, M.; Cunsolo, V.; Saletti, R.; Di Francesco, A.; Arizza, V.; et al. New Bioactive Peptides from the Mediterranean Seagrass *Posidonia oceanica* (L.) Delile and Their Impact on Antimicrobial Activity and Apoptosis of Human Cancer Cells. *Int. J. Mol. Sci.* **2023**, *24*, 5650. [CrossRef] [PubMed]
17. Luparello, C.; Branni, R.; Abruscato, G.; Lazzara, V.; Sugár, S.; Arizza, V.; Mauro, M.; Di Stefano, V.; Vazzana, M. Biological and Proteomic Characterization of the Anti-Cancer Potency of Aqueous Extracts from Cell-Free Coelomic Fluid of *Arbacia lixula* Sea Urchin an In Vitro Model of Human Hepatocellular Carcinoma. *J. Mar. Sci. Eng.* **2022**, *10*, 1292. [CrossRef]
18. Punginelli, D.; Schillaci, D.; Mauro, M.; Deidun, A.; Barone, G.; Arizza, V.; Vazzana, M. The potential of antimicrobial peptides isolated from freshwater crayfish species in new drug development: A review. *Dev. Comp. Immunol.* **2022**, *126*, 104258. [CrossRef]
19. Mauro, M.; Lazzara, V.; Punginelli, D.; Arizza, V.; Vazzana, M. Antitumoral compounds from vertebrate sister group: A review of *Mediterranean ascidians. Dev. Comp. Immunol.* **2020**, *108*, 103669. [CrossRef]
20. Mauro, M.; Pinto, P.; Settanni, L.; Puccio, V.; Vazzana, M.; Hornsby, B.L.; Fabbrizio, A.; Di Stefano, V.; Barone, G.; Arizza, V. Chitosan Film Functionalized with Grape Seed Oil—Preliminary Evaluation of Antimicrobial Activity. *Dev. Comp. Immunol.* **2022**, *14*, 5410. [CrossRef]
21. Kios, K.; Kakasis, S.; Syropoulou, F.; Boziaris, I.S. Chapter 10—Seafood and shellfish. In *Functional Foods and Their Implications for Health Promotion*; Zabetakis, I., Tsoupras, A., Lordan, R., Ramji, D., Eds.; Academic Press: Cambridge, MA, USA, 2023; pp. 281–302.
22. de Souza, M.M.M.; da Boa Morte, E.S.; Cardoso, L.G.; Nunes, D.V.; de Souza, C.O.; Druzian, J.I.; Cardoso, R.d.C.V. Nutritional contribution of shellfish from the biodiversity of Todos os Santos Bay, Brazil. *J. Food Compos. Anal.* **2021**, *102*, 103999. [CrossRef]
23. Botta, R.; Asche, F.; Borsum, J.S.; Camp, E.V. A review of global oyster aquaculture production and consumption. *Mar. Policy* **2020**, *117*, 103952. [CrossRef]
24. Lee, H.J.; Kim, D.; Do, K.; Yang, C.B.; Jeon, S.W.; Jang, A. Effects of Horse Meat Hydrolysate on Oxidative Stress, Proinflammatory Cytokines, and the Ubiquitin-Proteasomal System of C2C12 Cells. *Food Sci. Anim. Resour.* **2024**, *44*, 132–145. [CrossRef] [PubMed]
25. Hyun, J.; Kang, S.-I.; Lee, S.-W.; Amarasiri, R.P.G.S.K.; Nagahawatta, D.P.; Roh, Y.; Wang, L.; Ryu, B.; Jeon, Y.-J. Exploring the Potential of Olive Flounder Processing By-Products as a Source of Functional Ingredients for Muscle Enhancement. *Antioxidants* **2023**, *12*, 1755. [CrossRef] [PubMed]

5. Jeon, S.-H.; Choung, S.-Y. Oyster Hydrolysates Attenuate Muscle Atrophy via Regulating Protein Turnover and Mitochondria Biogenesis in C2C12 Cell and Immobilized Mice. *Nutrients* **2021**, *13*, 4385. [CrossRef] [PubMed]
7. Hu, Y.; Li, Q.; Xu, C.; Liu, S.; Kong, L.; Yu, H. Response to selection for growth in successive mass selected generations of Iwagaki oyster Crassostrea nippona. *Aquaculture* **2022**, *560*, 738575. [CrossRef]
3. Gong, J.; Li, Q.; Yu, H.; Liu, S.; Kong, L. First de novo transcriptome assembly of *Iwagaki oyster*, *Crassostrea nippona*, and comparative evolutionary analysis of salinity-stress response genes in *Crassostrea oysters*. *Mar. Genom.* **2021**, *56*, 100805. [CrossRef] [PubMed]
9. Gong, J.; Li, Q.; Yu, H.; Liu, S.; Kong, L. Effects of low salinity on hemolymph osmolality and transcriptome of the Iwagaki oyster, Crassostrea nippona. *Fish Shellfish Immunol.* **2022**, *126*, 211–216. [CrossRef]
0. Venugopal, V.; Gopakumar, K. Shellfish: Nutritive Value, Health Benefits, and Consumer Safety. *Compr. Rev. Food Sci. Food Saf.* **2017**, *16*, 1219–1242. [CrossRef]
1. Hong, Y.J.; Saroj, R.K.; Park, W.I.; Yi, G.-C. One-dimensional semiconductor nanostructures grown on two-dimensional nanomaterials for flexible device applications. *APL Mater.* **2021**, *9*, 060907. [CrossRef]
2. Ko, S.-C.; Kang, N.; Kim, E.-A.; Kang, M.C.; Lee, S.-H.; Kang, S.-M.; Lee, J.-B.; Jeon, B.-T.; Kim, S.-K.; Park, S.-J.; et al. A novel angiotensin I-converting enzyme (ACE) inhibitory peptide from a marine *Chlorella ellipsoidea* and its antihypertensive effect in spontaneously hypertensive rats. *Process. Biochem.* **2012**, *47*, 2005–2011. [CrossRef]
33. Ryan, J.T.; Ross, R.P.; Bolton, D.; Fitzgerald, G.F.; Stanton, C. Bioactive Peptides from Muscle Sources: Meat and Fish. *Nutrients* **2011**, *3*, 765–791. [CrossRef] [PubMed]
34. Zou, Y.; Robbens, J.; Heyndrickx, M.; Debode, J.; Raes, K. Bioprocessing of marine crustacean side-streams into bioactives: A review. *JCTB* **2021**, *96*, 1465–1474. [CrossRef]
35. Yoon, J.H.; Lee, S.-M.; Lee, Y.; Kim, M.J.; Yang, J.W.; Choi, J.Y.; Kwak, J.Y.; Lee, K.-P.; Yang, Y.R.; Kwon, K.-S. Alverine citrate promotes myogenic differentiation and ameliorates muscle atrophy. *Biochem. Biophys. Res. Commun.* **2022**, *586*, 157–162. [CrossRef] [PubMed]
36. Jayawardhana, H.H.A.C.K.; Liyanage, N.M.; Nagahawatta, D.P.; Lee, H.-G.; Jeon, Y.-J.; Kang, S.I. Pepsin Hydrolysate from Surimi Industry-Related Olive Flounder Head Byproducts Attenuates LPS-Induced Inflammation and Oxidative Stress in RAW 264.7 Macrophages and In Vivo Zebrafish Model. *Mar. Drugs* **2024**, *22*, 24. [CrossRef] [PubMed]
37. Wang, Y.-H.; Kuo, C.-H.; Lee, C.-L.; Kuo, W.-C.; Tsai, M.-L.; Sun, P.-P. Enzyme-Assisted Aqueous Extraction of Cobia Liver Oil and Protein Hydrolysates with Antioxidant Activity. *Catalysts* **2020**, *10*, 1323. [CrossRef]
38. Wang, L.; Hu, J.; Lv, W.; Lu, W.; Pei, D.; Lv, Y.; Wang, W.; Zhang, M.; Ding, R.; Lv, M. Optimized extraction of astaxanthin from shrimp shells treated by biological enzyme and its separation and purification using macroporous resin. *Food Chem.* **2021**, *363*, 130369. [CrossRef] [PubMed]
39. Nguyen, H.C.; Ngo, K.N.; Tran, H.K.; Barrow, C.J. Enzyme-Assisted Coextraction of Phenolics and Polysaccharides from *Padina gymnospora*. *Mar. Drugs* **2024**, *22*, 42. [CrossRef]
40. Ngo, D.-H.; Vo, T.-S.; Ngo, D.-N.; Wijesekara, I.; Kim, S.-K. Biological activities and potential health benefits of bioactive peptides derived from marine organisms. *Int. J. Biol. Macromol.* **2012**, *51*, 378–383. [CrossRef]
41. Liu, S.; Zhang, L.; Li, S. Advances in nutritional supplementation for sarcopenia management. *Front. Nutr.* **2023**, *10*, 1189522. [CrossRef]
42. Nagahawatta, D.; Sanjeewa, K.; Jayawardena, T.U.; Kim, H.-S.; Yang, H.-W.; Jiang, Y.; Je, J.-G.; Lee, T.-K.; Jeon, Y.-J. Drying seaweeds using hybrid hot water Goodle dryer (HHGD): Comparison with freeze-dryer in chemical composition and antioxidant activity. *Fish. Aquat. Sci.* **2021**, *24*, 19–31. [CrossRef]
43. Qiao, M.; Tu, M.; Wang, Z.; Mao, F.; Chen, H.; Qin, L.; Du, M. Identification and Antithrombotic Activity of Peptides from Blue Mussel (*Mytilus edulis*) Protein. *Int. J. Mol. Sci.* **2018**, *19*, 138. [CrossRef] [PubMed]
44. Tavano, O.L. Protein hydrolysis using proteases: An important tool for food biotechnology. *J. Mol. Catal. B Enzym.* **2013**, *90*, 1–11. [CrossRef]
45. Rodrigues, R.B.; Lui, T.A.; Neu, D.H.; Boscolo, W.R.; Bittencourt, F.J.R.C.A. Valine in diets for *Nile tilapia*. *Rev. Cienc. Agron.* **2018**, *49*, 467–474. [CrossRef]
46. Kamei, Y.; Hatazawa, Y.; Uchitomi, R.; Yoshimura, R.; Miura, S. Regulation of Skeletal Muscle Function by Amino Acids. *Nutrients* **2020**, *12*, 261. [CrossRef] [PubMed]
47. Bodine, S.C.; Latres, E.; Baumhueter, S.; Lai, V.K.-M.; Nunez, L.; Clarke, B.A.; Poueymirou, W.T.; Panaro, F.J.; Na, E.; Dharmarajan, K.; et al. Identification of Ubiquitin Ligases Required for Skeletal Muscle Atrophy. *Science* **2001**, *294*, 1704–1708. [CrossRef] [PubMed]
48. Cid-Díaz, T.; Santos-Zas, I.; González-Sánchez, J.; Gurriarán-Rodríguez, U.; Mosteiro, C.S.; Casabiell, X.; García-Caballero, T.; Mouly, V.; Pazos, Y.; Camiña, J.P. Obestatin controls the ubiquitin–proteasome and autophagy–lysosome systems in glucocorticoid-induced muscle cell atrophy. *J. Cachexia Sarcopenia Muscle* **2017**, *8*, 974–990. [CrossRef]
49. Torres-Velarde, J.; Llera-Herrera, R.; García-Gasca, T.; García-Gasca, A. Mechanisms of stress-related muscle atrophy in fish: An ex vivo approach. *Mech. Dev.* **2018**, *154*, 162–169. [CrossRef]
50. Otsuka, Y.; Egawa, K.; Kanzaki, N.; Izumo, T.; Rogi, T.; Shibata, H. Quercetin glycosides prevent dexamethasone-induced muscle atrophy in mice. *Biochem. Biophys. Rep.* **2019**, *18*, 100618. [CrossRef]

51. Chen, K.; Gao, P.; Li, Z.; Dai, A.; Yang, M.; Chen, S.; Su, J.; Deng, Z.; Li, L. Forkhead Box O Signaling Pathway in Skeletal Muscle Atrophy. *Am. J. Pathol.* **2022**, *192*, 1648–1657. [CrossRef]
52. Brunetti, A.; Goldfine, I.D. Role of myogenin myoblast differentiation and its regulation by fibroblast growth factor. *J. Biol. Chem* **1990**, *265*, 5960–5963. [CrossRef] [PubMed]
53. Legerlotz, K.; Smith, H.K. Role of MyoD in denervated, disused, and exercised muscle. *Muscle Nerve* **2008**, *38*, 1087–1100. [CrossRef] [PubMed]
54. Kim, J.; Guan, K.-L. mTOR as a central hub of nutrient signalling and cell growth. *Nat. Cell Biol.* **2019**, *21*, 63–71. [CrossRef] [PubMed]
55. Wells, L.; Edwards, K.A.; Bernstein, S.I. Myosin heavy chain isoforms regulate muscle function but not myofibril assembly. *EMBO J.* **1996**, *15*, 4454–4459. [CrossRef] [PubMed]
56. Marina, T.; Marija, C.; Ida, R. Functional Foods and the Young. *J. Food Prod. Mark.* **2014**, *20*, 441–451. [CrossRef]
57. Smith, G.I.; Julliand, S.; Reeds, D.N.; Sinacore, D.R.; Klein, S.; Mittendorfer, B. Fish oil–derived n-3 PUFA therapy increases muscle mass and function in healthy older adults. *Am. J. Clin. Nutr.* **2015**, *102*, 115–122. [CrossRef] [PubMed]
58. Pérez, M.; Dominguez-López, I.; Lamuela-Raventós, R.M. The Chemistry Behind the Folin–Ciocalteu Method for the Estimation of (Poly)phenol Content in Food: Total Phenolic Intake in a Mediterranean Dietary Pattern. *J. Agric. Food Chem.* **2023**, *71*, 17543–17553. [CrossRef] [PubMed]
59. Premarathna, A.D.; Ahmed, T.A.E.; Kulshreshtha, G.; Humayun, S.; Shormeh Darko, C.N.; Rjabovs, V.; Hammami, R.; Critchley, A.T.; Tuvikene, R.; Hincke, M.T. Polysaccharides from red seaweeds: Effect of extraction methods on physicochemical characteristics and antioxidant activities. *Food Hydrocoll.* **2024**, *147*, 109307. [CrossRef]
60. Kimira, Y.; Osawa, K.; Osawa, Y.; Mano, H. Preventive Effects of Collagen-Derived Dipeptide Prolyl-Hydroxyproline against Dexamethasone-Induced Muscle Atrophy in Mouse C2C12 Skeletal Myotubes. *Biomolecules* **2023**, *13*, 1617. [CrossRef]
61. Amarasiri, R.P.G.S.K.; Hyun, J.; Lee, S.-W.; Kim, J.; Jeon, Y.-J.; Lee, J.-S. Alcalase-Assisted Mytilus edulis Hydrolysate: A Nutritional Approach for Recovery from Muscle Atrophy. *Mar. Drugs* **2023**, *21*, 623. [CrossRef]
62. Nagahawatta, D.P.; Liyanage, N.M.; Jayawardhana, H.H.A.C.K.; Lee, H.-G.; Jayawardena, T.U.; Jeon, Y.-J. Anti-Fine Dust Effect of Fucoidan Extracted from *Ecklonia maxima* Leaves in Macrophages via Inhibiting Inflammatory Signaling Pathways. *Mar. Drugs* **2022**, *20*, 413. [CrossRef] [PubMed]
63. Liu, S.; Zhang, X.; Zang, X.; Liu, B.; Arunakumara, K.K.I.U.; Xu, D.; Zhang, X. Growth, feed efficiency, body muscle composition, and histology of flounder (*Paralichthys olivaceus*) fed GH transgenic Synechocystis. *Aquaculture* **2008**, *277*, 78–82. [CrossRef]
64. Cai, L.; Qin, X.; Xu, Z.; Song, Y.; Jiang, H.; Wu, Y.; Ruan, H.; Chen, J. Comparison of Cytotoxicity Evaluation of Anticancer Drugs between Real-Time Cell Analysis and CCK-8 Method. *ACS Omega* **2019**, *4*, 12036–12042. [CrossRef] [PubMed]
65. Jayawardena, T.U.; Kim, H.-S.; Sanjeewa, K.K.A.; Kim, S.-Y.; Rho, J.-R.; Jee, Y.; Ahn, G.; Jeon, Y.-J. *Sargassum horneri* and isolated 6-hydroxy-4,4,7a-trimethyl-5,6,7,7a-tetrahydrobenzofuran-2(4H)-one (HTT); LPS-induced inflammation attenuation via suppressing NF-κB, MAPK and oxidative stress through Nrf2/HO-1 pathways in RAW 264.7 macrophages. *Algal Res.* **2019**, *40*, 101513. [CrossRef]
66. Doyuk, F.; Dost, K. Simultaneous determination of six antibiotics belonging to four different classes in chicken meat BY HPLC/DAD and verification BY LC-MS/MS. *Food Chem.* **2023**, *426*, 136549. [CrossRef] [PubMed]
67. Ren, Q.; Boiani, M.; He, T.; Wichers, H.J.; Hettinga, K.A. Heating affects protein digestion of skimmed goat milk under simulated infant conditions. *Food Chem.* **2023**, *402*, 134261. [CrossRef] [PubMed]
68. Liu, W.-W.; Meng, J.; Cui, J.; Luan, Y.-S. Characterization and Function of MicroRNA∗s in Plants. *Front. Plant Sci.* **2017**, *8*, 2200. [CrossRef]

marine drugs

Article

Venomics Reveals the Venom Complexity of Sea Anemone *Heteractis magnifica*

Ming Li [1], Kailin Mao [1], Meiling Huang [1], Yanling Liao [1], Jinxing Fu [1], Kun Pan [1], Qiong Shi [2,3,*] and Bingmiao Gao [1,*]

1 Key Laboratory of Tropical Translational Medicine of Ministry of Education, Hainan Key Laboratory for Research and Development of Tropical Herbs, School of Pharmacy, Hainan Medical University, Haikou 571199, China; mingli@hainmc.edu.cn (M.L.); maokailin@hainmc.edu.cn (K.M.); mgling213@163.com (M.H.); liaoyanling@hainmc.edu.cn (Y.L.); fjx1900201062@163.com (J.F.); pankun001219@163.com (K.P.)
2 Laboratory of Aquatic Genomics, College of Life Sciences and Oceanography, Shenzhen University, Shenzhen 518057, China
3 Shenzhen Key Laboratory of Marine Genomics, Guangdong Provincial Key Laboratory of Molecular Breeding in Marine Economic Animals, BGI Academy of Marine Sciences, BGI Marine, Shenzhen 518081, China
* Correspondence: shiqiong@genomics.cn or shiqiong@szu.edu.cn (Q.S.); gaobingmiao@hainmc.edu.cn (B.G.)

Abstract: The venoms of various sea anemones are rich in diverse toxins, which usually play a dual role in capturing prey and deterring predators. However, the complex components of such venoms have not been well known yet. Here, venomics of integrating transcriptomic and proteomic technologies was applied for the first time to identify putative protein and peptide toxins from different tissues of the representative sea anemone, *Heteractis magnifica*. The transcriptomic analysis of *H. magnifica* identified 728 putative toxin sequences, including 442 and 381 from the tentacles and the column, respectively, and they were assigned to 68 gene superfamilies. The proteomic analysis confirmed 101 protein and peptide toxins in the venom, including 91 in the tentacles and 39 in the column. The integrated venomics also confirmed that some toxins such as the ShK-like peptides and defensins are co-expressed in both the tentacles and the column. Meanwhile, a homology analysis was conducted to predict the three-dimensional structures and potential activity of seven representative toxins. Altogether, this venomics study revealed the venom complexity of *H. magnifica*, which will help deepen our understanding of cnidarian toxins, thereby supporting the in-depth development of valuable marine drugs.

Keywords: sea anemone; *Heteractis magnifica*; transcriptomics; proteomics; venom; toxin

1. Introduction

Like other cnidarians, sea anemones release venom through specialized cells known as nematocysts, which are hollow tubular structures containing numerous toxins [1]. Upon contact with prey, the tubule undergoes an explosive eversion, penetrating the target organism and releasing its venom. Among Cnidaria, sea anemones (Actiniaria) represent one of the most extensively studied groups with diverse toxins. In 1903, physiologist Charles Richet conducted pioneering research on *Actinia equina* and *Anemonia sulcata*, where he undertook the partial purification and characterization of two active components from the tentacular extracts [2]. Even though his research methods and subsequent chromatography made it possible to separate sea anemone toxins, most marine toxicologists still studied crude extracts of whole animals or antennae. Until the 1970s, *Anemonia sulcata* was homogenized and extracted with ethanol to isolate two new toxins ATX I and ATX II [3–5]. Later, scientists began to study other species of sea anemones. According to literature and database searches, 14 peptide toxins have been isolated from *H. magnifica* (HM) using traditional isolation and molecular cloning methods, including voltage-gated channel toxins HmK, Rp II, Rp III, Hmg 1b-2 and Hmg 1b-5 [6–11], protease inhibitor toxins HMIQ3c1, HMGS1, HMGS2 and HMRG1 [12], pore-forming toxins HMg I, HMg II, HMg III, and HmT, and hemolytic toxins [13–16]. However, access to different databases shows that the number of recognized toxins in *H. magnifica* is still relatively far from enough.

Although the scientific community has devoted considerable attention and conducted extensive research on sea anemone toxins, the identification of individual toxins has proven challenging. This difficulty primarily arises from the substantial venom sample required for isolation and characterization using classical biochemical techniques as well as the intrinsic instability of sea anemone's venom proteins. The transcriptomic analysis of *Nematostella vectensis* was published in 2007 and marked the inspired beginning of an 'omics' era for sea anemone venom research [17]. This is the first time that transcriptome sequencing has been applied to sea anemones. So far, 16 sea anemones including *Cryptodendrum adhaesivum, Heterodactyla hemprichii, Heteractis crispa, Macrodactyla doreensis, Entacmaea quadricolor, Telmatactis stephensoni, Oulactis* sp, *Anemonia sulcata, Megalactis griffithsi, Exaiptasia diaphana, Pachycerianthus* cf, *Isarachnanthus nocturnus, Nematostella vectensis, Ceriantheomorphe brasiliensis, Pachycerianthus borealis* and *Stichodactyla helianthus* have been subjected to transcriptome sequencing [18–21]. The integration of transcriptome research into the investigation of sea anemone toxins has, to a certain extent, addressed the challenges associated with acquiring these venomous substances. However, owing to the scarcity of available reference genomes for cnidarians, the majority of these studies have adopted a strategy involving de novo transcriptome assembly from short reads. Therefore, transcriptome sequencing may produce false positive results that are difficult to distinguish. Recently, with the improvement in mass spectrometry instrumentation (for protein sequencing) and bioinformatic tools, proteomics has been used to identify sea anemone toxins. According to the literature, two sea anemones including *Bunodactis verrucosa* and *Nematostella vectensis* were subjected to proteomic sequencing [22,23]. In addition, some of the sea anemone toxins have been studied by venomics of integrating transcriptomic and proteomic technologies [24]. At present, three sea anemones (*Stichodactyla haddoni, Anthopleura dowii* and *Bunodosoma caissarum*) have been identified using integrated multi-omics technology for peptide discovery [24–26]. This approach can be complemented by the additional proteomic characterization of venom proteins, where the utilization of species-specific transcriptome annotation enhances protein identifications in comparison to searches in public databases [24,26]. These omics provide a detailed overview of the proteins and peptide toxins in sea anemones, revealing the high complexity and diversity of venom compared to traditional research results [19].

Many venomous species possess a central venom gland or duct through which toxic peptides and other biomolecules can be injected into predators and prey [27,28]. Cnidarians stand out as one of the few organisms without a centralized venom transmission system. Instead, their venom is transmitted at the cellular level through specialized stinging capsules known as nematocysts. In sea anemones, these nematocysts are distributed in various regions of polyps, including the actinopharynx, tentacles, column, acrorhagi, and mesenterial filaments. Attributed to differences in various physiological functions, there are also differences in the venom produced by nematode vesicles between tissues [29], providing more opportunities to regulate the composition of the venom and allowing the same organism to produce multiple types of venom [30].

Sea anemones lack active locomotion ability, and during their evolution, they choose to rely on toxins in their tentacles to cause paralysis, thereby fixing food and repelling predators. Therefore, the nematocyst in tentacles is the most abundant tissue and serves as the primary source for extracting toxin peptides [31–33]. In addition, the mesenterial filaments of sea anemones house venom components to play a crucial role in aiding digestion. However, these filaments can also be projected beyond the anemone's body, serving functions in external digestion, competition, and defense [29]. Accumulated evidence suggests that various species within the order Actiniaria (sea anemones) exhibit a differential expression of toxins during distinct life stages and in different anatomical regions [34,35]. In tentacles, mesenterial, filaments, and the body column, different toxin profiles have been observed [35]. However, there is currently no transcriptomic or proteomic study on the distribution, diversity, and potential medicinal value of peptide toxins in different tissues of *H. magnifica*, including tentacles and the column.

In this work, in order to explore the complexity and diversity of toxins in *H. magnifica*, integrated venomics including transcriptomics and proteomics sequencing was applied to study the tentacles and column of this sea anemone species. These findings will contribute to the expanding repertoire of *H. magnifica* venom proteins and peptides as well as aid in designing targeted experiments for further isolation and characterization of more valuable specific peptides. Moreover, the identification of putative venom peptides will offer new insights into the evolutionary diversification of toxins and various strategies for prey capture or predator defense.

2. Results

2.1. Summary of the Transcriptome Sequencing and Assembly

A total of 41,408,576 and 44,188,792 reads were generated using an Illumina sequencing platform for the tentacles (HM-T) and column (HM-C), respectively (Supplementary Table S1). After the removal of low-quality reads, we selected 207,530 clean reads for the tentacles (with an average length of 942 bp) and 148,422 clean reads for the column (with an average length of 900 bp) for further assembly and mapped the original sequences back to the assembled transcripts (Figure 1A). These tentacles and column short reads were then de novo assembled into 134,263 and 94,024 transcripts, respectively, by paired-end joining and gap filling with the Trinity software [36,37].

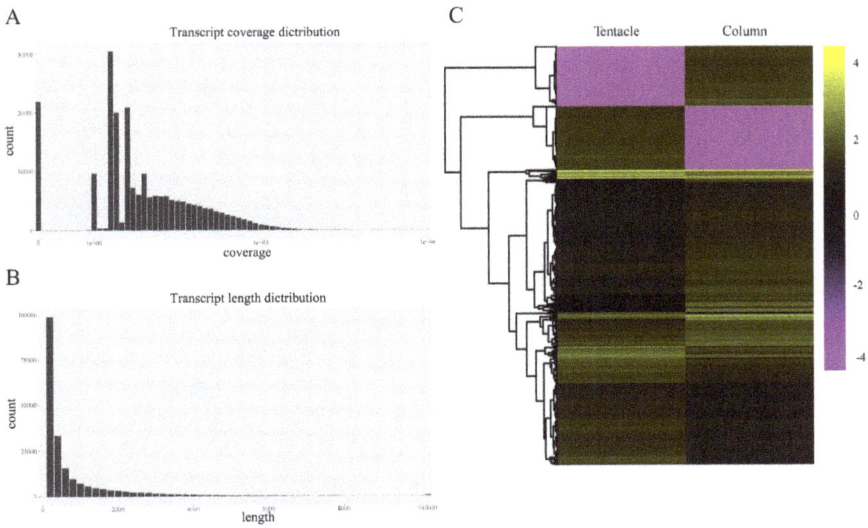

Figure 1. Summary of the *H. magnifica* transcriptome assembly. (**A**) The coverage of assembled transcripts after mapping of raw sequences back to the assembly; (**B**) the transcript length distribution; (**C**) a heatmap of differentially expressed genes (DEGs; colored for RPKM values) for the column and tentacles tissues in *H. magnifica*.

The average length of tentacles transcripts was 675 bp with an N50 of 1105 bp, and the average length of the column transcripts was 721 bp with an N50 of 1256 bp (Figure 1B and Supplementary Table S1). Gene transcription in *H. magnifica* was investigated through tissue-specific transcriptome sequencing (RNA-seq) conducted on two anatomically distinct regions including the tentacles and column. To estimate the relative abundance of each transcript, the paired trimmed reads were mapped back to the transcriptome assembly. The values of Fragments Per Kilobase of the exon model per Million mapped fragments (FPKMs) were then calculated using Stringtie (v1.3.3 b, Mihaela Pertea laboratory, Baltimore, MD, USA). Differential gene expression was assessed through tissue-specific RNA-seq in the two anatomically distinct regions (FDR ≤ 0.05) (Figure 1C).

2.2. *Statistics of Transcriptome Annotation*

To assess the functions of assembled transcripts, a total of 134,263 unigenes from the tentacles and 94,024 unigenes from the column were functionally annotated based on five public databases. These annotations predicted 44,831 genes in the tentacles (Supplementary Figure S1 and Supplementary Table S2) and 29,174 genes in the column (Supplementary Figure S2 and Supplementary Table S3). Among the annotated 44,831 tentacles genes, approximately 18,370 genes were categorized into five main processes (environmental information processing, cellular processes, genetic information processing, organic systems, and metabolism) according to the KEGG analysis (Supplementary Figure S3). The KOG database identified and annotated 17,778 genes, classifying them into 25 molecular families (Supplementary Figure S4). Moreover, the GO annotation revealed that 25,217 genes were associated with 29 functional annotations, which were further categorized into three distinct biological processes (Supplementary Figure S5).

Out of the 29,174 genes identified in the column, 11,964 genes were further classified into five primary categories (environmental information processing, cellular processes, genetic information processing, organic systems, and metabolism; see Supplementary Figure S6). A total of 11,868 unigenes were annotated with KOG classifications and grouped into 25 molecular families (Supplementary Figure S7). Additionally, another GO analysis revealed that 17,458 unigenes were associated with 29 functional annotations, which were further divided into three distinct biological processes (Supplementary Figure S8). These findings indicated that over 79.14% and 86.42% of reads from the tentacles (Supplementary Figure S9) and column (Supplementary Figure S10), respectively, corresponded to some sea anemones, with *Actinia tenebrosa* and *Exaiptasia diaphana* as the top two matched species. The annotated genes from the tentacles (44,831) and column (29,174) were deduced as reference protein databases for subsequent proteomic studies.

2.3. *Toxins Identified in the Transcriptomes of H. magnifica*

To identify candidate toxins in *H. magnifica*, the assembled transcripts were compared to the UniProt animal toxin database using BLAST [18]. A total of 630 protein sequences and 98 peptide sequences were identified in transcriptomics. A total of 442 sequences were identified in tentacles, of which 367 were proteins and 75 were peptides. Out of 381 sequences in the column, 332 protein sequences and 75 peptide sequences were identified (Figure 2). These transcripts were divided into 68 groups according to the predicted functions and then categorized into various biological functions, such as neurotoxin, hemostatic and hemorrhagic toxins, pore-forming protein, mixed-function enzymes, protease inhibitors, allergen, and innate immunity, and mixed-function protein (Figure 2).

The tentacles and column transcriptomes of *H. magnifica* were predicted with 442 and 381 toxin transcripts, respectively (Figure 3A and Supplementary Tables S4 and S5). In addition, this experiment also used FPKM values to represent the transcription levels of the genes encoding proteins or peptide toxins to evaluate the differential expression sequences between different tissues, including tentacles and columns. The top 10 protein or peptide toxins, with the highest FPKM values, were selected from each dataset (Figure 3B). The results showed that ShK-like peptides are the most abundant toxin transcripts in the tentacles, while collagen is the most abundant toxin transcript in the column. The tentacles of *H. magnifica* contained a significantly higher transcription of ShK-like peptides compared to the column. As one of the most crucial toxins present in sea anemone venoms, ShK-like peptides usually play a critical role in protecting sea anemones during defense and predation activities [38]. Furthermore, the collagen in the column exhibited a relatively higher transcription level compared to the tentacles.

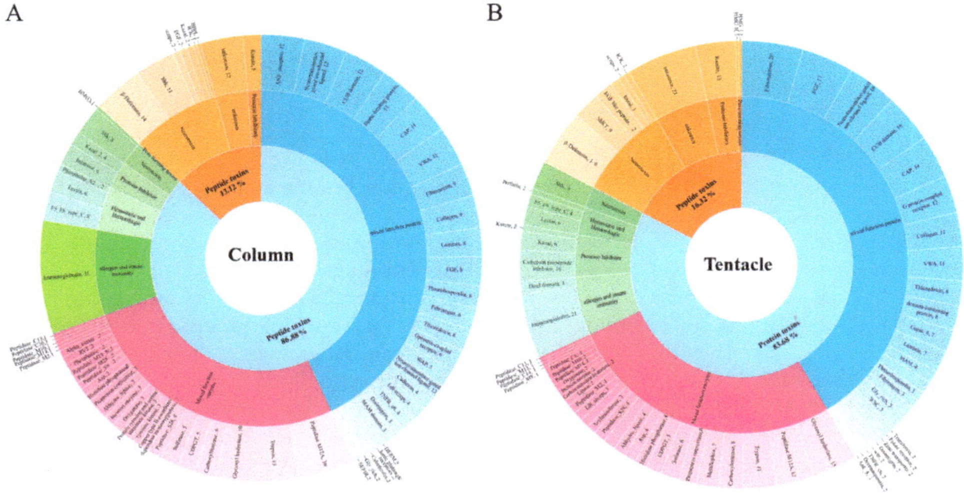

Figure 2. Putative toxin families identified in the transcriptomes of *H. magnifica*. The distinctive protein and peptide sequences from the tentacles (442) and the column (381) were compared by blasting the UniProt database (www.UniProt.org/program/toxins, Accessed on: 30 August 2023). They were then classified into toxin families based on their cysteine scaffolds and amino acid sequences. The pie charts illustrate the proportional contribution of each toxin family to the predicted venom proteome. The number of homologs identified for each protein and peptide family is displayed next to the family name. (**A**) Classification of the tentacles transcriptome; (**B**) classification of the column transcriptome.

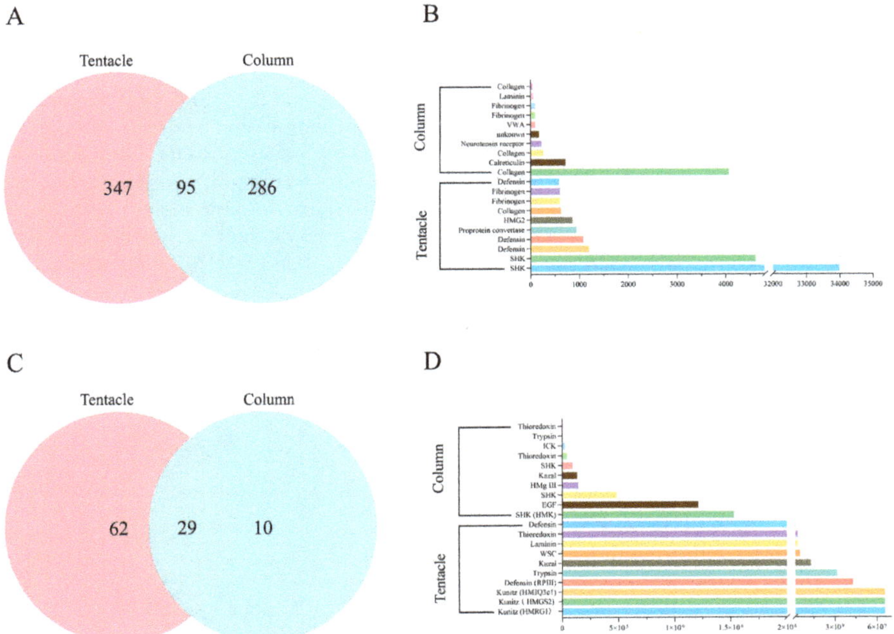

Figure 3. Comparison of proteome and transcriptome data in the *H. magnifica* tissues. (**A**) Relationship between the tentacles and the column data from the transcriptome datasets of *H. magnifica*; (**B**) ten

transcripts of proteins and peptides with the highest levels of transcription from the *H. magnifica* transcriptomes; (**C**) relationship between the tentacles and the column data from the proteomic datasets of *H. magnifica*; (**D**) ten proteins and peptides with the highest levels of translation from the *H. magnifica* proteomes.

2.4. Proteomic Analysis of *H. magnifica* Venom Proteins and Peptides

A total of 101 proteins and peptides were identified in the tentacles and column samples, of which 79 were protein sequences and 22 were peptide sequences, with the assistance of a local protein database generated by TransDecoder from our transcriptome data of the *H. magnifica* tentacles and column (Supplementary Table S6).

These identified proteins and peptides were further classified by family, class, or main molecular function according to the UniProt database and then grouped into seven main categories. They included 36 functional proteins, 26 proteases, 23 neurotoxins, 9 protease inhibitors, 5 allergen and innate immunity proteins, 1 pore-forming toxin, and 1 hemostatic and hemorrhagic protein. Among them, 91 were identified in the tentacles, and 39 were present in the column (Figure 3C). Furthermore, among the 29 co-expressed sequences, 22 were peptide toxins and 7 were protein toxins (Supplementary Table S6). The comparison of transcriptome and proteome expression levels showed significant variances in the highest expressed proteins. The proteome analysis showed that those peptides with the highest expression levels in the tentacles were HMRG1, HMGS2, HMIQ3c1, and RPIII, which have been studied in previous reports [6,12]. HmK, HMg III, and EGF-like peptides are highly expressed in the column. HM-052 is the same sequence as the reported HmK, and HM-028 is an EGF-like peptide with high similarity to Gigantoxin I (Figure 3D). The proteomic analysis detected a considerable abundance of toxins linked to various types, including Kunitz-type peptides, Anemone type 5 potassium channel toxins, ShK-like peptides, and β-defensin peptides. These toxin sequences also showed significant similarities to previously documented members of corresponding families (Supplementary Table S6).

2.4.1. ShK-like Peptides

ShK-like peptides are important toxins in sea anemones, and 13 ShK-like peptides were identified in the transcriptomes of *H. magnifica*. Subsequent proteomic analysis confirmed seven of these sequences, with HM-052 being exclusively expressed in the column and HM-062 only in the tentacles. The remaining five sequences were co-expressed in both the tentacles and the column (Figure 4A). Additionally, the transcriptomic analysis revealed 14 proteins containing ShK-like peptides, and the proteomic validation confirmed five of these proteins (Supplementary Table S6). Notably, HM-701 and HM-703 were uniquely expressed in the tentacles, while the remaining ShK-like peptides were identified in both the tentacles and the column (Supplementary Table S6). The results also showed that HM-052 has been reported in the literature as HmK, which is the first ShK-like peptide discovered in *H. magnifica* [39] (Figure 4B).

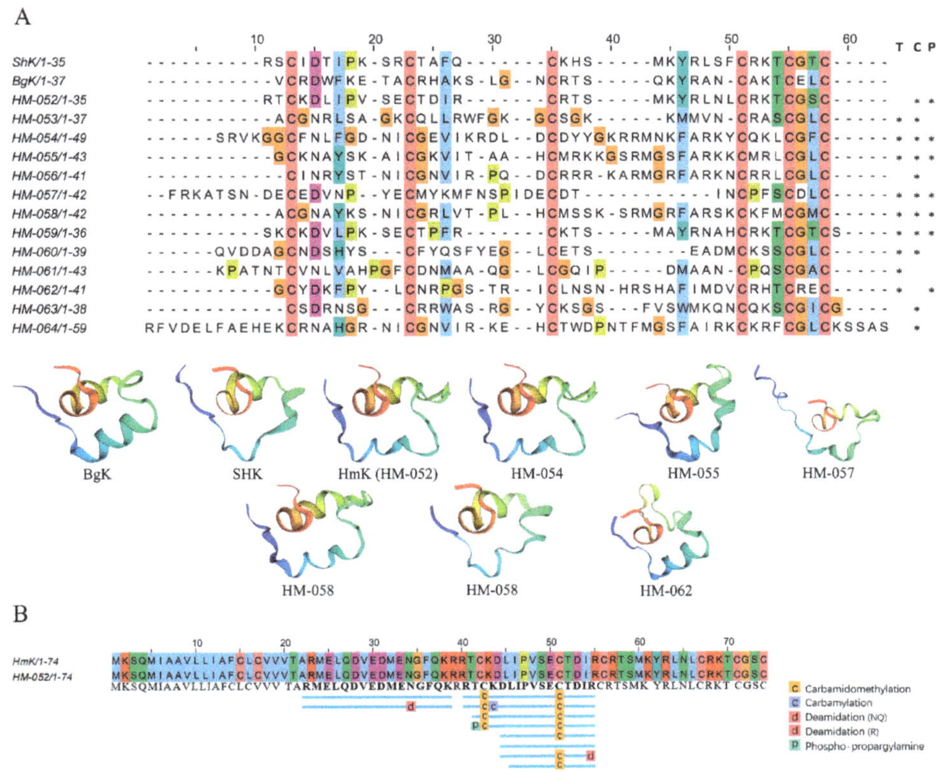

Figure 4. Deduced ShK domain transcripts from *H. magnifica*. (**A**) Peptide sequences from the mature region of representative sea anemones with the ShK domain were aligned for comparison. Residue shading was highlighted based on the Clustal color scheme (T: tentacles transcriptome, C: column transcriptome, P: proteomics; *: identification of sequences in corresponding omics data); (**B**) Peptide coverage.

2.4.2. β-Defensin Peptides

β-defensin peptides, secreted as components of the innate immune system in various organisms, are one of the best-described groups of antimicrobial peptides from a large number of plants, animals, and fungi for host defense [40,41]. However, the β-defensin peptides subfamily has been converted into neurotoxins in several animals to block ion channels [42]. There are 23 β-defensin peptides in the transcriptome, of which nine are only identified in tentacles, seven are only identified in column, and seven are co-expressed.

Among them, seven transcript sequences belonging to β-defensin peptides were confirmed in the proteome data using the NCBI blast tool (Supplementary Table S6). Specifically, HM-022 and HM-023 were identical to the reported toxins RP II and RP III, respectively (Figure 5A,C). These toxins can inhibit signal transduction during the inactivation of sodium channels. Furthermore, HM-009 identified by the proteome showed a high similarity (97%) with Pi-shtx-Hcr5d obtained by Kalina et al. from *Heteractis crispa* (Figure 5B) [43], and it showed a similarity (87%) with toxin Hmg 1b-5 from *H. magnifica*.

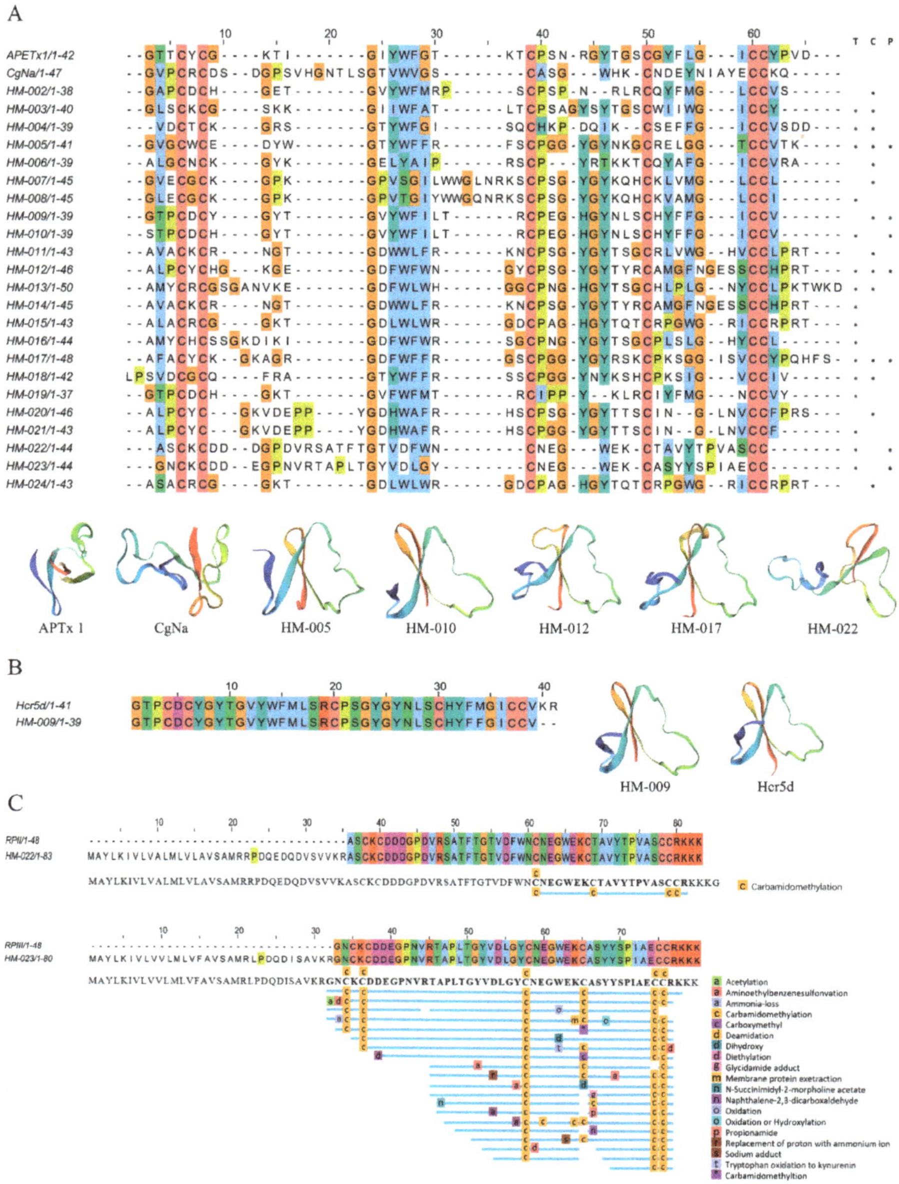

Figure 5. Deduced β-defensin transcripts from *H. magnifica*. (**A**) Representative mature region peptide sequences with β-defensins. These sequences were aligned using MUSCLE and visualized with Jalview, employing residue shading based on the Clustal protein color scheme (T: tentacles transcriptome, C: column transcriptome, P: proteomics; *: Identification of sequences in corresponding omics data); (**B**) similar to a sequence from *Heteractis crispa* (Hc) in transcriptome sequence alignment; (**C**) peptide coverage.

2.4.3. Inhibitor Cystine Knot Fold Peptides

Through transcriptomic studies, two inhibitor cystine knot fold (ICK) peptides were successfully identified, namely HM-034 and HM-035 (Figure 6A). HM-034 was only specifically identified in the tentacles transcriptome. It is worth noting that the HM-035 has been detected in both the tentacles and the column, and its existence has been confirmed by the proteomic analysis (Figure 6B and Supplementary Table S6). This cysteine pattern is also distinctive and aligns with only the voltage-gated potassium channel (Kv) type 5 toxin documented thus far, namely BcsTx3 (κ-actitoxin-Bcs4a) from *Bunodosoma caissarum* [44].

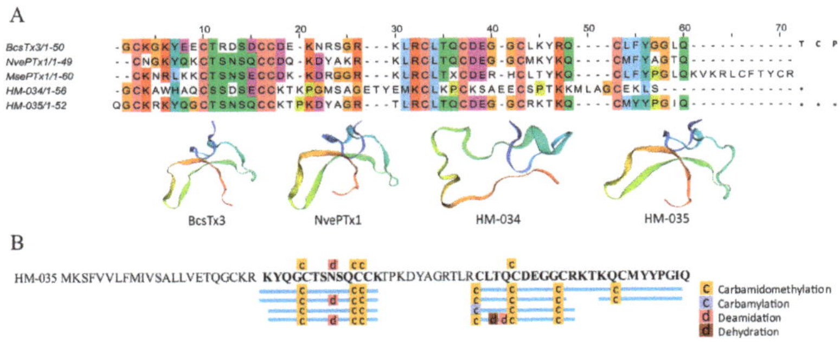

Figure 6. Deduced ICK transcripts from *H. magnifica*. (**A**) Representative peptide sequences with ICK domains. Sequences were aligned with residue shading, which is based on the Clustal protein color scheme (T: tentacles transcriptome, C: column transcriptome, P: proteomics; *: identification of sequences in corresponding omics data); (**B**) peptide coverage.

2.4.4. EGF-like Peptides

Several peptide factors bind to specific receptors on the cell surface to control cell growth, differentiation, and survival. Recent studies have revealed that diverse invertebrates also express a variety of growth factors and growth factor receptors [45]. By conducting a comprehensive transcriptome analysis of *H. magnifica*, a total of five peptide sequences (Figure 7A) and 23 protein sequences with EGF-like peptides were discovered (Supplementary Table S7). A further proteomic investigation supported the existence of three EGF-like peptide analogs and one EGF domain protein in the tentacles (Supplementary Table S6). Specifically, our experiments led to the identification of HM-025, HM-026, and HM-028 as EGF-like peptides, while the HM-028 exhibited a strikingly high sequence similarity (81%) with Gigantoxin I (Figure 7B). In fact, Gigantoxin I is one representative EGF-like peptide in various sea anemones; it was first isolated from *Stichodactyla gigantea*. Both EGF activity and toxicity are observed in this toxin, which causes tyrosine phosphorylation of the EGF receptor in A431 cells [46].

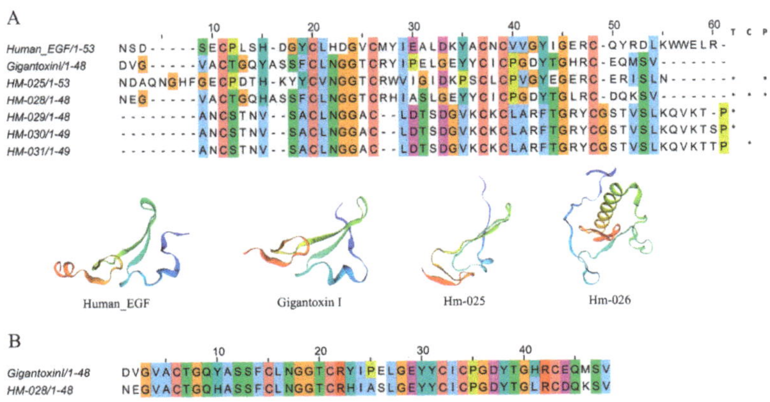

Figure 7. Deduced EGF domain transcripts from *H. magnifica*. (**A**) Representative mature peptide sequences with EGF domains. Residue shading was applied with the Clustal color scheme (T: tentacles transcriptome, C: column transcriptome, P: proteomics; *: identification of sequences in corresponding omics data); (**B**) similar sequence in *Stichodactyla gigantea*.

2.4.5. Kunitz-Type Peptides

The Kunitz-type protease inhibitors are widespread [47]. The transcriptomic analysis of *H. magnific* in this study revealed 12 homologous sequences, including eight peptides expressed in the tentacles and five sequences expressed in the column (Figure 8A), and the proteomic analysis supported the existence of three previously identified Kunitz-type peptides (Supplementary Table S6).

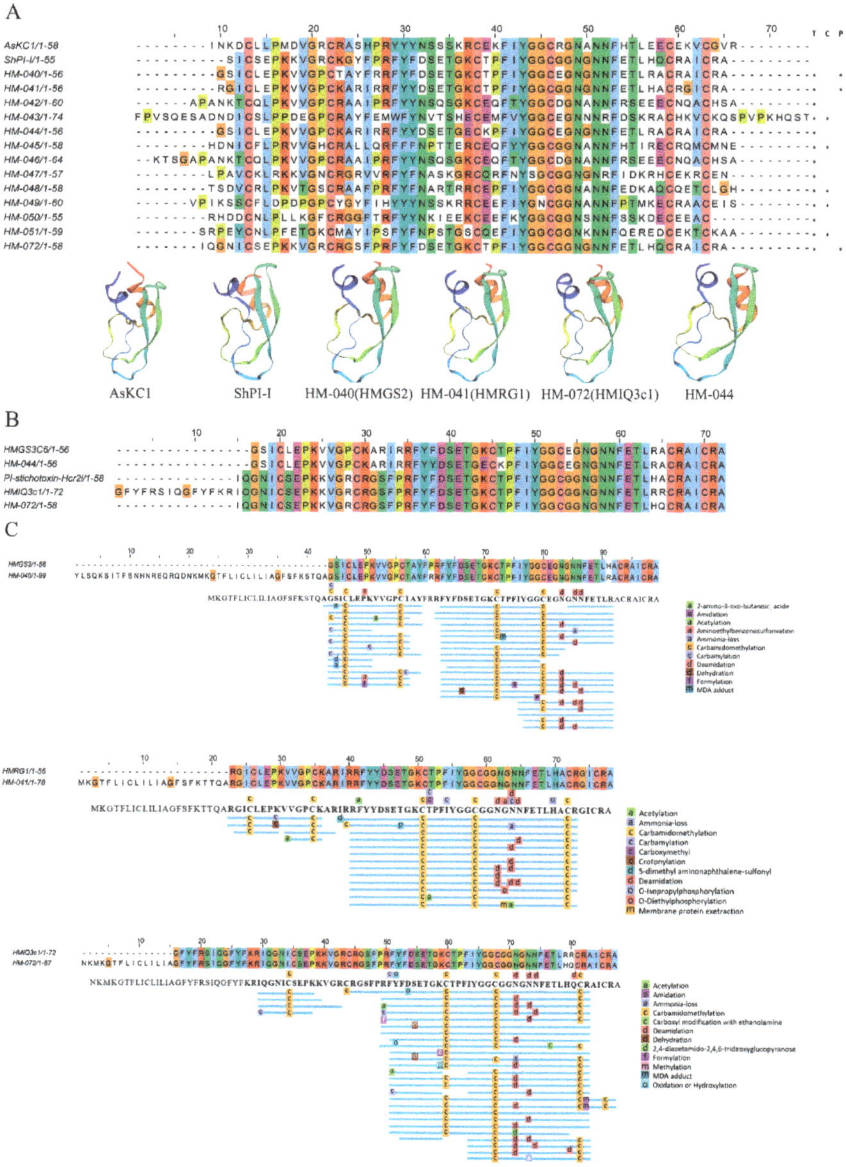

Figure 8. Deduced Kunitz-type transcripts from *H. magnifica*. (**A**) Representative mature peptide regions with Kunitz-type peptides. Residue shading was applied with the Clustal color scheme (T: tentacles transcriptome; C: column transcriptome, P: proteomics; *: identification of sequences in corresponding omics data); (**B**) similar sequence in the sea anemone *Heteractis crispa* (Hc); (**C**) peptide coverage.

Specifically, HM-040 and HM-041 share the same sequence as HMGS2 and HMRG1 toxins (Figure 8C) that were previously reported [12]. HM-072 is very similar to the previously reported HMIQ3c1 [48] (Figure 8B) with only arginine substituted for histidine and glutamine at positions 50 and 51. In addition, the HM-044 sequence exhibits a high similarity (95%) with the reported toxin HMGS3c6 (Figure 8B). The main difference is that phenylalanine replaces tyrosine at position 22 on HM-044 (Figure 8B). During the 1970s, the first report of protease inhibitors was published in sea anemones [49,50]. Subsequent studies have yielded additional information on Kunitz-type peptides, such as AsKC1 and ShPI-I toxins derived from the anemones *Anemonia sulcata* and *Stichodactyla haddoni*, respectively [51,52]. Previous findings proved that Kunitz-type peptides have dual functions as potassium channel blockers and serine protease inhibitors, enabling them to act as defense molecules and neurotoxins for immobilizing prey or predators [53].

2.4.6. Kazal Domain Peptides

Kazal-type serine proteinase inhibitors (KPIs) are characterized by the presence of one or more Kazal domains [54]. Here, a total of 12 Kazal domain peptides were identified through transcriptomic analysis, and four peptide sequences HM-036 to HM-039 exhibited the same cysteine pattern as the previously reported toxin, PI-atitoxin-Avd5a (Figure 9A). Among them, only the peptide HM-038 and two proteins HM-723 and HM-724 containing Kazal domains were validated by the proteomic analysis (Figure 9B and Supplementary Table S6). The Kazal domain peptides produced by the glandular cells of sea anemones are reported to exhibit antibacterial activity against *Streptomyces aureus*; furthermore, they are closely associated with the immune defense response of sea anemones [55].

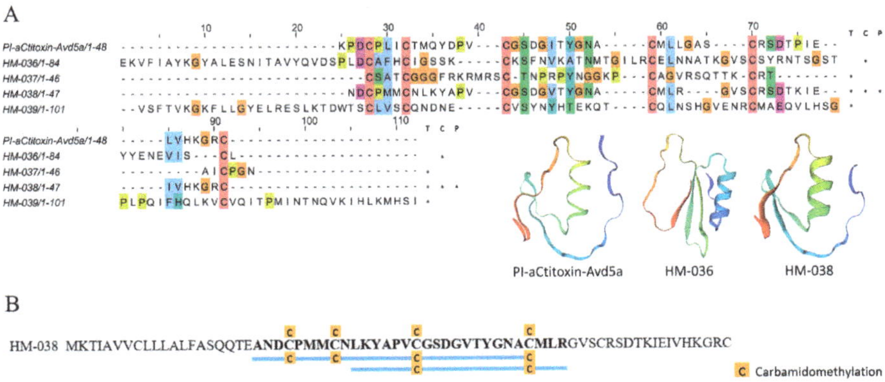

Figure 9. Deduced Kazal domain transcripts from *H. magnifica*. (**A**) Representative mature peptide sequences with Kazal domain peptides. Residue shading was applied with the Clustal color scheme (T: tentacles transcriptome, C: column transcriptome, P: proteomics; *: identification of sequences in corresponding omics data); (**B**) peptide coverage.

2.4.7. Pore-Forming Toxins

Pore-forming toxins (PFTs) are frequently reported components in cnidarian venoms [56]. Five PFTs including HMg I, HMg II, HMg III, HmT and hemolytic toxins have been reported in *H. magnifica* (Figure 10A) [13–16]. Here, three putative PFTs were identified in the tentacles transcriptome of *H. magnifica*. HM-164 was confirmed by proteomics, and its sequence is similar to HMg III. The peptides HM-032 and HM-033 are new pore-forming toxins. However, their results of sequence alignment showed that both peptides also contained HMg I and HMg II sequences, respectively (Figure 10B,C).

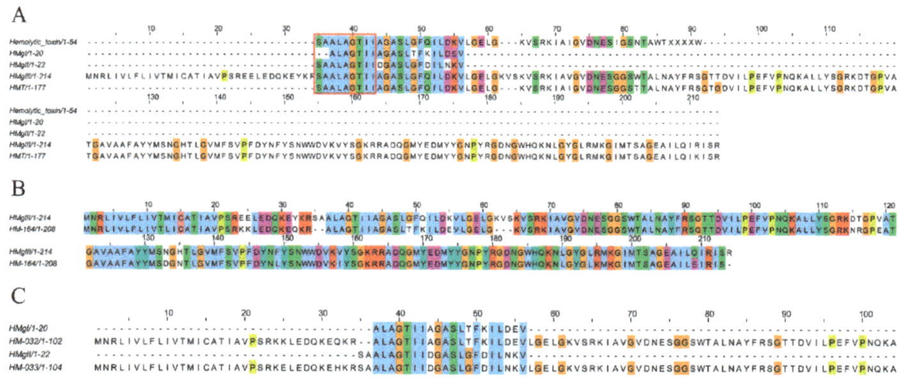

Figure 10. Deduced pore-forming transcripts from *H. magnifica*. (**A**) Representative sequences with pore-forming peptides. Residue shading was applied with the Clustal color scheme; (**B**) proteomics and transcriptomics identification of HMg III-like toxin; (**C**) transcriptomics identification of HMg I-like and HMg II-like toxins.

3. Discussion

Here, we for the first time reported the integrated transcriptome and proteome sequencing of *H. magnifica*. The transcriptomic analysis identified 728 putative toxin sequences, including 442 and 381 from the tentacles and the column, respectively, and they were assigned to 68 gene superfamilies. More importantly, the subsequent proteomic analysis confirmed 101 protein and peptide toxins in the venom, including functional proteins (such as pore-forming proteins, collagens, and G protein-coupled receptors, neurotoxins such as ShK-like peptides, ICK peptides, and β-defensin peptides) and proteases (such as metalloproteinases, lipases, and other enzymes).

ShK-like peptides were named after the first toxin identified in the sea anemone *Stichodactyla helianthus* [57], block Kv1.3 or Kv1.1 and suppressing the proliferation of effective memory T cells at picomolar concentrations [58–60]. Here, seven ShK-like peptides including HmK were identified in the tentacles and the column of *H. magnifica*, all of which were present in the transcriptome and proteomic data. These peptide toxins have therapeutic potential for autoimmune diseases involving effective memory T cells activated by Kv1.3 potassium channels [25]. Eventually, ShK-186 was developed, which was 100 times more selective for Kv1.3 than for Kv1.1, Kv1.4 and Kv1.6 [61–63]. These ShK-like peptides have not been reported in the venom proteomes of other sea anemone species. Therefore, the ShK-like peptide library from various sea anemones may be more diverse beyond our expectation. In addition, the proteomic results showed that ShK-like peptides are highly expressed in the column, and the column usually recognizes toxins more than tentacles. Previous studies have shown that the interstitial filaments in sea anemones contain venomous components to aid digestion, but they may also be projected outside the sea anemone for different purposes, such as external digestion, competition, and defense [64].

The most abundant toxin family identified in this study was β-defensin peptides with 22 transcriptome sequences, seven of which were validated by the proteomics analysis. These peptide members share a defensin domain. Two peptides matched the previously described toxins from *H. magnifica* (Rp II and Rp III), which can selectively inhibit voltage-gated sodium channels according to the UniProt database. β-defensin peptides are often crucial antimicrobial peptides, which are secreted as integral components of the innate immune system in various organisms [65,66]. In the sea anemone venoms, β-defensin peptides exhibit potential neurotoxic properties by blocking ligands and voltage-gated ion channels, including voltage-gated sodium channel (NaV) types 1, 2, and 4, Kv type 3, and acid-sensing ion channels (ASIC) [67–70]. Our multi-omics results show that more β-defensin peptides are recognized by tentacles, and the β-defensin peptides in sea anemone venoms may have evolved for predation and defense purposes with the ability to induce pain and prevent potential predators [38].

The Kunitz-type peptides have a highly conserved region, being characterized by serine protease inhibition, which are also observed among various sea anemones. Three peptides with protease inhibitory activity, such as HMIQ3c1, HMGS2 and HMRG1, were identified in this study. The physiological role of these inhibitors may be related to resistance to prey proteases from degrading their venom peptides [71]. Some of them have shown inhibitory potential against proteases of other classes or the ability to block voltage-gated potassium ion channels, which are crucial regulators of various physiological processes such as host defense, blood coagulation, platelet modulation, fibrinolysis, and action potential transduction [72–74].

In the transcriptome data, many transcripts from the tentacles of *H. magnifica* showed a high percentage of identity with reported transcripts from some closely related species. By comparing different sea anemone toxins, HM-009 and HM-072 in *H. magnifica* transcriptomes are similar in sequence to the Hcr5d and Hcr2i from *Heteractis crispa*. This indicates that these peptide toxins with important roles may be conserved somehow in various sea anemones, although there are certain significant differences in the identified peptide toxins among different species.

Transcriptomics is inherently constrained to identifying candidates with significant sequence homology to known toxins, and it is susceptible to generating false positives [75–77]. Therefore, in this study, proteomics was integrated to further verify the transcriptomic data of different tissues to obtain the complete venom composition. The significance of incorporating proteomic data becomes apparent when considering the discrepancy in both the number and type of putative toxin families during identification through our homology-based annotation compared to those findings derived from the combined transcriptome and proteome data. These data in this study reinforce the need to employ specific databases for further proteomics analyses in order to search for new peptide sequences with novel biological activities.

4. Materials and Methods

4.1. Sea Anemone Collection

Specimens of wild *H. magnifica*, 30–50 cm in length, were collected from offshore areas of Sanya City, Hainan Province, China. These specimens were kept alive in indoor aquaria with artificial seawater. After at least two weeks, they were transferred to smaller tanks for acclimation of 15 min, and then the tentacles and column tissues were cut using tweezers and a scalpel. Once these tissues were removed for storage in liquid nitrogen, the specimens were returned to their original aquarium for continuous aquaculture.

4.2. Transcriptome Construction and Quality Checking

For each tissue, total RNA was extracted using an RNeasy Mini Kit (Qiagen, Duesseldorf, North Rhine Westphalia, Germany). Then, the RNA extractions were quantified on a nanodrop spectrophotometer (Thermo Scientific, Waltham, MA, USA), and RNA integrity number (RIN) values were determined on an Agilent 4200 Bioanalyzer (Agilent, Santa Clara, CA, USA). Illumina sequencing libraries were constructed from total RNA using the NEBNext Ultra RNA Library Prep Kit (NEB, Beverly, MA, USA). The qualified library was sequenced on an Illumina Novaseq 6000 (Illumina, San Diego, CA, USA) high-throughput sequencing platform. The applied sequencing strategy is PE 150. Each sample contained at least 6 GB of sequencing data. Due to the importance of initial error correction on the transcriptome assembly, the initial data are screened for quality control. After filtration of those low-quality reads (containing more than 1% non-sequenced bases or more than 50% low-quality bases) as well as adapter sequences, paired-end sequences were de novo assembled into contigs using Trinity with default parameters [78,79].

4.3. Functional Annotation of the Assembled Transcriptome

To identify potential toxin-like transcripts, the translated transcriptome data were compared with putative animal toxin sequences, which were previously achieved by searching various public databases (with an E-value $\leq 10^{-5}$ as the threshold) [80]. Searches were conducted against five public databases, including (a) the Kyoto Encyclopedia of Genes and Genomes (KEGG) [81,82], (b) cnidarian protein sequences from the GenBank non-redundant (Nr) protein database [83], (c) UniProt animal toxin and venom database (UniProt) [84], (d) clusters of orthologous groups for eukaryotic complete genomes (KOG) [85], and (e) Gene Ontology (GO).

4.4. Candidate Toxin Gene Identification

A BLASTn (E-value < 10.0 and matching length > 60%) search was performed on the transcriptomes using toxin genes from cnidarians and other venomous lineages in the ToxProt dataset (http://www.uniprot.org/program/Toxins, accessed on 30 August 2023) and NCBI Nr Cnidarian Venomous Lineages to identify candidate toxin genes from the combined transcriptomes for each focal taxa [86]. Toxin-related genes from these sea anemone toxin candidates were recognized by SignalP 4.0 (http://www.cbs.dtu.dk/services/SignalP/, accessed on 15 September 2023) as signal peptides and mature peptide regions. Finally, only sequences containing the predicted signal peptides and mature peptides were considered as putative toxin candidates.

4.5. Identification of Protein and Peptide Toxins

A BLAST search of the UniProt database was performed manually to identify sea anemone proteins and peptide toxin families. Subsequent analysis and validation of the cysteine frameworks of peptides and protein toxins were conducted. Proteins and peptide toxins were assigned to known superfamilies in the BLAST database (in our present study, an amino acid number less than 80 was defined as "peptide").

4.6. Venom Sample Preparation for the Proteomic Analysis

Fresh tentacles and column tissues from *H. magnifica* were collected. Sample preparation includes protein extraction, denaturation, reduction, and alkylation as well as tryptic digestion and peptide cleanup. A commercially available iST Sample Preparation kit (PreOmics, Planegg, Germany) was applied according to the manufacturer's protocols. Briefly, 50 μL of Lyse buffer was added to each sample with heating to 95 °C for 10 min and rotation at 1000 rpm. After cooling down to room temperature, trypsin digestion buffer was added, and these samples were incubated at 37 °C for 2 h with shaking at 500 rpm. The digestion process was stopped with a stop buffer. Clean-up and desalting were carried out in the iST cartridge using the recommended washing buffers. Peptides were eluted with elution buffer (2 × 100 μL) and then dried in a speed vacuum concentrator.

4.7. Tandem Mass Spectrometry (MS/MS)

These peptide samples were subsequently re-dissolved in solvent A (0.1% formic acid in distilled water) and analyzed by Q-Exactive Plus coupled to an EASY-nanoLC 1200 system (Thermo Fisher Scientific, Waltham, MA, USA). Every 3 μL peptide sample was loaded onto a 25 cm analytical column (75 μm inner diameter, 1.9 μm resin (Dr. Maisch)) and separated with 60 min gradient starting at 2% buffer B (80% acetonitrile with 0.1% formic acid) followed by a stepwise increase to 35% in 47 min and then 100% in 1 min, which was held for 12 min. The column flow rate was maintained at 300 nL/min with the column temperature at 40 °C. The electrospray voltage was set to 2 kV. The mass spectrometer was run under a data-dependent acquisition (DDA) mode and automatically switched between MS and MS/MS modes. The survey of full-scan MS spectra (m/z 350–1800) was acquired in the Orbitrap with 70,000 resolutions. The automatic gain control (AGC) target of 3 e6 and the maximum injection time of 50 ms were set up. Then, the precursor ions were selected into the collision cell for fragmentation by higher-energy collision dissociation (HCD), and the normalized collection energy was set at 28. The MS/MS resolution was set at 17,500 with the automatic gain control (AGC) target of 1×10^5, the maximum injection time of 45 ms, and dynamic exclusion for 30 s.

4.8. Spectral Searches and Bioinformatics Analysis

Tandem mass spectra were processed by PEAKS Studio version 10.6 (Bioinformatics Solutions Inc., Waterloo, ON, Canada). PEAKS DB was set up to search the database of targets assuming trypsin as the digestion enzyme. The DB was searched with a fragment ion mass tolerance of 0.02 Da and a parent ion tolerance of 7 ppm. Carbamidomethyl on cysteine was specified as the fixed modification. Oxidation on methionine, deamination on asparagine and glutamine, and acetylation on the protein-N term on serine were specified as the variable modifications. Those peptides with no more than 1% FDR (false discovery rate) and those proteins with less than 1% FDR and containing at least 1 unique peptide were filtered for further analysis.

4.9. Alignment and Homology Modeling

We used the SWISS-Model and thread-based LOMETS on I-TASSER to predict three-dimensional structural models for putative toxin peptides based on their primary amino acid sequences [87,88]. In the SWISS-Model server, homologous sequences with over 50% sequence identity were used as examined templates.

5. Conclusions

In summary, this study reports the first integrated omics analysis to characterize the venom composition of tentacles and column for the sea anemone *H. magnifica*, which will help deepen our understanding of cnidarian toxins, thereby supporting the in-depth development of valuable marine drugs. A total of 728 protein and peptide sequences were identified in the transcriptomes, and 101 sequences were confirmed by proteomics to exist in the venom of tentacles and the column, most of which can be classified as β-defensin peptides, ShK-like peptides, and Kunitz-type peptides. These peptides were then used for homologous modeling and three-dimension structure prediction, and we found that these sea anemone peptides may act on voltage-gated potassium ion channels, voltage-gated sodium ion channels, and nicotinic acetylcholine receptors, potentially becoming leading molecules for innovative marine drugs.

Supplementary Materials: The following supporting information can be downloaded at https://www.mdpi.com/article/10.3390/md22020071/s1. Figure S1: The number of annotated genes in the tentacles transcriptome of *H. magnifica*; Figure S2: The number of annotated genes in the column transcriptome of *H. magnifica*; Figure S3: KEGG pathway annotation of the *H. magnifica* tentacles transcriptome; Figure S4: KOG analysis of the tentacles transcriptome of *H. magnifica*; Figure S5: GO analysis of the tentacles transcriptome of *H. magnifica*; Figure S6: KEGG pathway annotation of the *H. magnifica* column transcriptome; Figure S7: KOG analysis of the column transcriptome of *H. magnifica*; Figure S8: GO analysis of the column transcriptome of *H. magnifica*; Figure S9: Analysis of the tentacles transcriptome of *H. magnifica* through the NCBI Nr database; Figure S10: Analysis of the column transcriptome of *H. magnifica* through the NCBI Nr database; Table S1: Sequencing statistics and assembly summary of the sea anemones transcriptome; Table S2: Numbers of tentacles genes annotated by different databases in *H. magnifica*; Table S3: Number of column genes annotated by different databases in *H. magnifica*; Table S4: Annotation of protein and peptide sequences from tentacles of *H. magnifica*; Table S5: Annotation of protein and peptide sequences from column of *H. magnifica*; Table S6: Sequences identified by proteomics in tentacles and column transcriptome of *H. magnifica*; Table S7: Annotation of *H. magnifica* protein and peptide sequences.

Author Contributions: Conceptualization, B.G. and Q.S.; investigation, M.L., Y.L., J.F. and M.H.; data analysis, M.L. and K.P., and writing—original draft preparation, M.L., Q.S. and K.M. writing—review and editing, M.L. and B.G.; funding acquisition, B.G. All authors have read and agreed to the published version of the manuscript and agreed to the published version of the manuscript.

Funding: This research was supported by the National Natural Science Foundation of China (no. 82060686), Special scientific research project of Hainan academician innovation platform (no. YSPTZX202132), and Hainan Provincial Key Point Research and Invention Program (no. ZDYF2022SHFZ309), Hainan Medical University graduate innovation and entrepreneurship training program (no. HYYS2021B20).

Institutional Review Board Statement: Not applicable.

Informed Consent Statement: Not applicable.

Data Availability Statement: Raw sequencing reads and Trinity-assembled contigs have been deposited in the NCBI under accession number BioProject PRJNA1023041 (Column: SAMN37638800, Tentacle: SAMN37638799).

Conflicts of Interest: The authors declare no conflicts of interest.

References

1. Jouiaei, M.; Yanagihara, A.A.; Madio, B.; Nevalainen, T.J.; Alewood, P.F.; Fry, B.G. Ancient venom systems: A review on Cnidaria toxins. *Toxins* **2015**, *7*, 2251–2271. [CrossRef] [PubMed]
2. Richet, C. De l'action de la congestine (virus des Actinies) sur les lapins et de ses effects anaphylactiques. *CR Soc. Biol. Paris* **1905**, *58*, 109–112.
3. Martin, E.J. Toxicity of dialyzed extracts of some *California anemones* (Coelenterata). *Pac. Sci.* **1963**, *17*, 302–304.
4. Martinez, G.; Kopeyan, C.; Schweitz, H.; Lazdunski, M. Toxin III from *Anemonia sulcata*: Primary structure. *FEBS Lett.* **1977**, *84*, 247–252. [CrossRef]
5. Béress, L.; Béress, R. Reinigung zweier krabbenlähmender Toxine aus der Seeanemone *Anemonia sulcata*. *Kiel. Meeresforsch* **1971**, *27*, 117–127.
6. Schweitz, H.; Bidard, J.N.; Frelin, C.; Pauron, D.; Vijverberg, H.P.M.; Mahasneh, D.M.; Lazdunski, M.; Vilbois, F.; Tsugita, A. Purification, sequence, and pharmacological properties of sea anemone toxins from *Radianthus paumotensis*. A new class of sea anemone toxins acting on the sodium channel. *Biochemistry* **1985**, *24*, 3554–3561. [CrossRef]
7. Wemmer, D.E.; Kumar, N.V.; Metrione, R.M.; Lazdunski, M.; Drobny, G.; Kallenbach, N.R. NMR analysis and sequence of toxin II from the sea anemone *Radianthus paumotensis*. *Biochemistry* **1986**, *25*, 6842–6849. [CrossRef] [PubMed]
8. Metrione, R.M.; Schweitz, H.; Walsh, K.A. The amino acid sequence of toxin RpIII from the sea anemone, *Radianthus paumotensis*. *FEBS Lett.* **1987**, *218*, 59–62. [CrossRef]
9. Pease, J.H.; Kumar, N.V.; Schweitz, H.; Kallenbach, N.R.; Wemmer, D.E. NMR studies of toxin III from the sea anemone *Radianthus paumotensis* and comparison of its secondary structure with related toxins. *Biochemistry* **1989**, *28*, 2199–2204. [CrossRef]
10. Kalina, R.S.; Kasheverov, I.E.; Koshelev, S.G.; Sintsova, O.; Peigneur, S.; Pinheiro-Junior, E.L.; Popov, R.S.; Chausova, V.E.; Monastyrnaya, M.M.; Dmitrenok, P.S.; et al. Nicotinic acetylcholine receptors are Novel targets of APETx-like toxins from the sea anemone *Heteractis magnifica*. *Toxins* **2022**, *14*, 697. [CrossRef]
11. Gendeh, G.S.; Young, L.C.; de Medeiros, C.L.C.; Jeyaseelan, K.; Harvey, A.L.; Chung, M.C.M. A new potassium channel toxin from the sea anemone *Heteractis magnifica*: Isolation, cDNA cloning, and functional expression. *Biochemistry* **1997**, *36*, 11461–11471. [CrossRef]
12. Sintsova, O.; Gladkikh, I. Peptide fingerprinting of the sea anemone *Heteractis magnifica* mucus revealed neurotoxins, Kunitz-type proteinase inhibitors and a new β-defensin α-amylase inhibitor. *J. Proteom.* **2018**, *173*, 12–21. [CrossRef]
13. Khoo, K.S.; Kam, W.K.; Khoo, H.E.; Gopalakrishnakone, P.; Chung, M.C. Purification and partial characterization of two cytolysins from a tropical sea anemone, *Heteractis magnifica*. *Toxicon* **1993**, *31*, 1567–1579. [CrossRef]
14. Mebs, D.; Claus, I.; Schroter, A.; Takeya, H.; Iwanga, S. Haemolytic proteins from sea anemones. *Recent Adv. Toxinol. Res.* **1992**, *2*, 392–395. [CrossRef]
15. Wang, Y.; Chua, K.L.; Khoo, H.E. A new cytolysin from the sea anemone, *Heteractis magnifica*: Isolation, cDNA cloning and functional expression. *Biochim. Biophys. Acta (BBA)-Protein Struct. Mol. Enzymol.* **2000**, *1478*, 9–18. [CrossRef]
16. Samejima, Y.; Yanagisawa, M.; Aoki-Tomomatsu, Y.; Iwasaki, E.; Ando, J.; Mebs, D. Amino acid sequence studies on cytolytic toxins from sea anemone *Heteractis magnifica*, *Entacmaea quadricolor* and *Stichodactyla mertensii* (Anthozoa). *Toxicon* **2000**, *38*, 259–264. [CrossRef]
17. Putnam, N.H.; Srivastava, M.; Hellsten, U.; Dirks, B.; Chapman, J.; Salamov, A.; Terry, A.; Shapiro, H.; Lindquist, E.; Kapitonov, V.V. Sea anemone genome reveals ancestral eumetazoan gene repertoire and genomic organization. *Science* **2007**, *317*, 86–94. [CrossRef] [PubMed]
18. Fu, J.X.; He, Y.B.; Peng, C.; Tang, T.L.; Jin, A.H.; Liao, Y.L.; Shi, Q.; Gao, B.M. Transcriptome sequencing of the pale anemones (*Exaiptasia diaphana*) revealed functional peptide gene resources of sea anemone. *Front. Mar. Sci.* **2022**, *9*, 15. [CrossRef]
19. Macrander, J.; Broe, M.; Daly, M. Tissue-specific venom composition and differential gene expression in sea anemones. *Genome Biol. Evol.* **2016**, *8*, 2358–2375. [CrossRef]
20. Delgado, A.; Benedict, C.; Macrander, J.; Daly, M. Never, nver make an enemy horizontal ellipsis Out of an Anemone: Transcriptomic comparison of Clownfish hosting sea anemone venoms. *Mar. Drugs* **2022**, *20*, 730. [CrossRef]

21. Klompen, A.M.; Macrander, J.; Reitzel, A.M.; Stampar, S.N. Transcriptomic analysis of four cerianthid (Cnidaria, Ceriantharia) venoms. *Mar. Drugs* **2020**, *18*, 413. [CrossRef] [PubMed]
22. Domínguez-Pérez, D.; Campos, A.; Alexei Rodríguez, A.; Turkina, M.V.; Ribeiro, T.; Osorio, H.; Vasconcelos, V.; Antunes, A. Proteomic analyses of the unexplored sea anemone *Bunodactis verrucosa*. *Mar. Drugs* **2018**, *16*, 42. [CrossRef] [PubMed]
23. Levitan, S.; Sher, N.; Brekhman, V.; Ziv, T.; Lubzens, E.; Lotan, T. The making of an embryo in a basal metazoan: Proteomic analysis in the sea anemone *Nematostella vectensis*. *Proteomics* **2015**, *15*, 4096–4104. [CrossRef]
24. Mazzi Esquinca, M.E.; Correa, C.N.; Marques de Barros, G.; Montenegro, H.; Mantovani de Castro, L. Multiomic approach for bioprospection: Investigation of toxins and peptides of Brazilian sea anemone *Bunodosoma caissarum*. *Mar. Drugs* **2023**, *21*, 197. [CrossRef]
25. Madio, B.; Undheim, E.A.; King, G.F. Revisiting venom of the sea anemone *Stichodactyla haddoni*: Omics techniques reveal the complete toxin arsenal of a well-studied sea anemone genus. *J. Proteom.* **2017**, *166*, 83–92. [CrossRef]
26. Ramírez-Carreto, S.; Vera-Estrella, R.; Portillo-Bobadilla, T.; Licea-Navarro, A.; Bernaldez-Sarabia, J.; Rudiño-Piñera, E.; Verleyen, J.J.; Rodríguez, E.; Rodríguez-Almazán, C. Transcriptomic and proteomic analysis of the tentacles and mucus of *Anthopleura dowii* Verrill, 1869. *Mar. Drugs* **2019**, *17*, 436. [CrossRef]
27. King, G.F.; Hardy, M.C. Spider-venom peptides: Structure, pharmacology, and potential for control of insect pests. *Annu. Rev. Entomol.* **2013**, *58*, 475–496. [CrossRef]
28. Guiguet, A.; Tooker, J.F.; Deans, A.R.; Mikó, I.; Ning, G.; Schwéger, S.; Hines, H.M. Comparative anatomy of venom glands suggests a role of maternal secretions in gall induction by cynipid wasps (Hymenoptera: Cynipidae). *Insect Syst. Divers.* **2023**, *7*, 3. [CrossRef]
29. Ashwood, L.M.; Undheim, E.A.; Madio, B.; Hamilton, B.R.; Daly, M.; Hurwood, D.A.; King, G.F.; Prentis, P.J. Venoms for all occasions: The functional toxin profiles of different anatomical regions in sea anemones are related to their ecological function. *Mol. Ecol.* **2022**, *31*, 866–883. [CrossRef] [PubMed]
30. Ashwood, L.M.; Mitchell, M.L.; Madio, B.; Hurwood, D.A.; King, G.F.; Undheim, E.A.; Norton, R.S.; Prentis, P.J. Tentacle morphological variation coincides with differential expression of toxins in sea anemones. *Toxins* **2021**, *13*, 452. [CrossRef] [PubMed]
31. Fautin, D.G. Structural diversity, systematics, and evolution of cnidae. *Toxicon* **2009**, *54*, 1054–1064. [CrossRef]
32. Madio, B.; Peigneur, S.; Chin, Y.K.; Hamilton, B.R.; Henriques, S.T.; Smith, J.J.; Cristofori-Armstrong, B.; Dekan, Z.; Boughton, B.A.; Alewood, P.F. PHAB toxins: A unique family of predatory sea anemone toxins evolving via intra-gene concerted evolution defines a new peptide fold. *Cell Mol. Life Sci.* **2018**, *75*, 4511–4524. [CrossRef]
33. Frazão, B.; Vasconcelos, V.; Antunes, A. Sea anemone (Cnidaria, Anthozoa, Actiniaria) toxins: An overview. *Mar. Drugs* **2012**, *10*, 1812–1851. [CrossRef]
34. Columbus-Shenkar, Y.Y.; Sachkova, M.Y.; Macrander, J.; Fridrich, A.; Modepalli, V.; Reitzel, A.M.; Sunagar, K.; Moran, Y. Dynamics of venom composition across a complex life cycle. *eLife* **2018**, *7*, e35014. [CrossRef]
35. Surm, J.M.; Smith, H.L.; Madio, B.; Undheim, E.A.; King, G.F.; Hamilton, B.R.; van der Burg, C.A.; Pavasovic, A.; Prentis, P.J. A process of convergent amplification and tissue-specific expression dominates the evolution of toxin and toxin-like genes in sea anemones. *Mol. Ecol.* **2019**, *28*, 2272–2289. [CrossRef] [PubMed]
36. Smith, E.G.; Surm, J.M.; Macrander, J.; Simhi, A.; Amir, G.; Sachkova, M.Y.; Lewandowska, M.; Reitzel, A.M.; Moran, Y. Micro and macroevolution of sea anemone venom phenotype. *Nat. Commun.* **2023**, *14*, 249. [CrossRef]
37. Shao, P.-J.; Chiu, Y.-L.; Tsai, P.-H.; Shikina, S. Possible germline progenitor cells in extra-gonadal tissues of the sea anemone, *Exaiptasia diaphana*. *Front. Mar. Sci.* **2023**, *10*, 1278022. [CrossRef]
38. Menezes, C.; Thakur, N.L. Sea anemone venom: Ecological interactions and bioactive potential. *Toxicon* **2022**, *208*, 31–46. [CrossRef] [PubMed]
39. Gendeh, G.S.; Chung, M.C.; Jeyaseelan, K. Genomic structure of a potassium channel toxin from *Heteractis magnifica*. *FEBS Lett.* **1997**, *418*, 183–188. [CrossRef]
40. Santana, F.L.; Estrada, K.; Ortiz, E.; Corzo, G. Reptilian β-defensins: Expanding the repertoire of known crocodylian peptides. *Peptides* **2021**, *136*, 170473. [CrossRef]
41. Parisi, K.; Shafee, T.M.A.; Quimbar, P.; van der Weerden, N.L.; Bleackley, M.R.; Anderson, M.A. The evolution, function and mechanisms of action for plant defensins. *Semin. Cell Dev. Biol.* **2019**, *88*, 107–118. [CrossRef]
42. Kim, C.H.; Lee, Y.J.; Go, H.J.; Oh, H.Y.; Lee, T.K.; Park, J.B.; Park, N.G. Defensin-neurotoxin dyad in a basally branching metazoan sea anemone. *FEBS J.* **2017**, *284*, 3320–3338. [CrossRef]
43. Kalina, R.; Gladkikh, I.; Dmitrenok, P.; Chernikov, O.; Koshelev, S.; Kvetkina, A.; Kozlov, S.; Kozlovskaya, E.; Monastyrnaya, M. New APETx-like peptides from sea anemone *Heteractis crispa* modulate ASIC1a channels. *Peptides* **2018**, *104*, 41–49. [CrossRef]
44. Orts, D.J.; Moran, Y.; Cologna, C.T.; Peigneur, S.; Madio, B.; Praher, D.; Quinton, L.; De Pauw, E.; Bicudo, J.E.; Tytgat, J. Bcs Tx3 is a founder of a novel sea anemone toxin family of potassium channel blocker. *FEBS J.* **2013**, *280*, 4839–4852. [CrossRef]
45. Muskavitch, M.A.; Hoffmann, F.M. Homologs of vertebrate growth factors in *Drosophila melanogaster* and other invertebrates. *Curr. Top. Dev. Biol.* **1990**, *24*, 289–328. [CrossRef]
46. Honma, T.; Nagai, H.; Nagashima, Y.; Shiomi, K. Molecular cloning of an epidermal growth factor-like toxin and two sodium channel toxins from the sea anemone *Stichodactyla gigantea*. *Biochim Biophys. Acta* **2003**, *1652*, 103–106. [CrossRef] [PubMed]

47. Rawlings, N.D.; Tolle, D.P.; Barrett, A.J. Evolutionary families of peptidase inhibitors. *Biochem. J.* **2004**, *378*, 705–716. [CrossRef] [PubMed]
48. Kvetkina, A.; Leychenko, E.; Yurchenko, E.; Pislyagin, E.; Peigneur, S.; Tytgat, Y.; Isaeva, M.; Aminin, D.; Kozlovskaya, E. A new Iq-peptide of the Kunitz type from the *Heteractis magnifica* sea anemone exhibits neuroprotective activity in a model of Alzheimer's disease. *Russ. J. Bioorg. Chem.* **2018**, *44*, 416–423. [CrossRef]
49. Fritz, H.; Brey, B.; Béress, L. Polyvalent isoinhibitors for trypsin, chymotrypsin, plasmin and kallikreins of sea anemones (*Anemonia sulcata*), isolation, inhibitory behavior and amino acid composition. *Hoppe-Seyler's Z. Fur Physiol. Chem.* **1972**, *353*, 19–30. [CrossRef] [PubMed]
50. Wunderer, G.; Kummer, K.; Fritz, H.; Béress, L.; Machleidt, W. Broad specificity inhibitors from sea anemones. In Proceedings of the Proteinase Inhibitors, Berlin, Germany, 16–20 October 1973; pp. 277–281.
51. Schweitz, H.; Bruhn, T.; Guillemare, E.; Moinier, D.; Lancelin, J.-M.; Béress, L.; Lazdunski, M. Kalicludines and kaliseptine: Two different classes of sea anemone toxins for voltage-sensitive K+ channels. *J. Biol. Chem.* **1995**, *270*, 25121–25126. [CrossRef] [PubMed]
52. García-Fernández, R.; Peigneur, S.; Pons, T.; Alvarez, C.; González, L.; Chávez, M.A.; Tytgat, J. The Kunitz-type protein ShPI-1 inhibits serine proteases and voltage-gated potassium channels. *Toxins* **2016**, *8*, 110. [CrossRef]
53. Mishra, M. Evolutionary aspects of the structural convergence and functional diversification of Kunitz-domain inhibitors. *J. Mol. Evol.* **2020**, *88*, 537–548. [CrossRef] [PubMed]
54. Rimphanitchayakit, V.; Tassanakajon, A. Structure and function of invertebrate Kazal-type serine proteinase inhibitors. *Dev. Comp. Immunol.* **2010**, *34*, 377–386. [CrossRef]
55. Tschesche, H.; Kolkenbrock, H.; Bode, W. The covalent structure of the elastase inhibitor from *Anemonia sulcata*-a "non-classical" Kazal-type protein. *Biol. Chem. Hoppe-Seyler* **1987**, *368*, 1297–1304. [CrossRef]
56. Podobnik, M.; Anderluh, G. Pore-forming toxins in Cnidaria. *Semin. Cell Dev. Biol.* **2017**, *72*, 133–141. [CrossRef]
57. Castañeda, O.; Sotolongo, V.; Amor, A.M.; Stöcklin, R.; Anderson, A.J.; Harvey, A.L.; Engström, Å.; Wernstedt, C.; Karlsson, E. Characterization of a potassium channel toxin from the Caribbean sea anemone *Stichodactyla helianthus*. *Toxicon* **1995**, *33*, 603–613. [CrossRef] [PubMed]
58. Chi, V.; Pennington, M.W.; Norton, R.S.; Tarcha, E.J.; Londono, L.M.; Sims-Fahey, B.; Upadhyay, S.K.; Lakey, J.T.; Iadonato, S.; Wulff, H.; et al. Development of a sea anemone toxin as an immunomodulator for therapy of autoimmune diseases. *Toxicon* **2012**, *59*, 529–546. [CrossRef]
59. Wang, Z.L.; Zhang, S.Y.; Hao, S.L.; Yang, W.X. Neurotoxins and pore forming toxins in sea anemones: Potential candidates for new drug development. *Histol. Histopathol.* **2023**, *38*, 9–28. [CrossRef]
60. Wang, X.L.; Li, G.Y.; Guo, J.K.; Zhang, Z.P.; Zhang, S.Z.; Zhu, Y.D.; Cheng, J.W.; Yu, L.; Ji, Y.H.; Tao, J. Kv1.3 channel as a key therapeutic target for neuroinflammatory diseases: State of the art and beyond. *Front. Neurosci.* **2020**, *13*, 11. [CrossRef] [PubMed]
61. Chandy, K.G.; Sanches, K.; Norton, R.S. Structure of the voltage-gated potassium channel Kv1.3: Insights into the inactivated conformation and binding to therapeutic leads. *Channels* **2023**, *17*, 2253104. [CrossRef]
62. Varanita, T.; Angi, B.; Scattolini, V.; Szabo, I. Kv1.3 K+ channel physiology assessed by genetic and pharmacological modulation. *Physiology* **2023**, *38*, 25–41. [CrossRef]
63. Manolios, N.; Papaemmanouil, J.; Adams, D. The role of ion channels in T cell function and disease. *Front. Immunol.* **2023**, *14*, 14. [CrossRef]
64. Basulto, A.; Pérez, V.M.; Noa, Y.; Varela, C.; Otero, A.J.; Pico, M.C. Immunohistochemical targeting of sea anemone cytolysins on tentacles, mesenteric filaments and isolated nematocysts of *Stichodactyla helianthus*. *J. Exp. Zool. Part A Comp. Exp. Biol.* **2006**, *305*, 253–258. [CrossRef]
65. Suarez-Carmona, M.; Hubert, P.; Delvenne, P.; Herfs, M. Defensins:"Simple" antimicrobial peptides or broad-spectrum molecules? *Cytokine Growth Factor Rev.* **2015**, *26*, 361–370. [CrossRef]
66. Shafee, T.M.; Lay, F.T.; Hulett, M.D.; Anderson, M.A. The defensins consist of two independent, convergent protein superfamilies. *Mol. Biol. Evol.* **2016**, *33*, 2345–2356. [CrossRef]
67. Chagot, B.; Diochot, S.; Pimentel, C.; Lazdunski, M.; Darbon, H. Solution structure of APETx1 from the sea anemone Anthopleura elegantissima: A new fold for an HERG toxin. *Proteins Struct. Funct. Bioinform.* **2005**, *59*, 380–386. [CrossRef]
68. Chagot, B.; Escoubas, P.; Diochot, S.; Bernard, C.; Lazdunski, M.; Darbon, H. Solution structure of APETx2, a specific peptide inhibitor of ASIC3 proton-gated channels. *Protein Sci.* **2005**, *14*, 2003–2010. [CrossRef] [PubMed]
69. Diochot, S.; Baron, A.; Rash, L.D.; Deval, E.; Escoubas, P.; Scarzello, S.; Salinas, M.; Lazdunski, M. A new sea anemone peptide, APETx2, inhibits ASIC3, a major acid-sensitive channel in sensory neurons. *EMBO J.* **2004**, *23*, 1516–1525. [CrossRef] [PubMed]
70. Smith, J.J.; Blumenthal, K.M. Site-3 sea anemone toxins: Molecular probes of gating mechanisms in voltage-dependent sodium channels. *Toxicon* **2007**, *49*, 159–170. [CrossRef] [PubMed]
71. Župunski, V.; Kordiš, D.; Gubenšek, F. Adaptive evolution in the snake venom Kunitz/BPTI protein family. *FEBS Lett.* **2003**, *547*, 131–136. [CrossRef] [PubMed]
72. An, D.; Pinheiro-Junior, E.L.; Béress, L.; Gladkikh, I.; Leychenko, E.; Undheim, E.A.B.; Peigneur, S.; Tytgat, J. AsKC11, a aunitz peptide from *Anemonia sulcata*, Is a novel activator of G Protein-coupled inward-rectifier potassium channels. *Mar. Drugs* **2022**, *20*, 140. [CrossRef] [PubMed]

73. Gladkikh, I.; Monastyrnaya, M.; Leychenko, E.; Zelepuga, E.; Chausova, V.; Isaeva, M.; Anastyuk, S.; Andreev, Y.; Peigneur, S.; Tytgat, J.; et al. Atypical Reactive Center Kunitz-Type Inhibitor from the Sea Anemone *Heteractis crispa*. *Marine Drugs* **2012**, *10*, 1545–1565. [CrossRef] [PubMed]
74. Kalinovskii, A.P.; Sintsova, O.V.; Gladkikh, I.N.; Leychenko, E.V. Natural inhibitors of mammalian α-Amylases as promising drugs for the treatment of metabolic diseases. *Int. J. Mol. Sci.* **2023**, *24*, 16514. [CrossRef] [PubMed]
75. Mouchbahani-Constance, S.; Sharif-Naeini, R. Proteomic and transcriptomic techniques to decipher the molecular evolution of venoms. *Toxins* **2021**, *13*, 154. [CrossRef] [PubMed]
76. Sunagar, K.; Morgenstern, D.; Reitzel, A.M.; Moran, Y. Ecological venomics: How genomics, transcriptomics and proteomics can shed new light on the ecology and evolution of venom. *J. Proteom.* **2016**, *135*, 62–72. [CrossRef] [PubMed]
77. von Reumont, B.M.; Anderluh, G.; Antunes, A.; Ayvazyan, N.; Beis, D.; Caliskan, F.; Crnković, A.; Damm, M.; Dutertre, S.; Ellgaard, L. Modern venomics—Current insights, novel methods, and future perspectives in biological and applied animal venom research. *GigaScience* **2022**, *11*, 27. [CrossRef]
78. Macrander, J.; Brugler, M.R.; Daly, M. A RNA-seq approach to identify putative toxins from acrorhagi in aggressive and non-aggressive *Anthopleura elegantissima* polyps. *BMC Genom.* **2015**, *16*, 1–19. [CrossRef]
79. Grabherr, M.G.; Haas, B.J.; Yassour, M.; Levin, J.Z.; Thompson, D.A.; Amit, I.; Adiconis, X.; Fan, L.; Raychowdhury, R.; Zeng, Q. Full-length transcriptome assembly from RNA-Seq data without a reference genome. *Nat. Biotechnol.* **2011**, *29*, 644–652. [CrossRef]
80. Altschul, S.F.; Gish, W.; Miller, W.; Myers, E.W.; Lipman, D.J. Basic local alignment search tool. *J. Mol. Biol.* **1990**, *215*, 403–410. [CrossRef]
81. Xie, C.; Mao, X.; Huang, J.; Ding, Y.; Wu, J.; Dong, S.; Kong, L.; Gao, G.; Li, C.-Y.; Wei, L. KOBAS 2.0: A web server for annotation and identification of enriched pathways and diseases. *Nucleic Acids Res.* **2011**, *39*, W316–W322. [CrossRef]
82. Law, S.T.S.; Yu, Y.; Nong, W.; So, W.L.; Li, Y.; Swale, T.; Ferrier, D.E.; Qiu, J.; Qian, P.; Hui, J.H.L. The genome of the deep-sea anemone *Actinernus* sp. contains a mega-array of ANTP-class homeobox genes. *Proc. R. Soc. B-Biol. Sci.* **2023**, *290*, 20231563. [CrossRef]
83. Liao, Q.; Gong, G.; Poon, T.C.; Ang, I.L.; Lei, K.M.; Siu, S.W.I.; Wong, C.T.T.; Rádis-Baptista, G.; Lee, S.M.-Y. Combined transcriptomic and proteomic analysis reveals a diversity of venom-related and toxin-like peptides expressed in the mat anemone *Zoanthus natalensis* (Cnidaria, Hexacorallia). *Arch. Toxicol.* **2019**, *93*, 1745–1767. [CrossRef] [PubMed]
84. Jungo, F.; Bougueleret, L.; Xenarios, I.; Poux, S. The UniProtKB/Swiss-Prot Tox-Prot program: A central hub of integrated venom protein data. *Toxicon* **2012**, *60*, 551–557. [CrossRef] [PubMed]
85. Liu, Y.; Liao, X.; Han, T.; Su, A.; Guo, Z.; Lu, N.; He, C.; Lu, Z. Full-length transcriptome sequencing of the scleractinian Coral *Montipora foliosa* reveals the gene expression profile of Coral–Zooxanthellae Holobiont. *Biology* **2021**, *10*, 1274. [CrossRef]
86. Jungo, F.; Bairoch, A. Tox-Prot, the toxin protein annotation program of the Swiss-Prot protein knowledgebase. *Toxicon* **2005**, *45*, 293–301. [CrossRef]
87. Waterhouse, A.; Bertoni, M.; Bienert, S.; Studer, G.; Tauriello, G.; Gumienny, R.; Heer, F.T.; de Beer, T.A.P.; Rempfer, C.; Bordoli, L. SWISS-MODEL: Homology modelling of protein structures and complexes. *Nucleic Acids Res.* **2018**, *46*, W296–W303. [CrossRef]
88. Zheng, W.; Zhang, C.; Bell, E.W.; Zhang, Y. I-TASSER gateway: A protein structure and function prediction server powered by XSEDE. *Future Gener. Comput. Syst.* **2019**, *99*, 73–85. [CrossRef] [PubMed]

marine drugs

MDPI

Article

Marine Bioprospecting: Enzymes and Stress Proteins from the Sea Anemones *Anthopleura dowii* and *Lebrunia neglecta*

Santos Ramírez-Carreto [1], Beatriz Miranda-Zaragoza [2], Nuno Simões [3,4,5], Ricardo González-Muñoz [6] and Claudia Rodríguez-Almazán [2,*]

1 Instituto Nacional de Salud Pública, Centro de Investigación Sobre Enfermedades Infecciosas, Av. Universidad #655, Santa María Ahuacatitlan, Cuernavaca C.P. 62100, Mexico; santos.ramirez@ibt.unam.mx

2 Departamento de Micro y Nanotecnologías, Instituto de Ciencias Aplicadas y Tecnología, Universidad Nacional Autónoma de México, Cto. Exterior S/N, C.U., Coyoacán, Ciudad de México C.P. 04510, Mexico; beatriz.miranda@icat.unam.mx

3 Unidad Multidisciplinaria de Docencia e Investigación en Sisal, Facultad de Ciencias, Universidad Nacional Autónoma de México, Puerto Abrigo s/n, Sisal C.P. 97356, Mexico; ns@ciencias.unam.mx

4 International Chair for Coastal and Marine Studies, Harte Research Institute for Gulf of Mexico Studies, Texas A and M University-Corpus Christi, Corpus Christi, TX 78412, USA

5 Laboratorio Nacional de Resiliencia Costera (LANRESC), Laboratorios Nacionales, CONACYT, Sisal C.P. 97356, Mexico

6 Instituto de Investigaciones Marinas y Costeras, CONICET, Facultad de Ciencias Exactas y Naturales, Universidad Nacional de Mar del Plata, Dean Funes 3350, Mar del Plata C.P. 7600, Argentina; ricordea.gonzalez@gmail.com

* Correspondence: claudiar@comunidad.unam.mx

Abstract: The bioprospecting of sea anemone tissues and secretions has revealed that they are natural libraries of polypeptides with diverse biological activities that can be utilized to develop of biotechnological tools with potential medical and industrial applications. This study conducted a proteomic analysis of crude venom extracts from *Anthopleura dowii* Verrill, 1869, and *Lebrunia neglecta* Duchassaing & Michelotti, 1860. The obtained data allowed us to identify 201 polypeptides, of which 39% were present in both extracts. Among the obtained sequences, hydrolase-type enzymes, oxidoreductases, transferases, heat shock proteins, adhesion proteins, and protease inhibitors, among others, were identified. Interaction analysis and functional annotation indicated that these proteins are primarily involved in endoplasmic reticulum metabolic processes such as carbon metabolism and protein processing. In addition, several proteins related to oxidative stress were identified, including superoxide dismutase, peroxiredoxins, thioredoxin, and glutathione oxidase. Our results provide novel information on the polypeptide composition of the crude venom extract from sea anemones, which can be utilized to develop molecules for therapeutic tools and industrial applications.

Keywords: *Lebrunia neglecta*; *Anthopleura dowii*; proteome; crude venom extract; enzymes; sea anemone

1. Introduction

The ocean is an ecosystem that is rich in biodiversity, serving as a leading source of natural compounds that have been explored for decades. The marine environment offers excellent economic opportunities for the development of new natural products due to its biodiversity. Compounds of marine origin have become base scaffolds to generate new molecules with therapeutic potential for various diseases or industrial applications [1,2]. There is a continuous search for natural compounds because not all end up having an application. Continued marine bioprospecting efforts allow for the sustainable exploration of new species of marine organisms to generate alternatives to using marine resources [3,4]. There is a particular interest in species that live in extreme environments or undergo significant physical–chemical and biological changes that promote physiological modifications in the organisms [5–7]. Species inhabiting highly variable marine zones have adapted their molecular machinery to different and varying conditions of salinity, light, wave action, ultraviolet radiation, pressure, temperature, and biological and chemical interaction, leading to the production of metabolites with biotechnological potential [2,3,8–10].

Some marine invertebrates live in environments with significant physical and biological variations, generating natural products with new biochemical characteristics to adapt to these environments [2,3,5,11]. These organisms are fascinating for marine bioprospecting due to the potential for novel natural marine products to be found. Marine invertebrate populations that inhabit rocky coastlines increase their energy costs because they face variations in the type of substrate and an extensive diversity of physical and biological factors, such as exposure to waves, temperature, desiccation, salinity, oxygen, light, fixation surface, competition, and predation [12–15]. In addition, the interaction of the tides with the waves is part of a scenario of manifestations of various adaptations and a peculiar diversity of species. In contrast, protected habitats such as reefs are characterized by low hydrodynamic stress, sedimentation, and the stratification of coastal water, producing marked daily or seasonal changes in temperature, salinity, and nutrient concentrations; metabolic adjustments are minor [16–18]. However, climate change is a factor that adds to the stress faced by marine organisms, generating metabolic adjustments in response to temperature [19,20].

Numerous reports address the issue of the physiological changes that marine organisms can undergo with the physical, chemical, and biological changes that can occur in the ocean. When ocean temperature increases, the metabolic activity of invertebrates can increase two- to three-fold. In a parallel manner, stress due to environmental warming inhibits oxygen consumption and leads to significant changes in metabolic processes, stimulating CoxI (aerobic metabolism), HSP70 (stress response), and SOD (antioxidant molecule), and organisms such as the pelagic bigfin reef squid, *Sepioteuthis lessoniana,* respond rapidly to cellular and metabolic adjustments of these proteins [21]. Organisms that live in hydrothermal vents increase the production of superoxide dismutase and catalase [22]. The reduction in oxygen in water bodies globally due to climate change is a factor that generates hypoxic conditions, which reduces locomotion and metabolic rate and increases oxidative stress, affecting other physiological processes of organisms that must be adjusted to quickly adapt to those conditions [23]. Organisms such as crustaceans, which face changes in salinity, generate a stress response and, at the same time, metabolic disorders, affecting the production of chitin and the enzymes required to produce this [24,25].

These suggest that marine enzymes have evolved to adapt to diverse conditions, potentially resulting in better stability and specificity for substrates and presenting advantages over their counterparts. In recent decades, interest has been generated in studying polypeptides, including enzymes, from marine organisms with applications in various areas, such as in the treatment of diabetes [14], like anticoagulants, and in the treatment of metabolic disorders, among other clinical and industrial applications [26].

In recent decades, peptides and proteins from cnidarians, particularly sea anemones, have been shown to have important structural and functional characteristics that benefit human health [14,27,28]. Approximately 1200 sea anemone species are distributed across marine habitats worldwide [29,30]. Sea anemones are marine invertebrates living in various environments, such as hydrothermal vents, reefs, and intertidal zones. Sea anemones are animals with a simple anatomy and sedentary lifestyles, producing polypeptides with functions in fundamental processes like feeding, defense, protection, reproduction, symbiosis, and growth [29]. While sea anemones have long been recognized as rich sources of polypeptides with various pharmacological activities, it was not until a few years ago that proteomic and transcriptomic methods were implemented to explore the proteins in their secretions and tissues [29,31,32]. Bioprospecting in these venomous marine invertebrates has traditionally focused on polypeptides with neurotoxic [33,34] and cytotoxic properties [35], with little attention given to other components of biotechnological interest, such as enzymes, amylases [36], chitinases [37], proteases [29,31], oxidoreductases [38,39], heat shock proteins [40], and structural proteins [41], which have been identified in their crude venom extracts (CVEs) and secretions. These enzymes can potentially be used to develop tools for medical and industrial applications [42,43].

In this study, marine bioprospecting was carried out involving proteomics on two sea anemones, *Anthopleura dowii,* located in the intertidal zone in Baja California Norte, Mexico, and *Lebrunia neglecta,* found in the reef area of Puerto Morelos, Quintana Roo, Mexico (Figure 1). The main goal of this study was to identify polypeptides (non-toxins) and provide new insights into engineering proteins for biotechnological applications with potential biomedical and industrial applications.

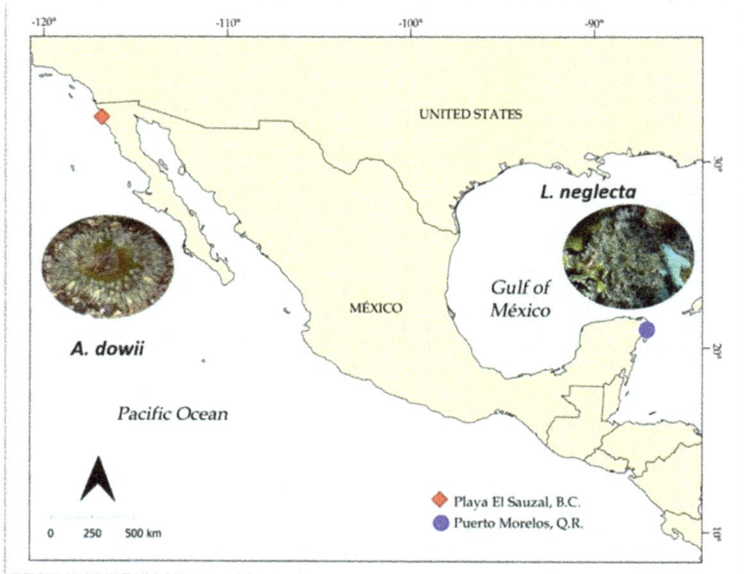

Figure 1. Sea anemone collection area. The red diamond indicates El Sauzal Beach, Baja California Norte, Mexico, where the *A. dowii* specimens were collected. The blue circle indicates the area where the Puerto Morelos reef is located, Quintana Roo, Mexico, where *L. neglecta* was collected.

2. Results

The bioprospecting of *A. dowii* and *L. neglecta* sea anemones was carried out using proteomics of crude venom extracts, aiming to identify (non-toxic) proteins that the anemones produce and may be used in biotechnological applications.

2.1. Proteomic Analysis of Crude Venom Extracts (CVEs)

Specimens of *A. dowii* and *L. neglecta* were collected, and CVEs of both species were obtained, as indicated in Section 4.2. The electrophoretic profiles of the CVEs from *A. dowii* and *L. neglecta* exhibited differences in the region from 10 to 250 kDa (Figure 2A). These differences suggested the possibility of identifying disparities and similarities in the polypeptide composition between both CVEs when analyzed using Shotgun proteomics. Of the 8554 total spectra obtained through an MS/MS analysis of the three biological replicates for each CVE, 120 and 159 proteins were identified in *A. dowii* and *L. neglecta,* respectively (Supplementary Material S1).

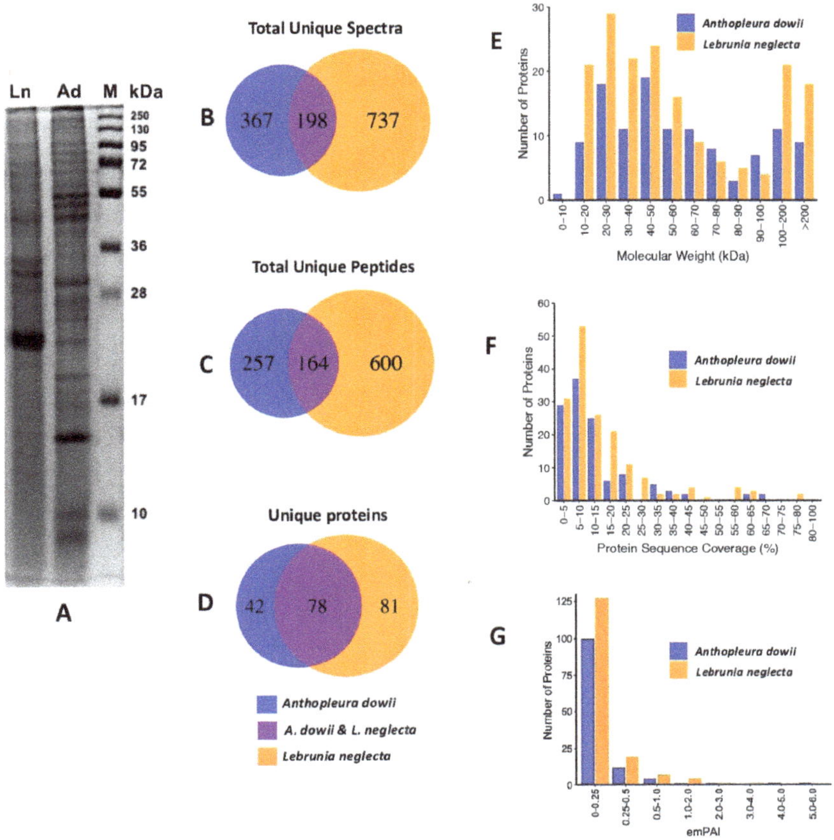

Figure 2. Proteomic comparison of crude venom extract from *Anthopleura dowii* and *Lebrunia neglecta*. (**A**) CVE electrophoretic profiles from *A. dowii* (Ad) and *L. neglecta* (Ln) samples analyzed with SDS–PAGE gel electrophoresis, 15% polyacrylamide, and Coomassie blue staining. (**B**–**D**) show Venn diagrams corresponding to the distribution of the number of unique spectra, unique peptides, and unique proteins, and the overlap between both samples. The distribution of the identified proteins with respect to their molecular weight, percentage of coverage, and the exponentially modified protein abundance index (emPAI) are shown in (**E**–**G**), respectively.

In order to identify the proteins obtained in both CVEs, the Refseq_Cnidaria database was used, and the parameters are described in Section 4.5. Tryptic peptides obtained from the CVE from *A. dowii* were also aligned with the polypeptide sequences generated from the tentacle transcriptome [31,44], increasing the number of proteins identified in *A. dowii* and improving the subsequent annotation process. Among the two groups of proteins, we identified 201 unique proteins. Of these, 21% (42 proteins) were exclusively found in the *A. dowii* CVE, 40% (81 proteins) were exclusive to the *L. neglecta* CVE, and 39% (78 proteins) were found in both CVEs (Figure 2B–D). Using the Scaffold program, the proteomic indices of both species were obtained and compared to estimate differences and similarities. Our results show, in Venn diagrams, that the totals of unique spectra and peptides are more than double for the *L. neglecta* samples (Figure 2B,C), which coincides with the more significant number of unique proteins identified in *L. neglecta* compared to *A. dowii* (Figure 2D). The distribution of molecular weights, imPAI, and sequence coverage are similar between both samples (Figure 2E–G). Although the number of identified proteins is like that reported in other studies on sea anemones' proteomics [31,45,46], we estimated that the identification rate was low, considering the complexity observed in the electrophoretic profiles of both CVEs. We speculate that the low identification and coverage rate could be due to the intrinsic characteristics of the sample that could complicate the identification process, as suggested by other research groups [45]. Therefore, we suggest that this type of sample may require a strategy that allows a larger amount of proteolytic peptides to be obtained for MS/MS sequencing analysis [47,48].

The identified proteins were annotated following the parameters and process indicated in Section 4.6. The information obtained from the automatic annotation using the Refseq_Cnidaria database was used and aligned with a homemade database generated from the *A. dowii* tentacle transcriptome, allowing, in some cases, for the complete sequence of the protein to be obtained and improving the annotation by submitting the sequence to the UniProt database. UniProt access codes were then subjected to GO analysis to classify proteins from both CVEs into the three main GO categories: "cellular components" (CC), "biological processes" (BP), and "molecular function" (MF) (Figure 3). The GO terms with the highest representation in the MF category were protein binding (GO:0005515), hydrolase activity (GO:0016787), and nucleotide binding (GO:0000166). In the case of BP, the most common GO terms shared between the samples from both CVEs were those related to metabolic processes (GO:0008152), cellular processes (GO:0009987), proteolysis (GO:0006508), and cell adhesion (GO:0007155). Our results predicted that most of the identified polypeptides are in the cytoplasm, extracellular region, and cytoskeleton (Figure 3). A Gene Ontology (GO) analysis of the *A. dowii* and *L. neglecta* proteomes revealed the presence of proteins previously reported in cnidarian secretions and found to be overrepresented under stress conditions, suggesting their role in protective mechanisms [40,49]. These proteins have also been documented in the sea anemone proteome of *Exaiptasia pallida* in the presence of the endosymbiont *Durusdinium trenchii* [50] and in the proteome of *Nematostella vectensis* under tissue damage conditions [40].

In order to identify the biological processes or metabolic pathways in which the protein components identified in both proteomes are participating, interaction maps were constructed for each group of proteins using the String v11.5 program, and the genome database of *N. vectensis* was used as a template [40]. Of the 119 and 158 proteins identified in *A. dowii* and *L. neglecta*, respectively, the amino acid sequences of 103 and 142 proteins were aligned with the *N. vectensis* database. Six and seven interaction groups were obtained in the proteomes of *A. dowii* and *L. neglecta*, respectively (Supplementary Material S2–S6). Group A in the *A. dowii* interactome comprises proteins involved in oxidation–reduction processes, such as mitochondrial superoxide dismutases (SODs), SODs of the Cu-Zn family, and peroxiredoxins-1. These proteins play a crucial role in cellular protection against the toxic free radicals generated within the cell [41]. Group B consists of proteins involved in protein processing and folding in the endoplasmic reticulum, including chaperones, thioredoxins, and heat shock proteins. Group C encompasses proteins related to metabolic processes such as glycolysis, gluconeogenesis, pyruvate metabolism, amino acid biosynthesis, and the pentose–phosphate pathway. Groups D and E include chitinase-type glycosidases and structural and adhesion proteins, respectively. Group F consists of proteins involved in signal transduction mediated by small GTPases. These proteins are predicted to participate in vesicular trafficking and are part of endocytosis, autophagy, and mitophagy.

In the case of *L. neglecta*, groups A and C comprise structural proteins and proteins related to cell adhesion. In Group B, we found that proteins participate in vesicular trafficking, while in Group D, the proteins are involved in the stress response, oxidation–reduction processes, and carbon metabolism. Group E has proteins that participate in translation, while group F consists of proteins related to fructose and mannose metabolism, the pentose phosphate pathway, and glycolysis/gluconeogenesis. Finally, group G includes proteins related to redox metabolism (Supplementary Material S6).

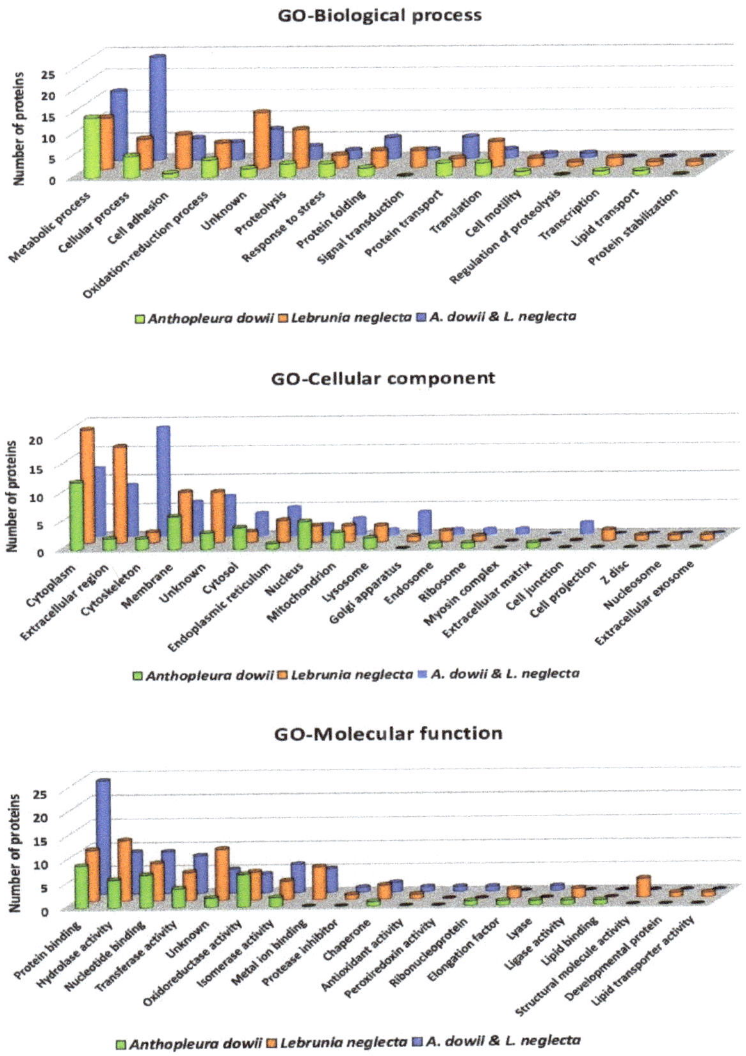

Figure 3. Annotation of the proteins identified in the crude venom extracts from *A. dowii* and *L. neglecta* by shotgun proteomics. They are classified into three main Gene Ontology (GO) categories (Supplementary Material S1): biological process, cellular component, and molecular function. The graphs show, on the *x*-axis, the sub-category of the identified proteins and, on the *y*-axis, the number of proteins identified for each sub-category. The annotation is based on amino acid sequence homology with respect to the proteins annotated in the UniProtKB database using BLASTP and QuickGO tools.

Although differences were observed between the number of unique proteins present in both CVEsamples (Figure 2D), the information generated from the annotation and the interactomes suggests that both samples are very homogeneous in terms of the presence of proteins with the same molecular function and their participation in biological processes. The most significant differences, attributed to the absence or presence of proteins with a specific function, are observed in those related to structural roles, lipid binding/transport, and development-related activities. Interestingly, some proteins identified only in *L. neglecta* were also found in the *A. dowii* tentacle transcriptome, suggesting that these proteins are present in *A. dowii* but were not identified in our proteomic analysis (see Supplementary Material S1).

2.2. Differentially Expressed Cellular Proteins

The results of the identification and annotation of the proteomes provided sufficient information to determine the type of proteins that make up the CVEs of both anemone species. However, to estimate the differences between the proteomes related to the different ecological factors found in the environments where *A. dowii* and *L. neglecta* live, it was necessary to identify the differentially expressed proteins in both species. This information, in turn, allowed for the identification of metabolic pathways in which up- or downregulated proteins contribute to the knowledge of the physiological response of sea anemones to physical, chemical, and biological factors. When a double change-fold restriction was applied (see Section 4.5), 59 proteins showed significant changes in expression levels between both proteomes. A volcano plot was made considering up- and downregulated proteins (Figure 4), which allows us to quickly identify changes in large datasets consisting of experimental replicates [51]. In the volcano plot, each dot represents a protein. Proteins with positive and negative log2 (fold change) values, derived from the ratio (*Lebrunia neglecta/Anthopleura dowii*) between both samples, are averaged in triplicate; this corresponds to the value on the x-axis, and p-log10 values are indicated on the y-axis (proteins with high statistical significance). The cutoff line on the x-axis indicates a two-fold change in the intensity of the MS signals. This analysis led us to identify that, of the differentially expressed proteins, 29 were downregulated and 30 were upregulated.

The UniProt database and the Quick GO program classified these proteins based on their molecular function. The main molecular functional groups of proteins that significantly lowered their expression levels included catalytic proteins (45%), nucleotide binding (17%), protein binding (17%), unfolded protein binding (14%), and proteins with unknown function (7%) (Figure 3B). Catalytic proteins were also the leading upregulated group (47%), followed by protein binding (23%), nucleotide binding (10%), metal-ion binding (10%), function unknown (7%), and binding to unfolded proteins (3%) (Figure 4C).

In order to determine the types of protein–protein interactions among the 59 differentially expressed proteins, an interaction map was generated with String v11.5 software, as indicated in Section 4.6. Interaction networks showed thirty-seven interactions obtained from 51 altered proteins that matched the *N. vectensis* database using confidence mode (score > 0.7). We identified six interaction groups (Figure 4D). Group (A), indicated in yellow, comprises proteins related to sugar metabolism, and group (B), marked in lime green, comprises proteins related to the response to stress and metabolism of reactive oxygen species. Group (C), in purple, includes structural proteins with actin domains; group (D), indicated in red, includes endoplasmic reticulum proteins that participate in protein folding and cytoplasmic proteins that respond to cell stress. Groups (E) and (F), marked in blue and green, are enzymes that participate in the modification of proteins and extracellular enzymes that act in the degradation of chitin, respectively (see Supplementary Material S5). Our analysis shows an increase in the expression of oxidoreductase enzymes such as superoxide dismutase [Cu-Zn], peroxiredoxin-1 isoform X2, superoxide dismutase [mn], and endoplasmic reticulum proteins involved in protein folding and redox metabolisms, such as calreticulin and peptidyl–prolyl cis–trans isomerases, suggesting more significant activity of redox metabolism in *L. neglecta* compared to *A. dowii*. There is a decrease in the expression of proteins related to stress belonging to the heat shock protein 70 family in *L. neglecta* compared to *A. dowii*. A marked decrease in the expression of hydrolase-type enzymes was observed in the samples of *L. neglecta* compared to *A. dowii* (Figure 4B–D).

The results show the enrichment of proteins with enzymatic activities differentially regulated in both anemone species. This regulation could be related to the response of the anemone to exposure to environmental stressors that could be occurring in its habitats, such as temperature, ultraviolet radiation, changes in salinity, the availability of nutrients, and the accumulation of pollutants, among others.

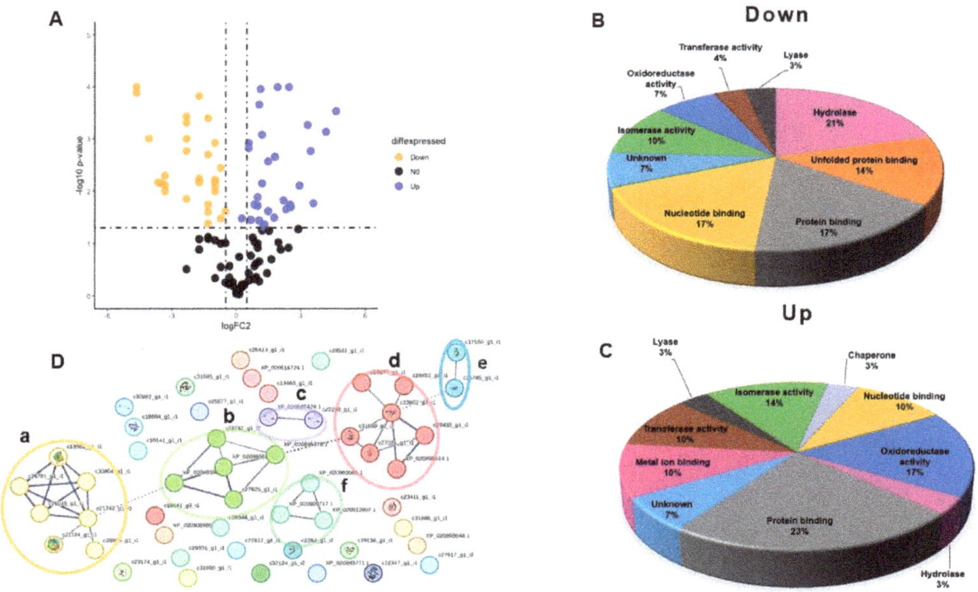

Figure 4. Differentially expressed proteins from proteomes of *A. dowii* and *L. neglecta*. (**A**) shows the volcano graph that represents the changes in the expression of 59 proteins present in both samples. The x-axis shows the log fold change of the *L. neglecta/A. dowii* ratio, while the y-axis shows the −log probability that was computed (p-value), associated with Student's t-test. Orange dots indicate significantly downregulated proteins and blue dots indicate upregulated proteins. In (**B**,**C**), pie diagrams are shown with the lower and higher functional annotations, respectively, showing the differentially expressed proteins. In (**D**), protein–protein interaction network is shown. The circles indicate the interaction nodes and these are also indicated with parenthesis. a, proteins of sugar metabolism; b, proteins related to the stress response and the metabolism of reactive oxygen species; c, structural proteins; d, groups proteins that participate in protein folding; e and f, groups enzymes that participate in the modification of proteins and extracellular enzymes that act in the degradation of chitin, respectively.

2.3. Proteins with Many Biotechnology Applications

Some members of the protein families identified here have been isolated and functionally characterized from other organisms, including humans, and have shown potential as molecular tools for medical and industrial use. In order to gain more information about the observed proteins, a comparative analysis was carried out through the alignments of their amino acid sequences and homology models of each of them.

2.3.1. Peptidases

Peptidases have proven helpful as additives in detergents, in the generation of products implemented in the pharmaceutical and cosmetic industries, in analytical processes such as treating samples for proteomic analysis, and in the food industry [42–44]. In the CVE of *A. dowii*, we identified a probable cysteine protease with 16% coverage in its sequence. The tryptic peptides matched the precursor sequence of a likely cysteine protease from our tentacle transcriptome database, and we named it Ad_Cathepsin (Id code: c29945_g1_i1. Supplementary S1). The precursor sequence showed 91.3% identity with a probable Cathepsin from the sea anemone *Actinia tenebrosa*, 58.3% with Cathepsin L2 from humans, and 58.2% identity with Cathepsin L from dogs (Figure 5A). In addition, it shares between 62.8 and 44.9% identity with proteases isolated from flour beetle genera, *Tribolium* and *Tenebrio*, which have been used in the development of formulations for the degradation of gluten present in some foods such as cereals [52,53]. Gluten triggers dermatitis herpetiformis and celiac disease in people intolerant to gluten. Because gluten is a common protein in foods, it is difficult for intolerant people to consume it, and they often suffer relapses due to its inadvertent intake [52,53]. The pre-treatment of food with proteases has proven not to be sufficient, which is why formulations that can be administered orally to patients are sought. However, exogenously administered proteases must possess a wide range of stability and activity to be useful at a gastric pH, show resistance to degradation by gastrointestinal tract proteases, and have favorable kinetics to allow for gluten degradation before gastric emptying [54]. Therefore, searching for new proteases that can function under certain environmental conditions is still necessary. The structural alignment presented in Figure 5B shows that Ad_Cathepsin maintains folding, made up of the characteristic domains of the papain superfamily [52]. The structural alignment shows that the most significant differences between these proteins are found in the region that joins the domain; on the contrary, the regions that bind to the substrate, S2 and P2, are highly conserved [55] and maintain residual Cys25 and His163 (Figure 5A,B).

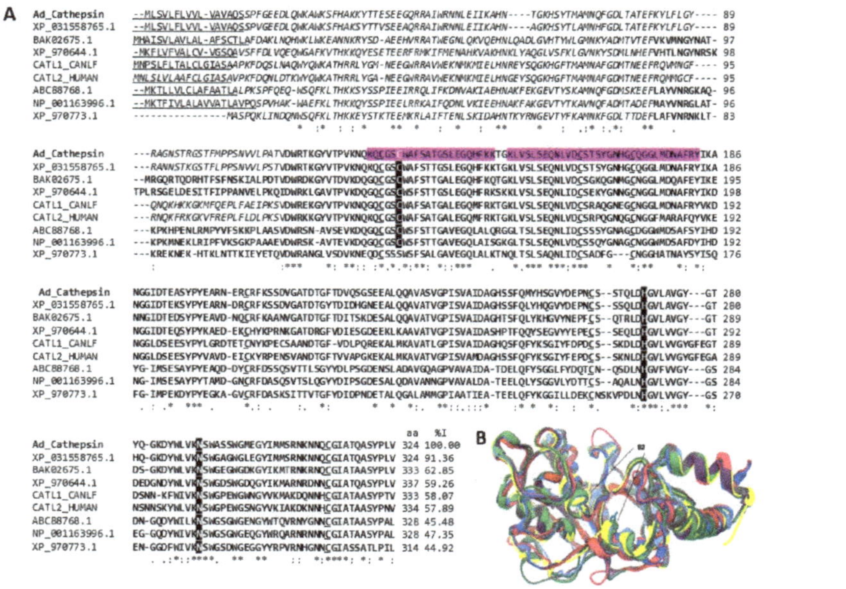

Figure 5. Cathepsins-L. (**A**) Alignment of amino acid sequences of putative Cathepsin-L from *A. dowii* (Ad_Cathepsin (c29945_g1_i1)) with Cathepsins-L from humans (Uni-ProtKB: CATL2_HUMAN), dogs (UniProtKB: CATL1_CANLF) and flour beetles (REF-SEQ: XP_0315589703_1, NP_001163996.1. GenBank: ABC88768.1, ABC88768.1, BAK02675.1). The region corresponding to the signal peptide is underlined, the propeptide is indicated in italics, and the region of the mature protein is indicated in bold. The residues that make up the active site are shaded black and highlighted with white letters. The cysteines that participate in the formation of disulfide bridges are underlined, and the region covered by the tryptic peptides in the Ad_Cathepsin sequence is highlighted by a magenta background. (**B**) Structure models of Cathepsin. Ad_Cathepsin is blue, XP_031558765.1 is yellow, BAK02675.1 is navy blue, XP_970644.1 is green, CATL1-CANLF is medium gray, CATL2_HUMAN is purple, ABC88768.1 is light gray, NP_001163996.1 is red, and XP_970773.1 is pink. The comparison of the Cα chains presented RMSD values that oscillate between 0.483 and 0.558.

2.3.2. Chitinases

Chitinases catalyze the cleavage of β-1,4-O-glycosidic bonds in chitin, a polysaccharide that is abundant in nature and present in various pathogenic organisms for man, such as fungi, protozoa, and helminths [56–58]. Chitinases have anti-inflammatory properties [59] and can help heal wounds [57]. Since the cuticle of arbovirus vector mosquitoes is made of chitin, chitinases can also be used as bioinsecticides for the biological control of *Aedes aegypti* mosquitoes [56]. In addition, they can be used to control phytopathogens, and transgenic plants and recombinant strains of *Bacillus* can be generated that allow for the control of insects and fungi [57].

The oligosaccharides generated from the degradation of chitin, such as N-acetylglucosamine, chitooligosaccharides, glucosamine, and chitosan, may have applications in the pharmaceutical and food industries [60]. Chitooligosaccharides have even been reported to have antitumor activity [57]. In the extracts of both sea anemone species, six probable chitinases were identified (Supplementary Material S1). The chitinase identified in the *A. dowii* tentacle transcriptome does not correspond to the amino acid sequences identified in the proteome of the *A. dowii* tentacle transcriptome; this suggests that the chitinases identified here are produced in anatomical regions other than the tentacles [38]. With 32% coverage, a sequence like Chitotriosidase-1-like from *Exaiptasia pallida* (XP_020909717.1) was identified (Figure 6A), a good candidate for intervention in infectious diseases caused by chitin-containing pathogens [61]. Chitotriosidases comprise a catalytic domain and a chitin-binding domain, with distorted folding of the b-sandwich [62]. Additionally, these proteins retain six catalytic cysteines, which are essential for maintaining conformation and disulfide bonds [62]. The structural overlap of the homology model of Chitotriosidase-1-like from *Exaiptasia pallida* (XP_020909717.1) presented RMSD values between 0.127 and 0.616 (Figure 6B).

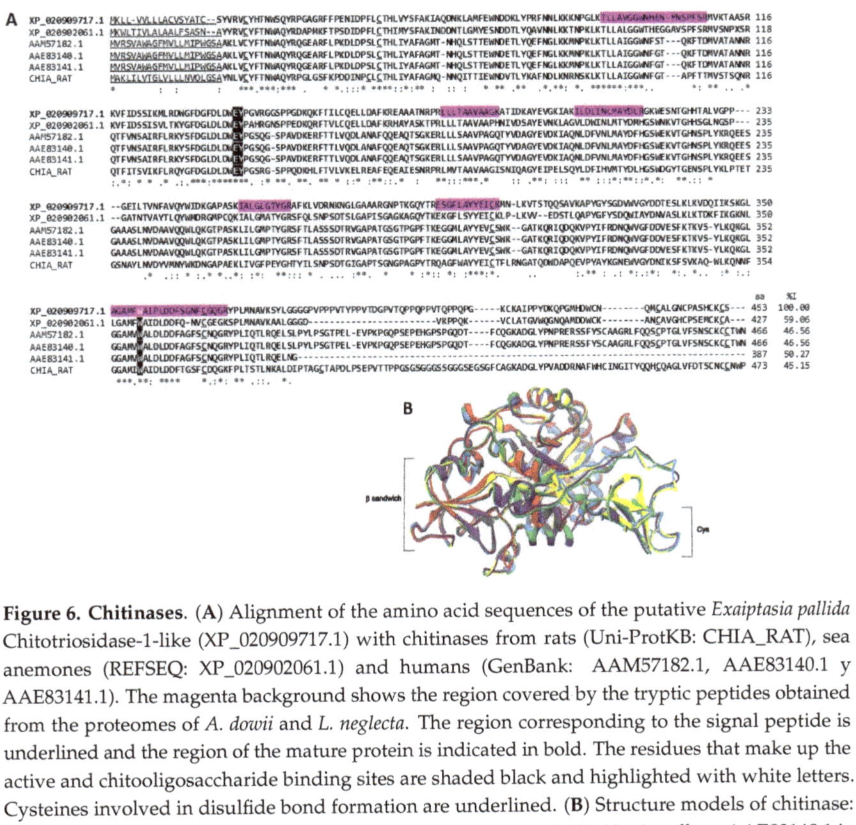

Figure 6. Chitinases. (**A**) Alignment of the amino acid sequences of the putative *Exaiptasia pallida* Chitotriosidase-1-like (XP_020909717.1) with chitinases from rats (Uni-ProtKB: CHIA_RAT), sea anemones (REFSEQ: XP_020902061.1) and humans (GenBank: AAM57182.1, AAE83140.1 y AAE83141.1). The magenta background shows the region covered by the tryptic peptides obtained from the proteomes of *A. dowii* and *L. neglecta*. The region corresponding to the signal peptide is underlined and the region of the mature protein is indicated in bold. The residues that make up the active and chitooligosaccharide binding sites are shaded black and highlighted with white letters. Cysteines involved in disulfide bond formation are underlined. (**B**) Structure models of chitinase: XP_020909717.1 is shown in green, XP_02092061.1 is in red, AAM57182.1 is yellow, AAE83140.1 is represented in gray, AAE83141.1 is blue, and CHIA_RAT is in purple. The comparison of the Cα chains presented RMSD values that oscillate between 0.127 and 0.616.

2.3.3. Antioxidant Enzymes

Reactive oxygen species (ROS), such as hydrogen peroxide, superoxide anions, and hydroxyl radicals, are generated under physiological conditions because of cellular metabolism [61]. Although ROS play an essential role in cellular processes such as signal transduction, gene expression, enzyme regulation, and immune responses, when these oxygen species increase beyond equilibrium, cells undergo oxidative stress, leading to damage to proteins, lipids, and nucleic acids (DNA and RNA) [62]. This cell damage is related to various pathologies, such as cancer, diabetes, Alzheimer's disease, Parkinson's disease, and Amyotrophic Lateral Sclerosis [62,63]. To conserve balanced ROS levels, organisms are endowed with various enzymes that function as antioxidants in cells; among these are peroxiredoxins, thioredoxins, catalases, and peroxide dismutases [63,64]. Sequences encoding probable antioxidant enzymes, such as those mentioned above, were identified in the proteomes of *A. dowii* and *L. neglecta* (Supplementary Material S1).

Peroxiredoxins (Prxs 1-6) can catalyze the reduction of H_2O_2, peroxynitrite, and alkyl hydroperoxides [61,65]. Their active sites contain a Cys (Cys peroxidative, CP-SH) and function as a site of oxidation via peroxides generating sulfenic acid (CP-S-OH), which reacts with another Cys residue, called cysteine resolution (CR), to form a water molecule and a disulfide bond that can be further reduced through an electron donor, completing a catalytic cycle [61]. There are two groups of Prxs in mammals, 2-Cys and 1-Cys; this classification is according to the requirement for the Cys residue to complete the catalytic cycle [65]. However, Prx6 is the only one that does not have a resolution Cys and uses glutathione (GSH) to complete its catalytic peroxidative reaction. Therefore, it is a Prx 1-Cys. Furthermore, Prx6 is a moonlighting enzyme as it also possesses calcium-independent phospholipase A2 and lysophosphatidylcholine acyltransferase activity [66].

A probable Prx6 with sequence similarity to human and chicken Prx6 was identified in the *A. dowii* proteome and transcriptome and was named Ad_Prx6 (Figure 7). The precursor sequence of Dw_Prx6 also showed about 95% identity with *Actinia tenebrosa* peroxiredoxin-6-like (Figure 7A). Analyzing peroxiredoxin models, we observe that Dw_Prx6 contains a thioredoxin fold conserving the active site and the conserved Thr-Pro-Val-Cys-Thr-Thr motif (Figure 7B) [67]. An experimental evaluation of exogenous Prx6 as a radioprotective agent was carried out in mice, reducing damage by ionizing radiation, generating an increase in the survival rate of animals exposed to lethal and sub-lethal doses [68,69], and establishing the possibility of being able to design Prx-based preparations to protect patients undergoing radiotherapy.

Another type of enzyme with antioxidant properties is superoxide dismutase (SOD). These proteins are metalloenzymes that catalyze the conversion of the free radical superoxide anion ($\bullet O^{2-}$) into hydrogen peroxide (H_2O_2) and molecular oxygen (O_2). Depending on the cofactor metal present in the active site, they are classified into four groups: Copper–Zinc-SOD (Cu, Zn-SOD), Iron SOD (Fe-SOD), Manganese SOD (Mn-SOD), and Nickel SOD. The therapeutic potential of SOD as an anti-inflammatory and anticancer agent has been suggested [70,71]. We identified a Copper–Zinc-SOD and a Mn-SOD in the *A. dowii* and *L. neglecta* proteomes. Mn-SOD was also identified in the *A. dowii* transcriptome and was named Ad_SOD [Mn].

Ad_SOD [Mn] presented between 70.2 and 69.8% identity in its amino acid sequence with the SOD2 sequences of humans (UniProtKB/Swiss-Prot: P04179.1) and the macaque *Macaca nemestrina*, respectively (Figure 8A). Furthermore, it showed 70.7 and 70.2% identity with AWU17515.1 and AAE36440.1. The structural analysis showed that the SODs displayed manganese SOD folding [72]. SODs present the highest structural homology between each other, showing the lowest RMSD values (0.014–0.536), conserving the structure and geometry of the active site (Figure 8B). The latter is human SOD2, differing from SOD2 P04179.1, and their recombinant forms can internalize leukemic T cells, generating apoptosis in 99% of them without causing toxic effects in healthy cells [70]. SOD enzymes have shown good potential for use in the design of therapeutic molecules for the treatment of cancer and other pathologies related to oxidative stress; however, exogenous

SOD has shown low bioavailability, mainly when applied orally, since the acidic pH and the proteases of the stomach cause its denaturation and degradation [71]. That is why the bioprospecting of SOD enzymes from different natural sources is essential to obtain possible candidates that are more stable at certain pH conditions and less vulnerable to proteolysis. This type of enzyme is used for veterinary use and in food supplements [71].

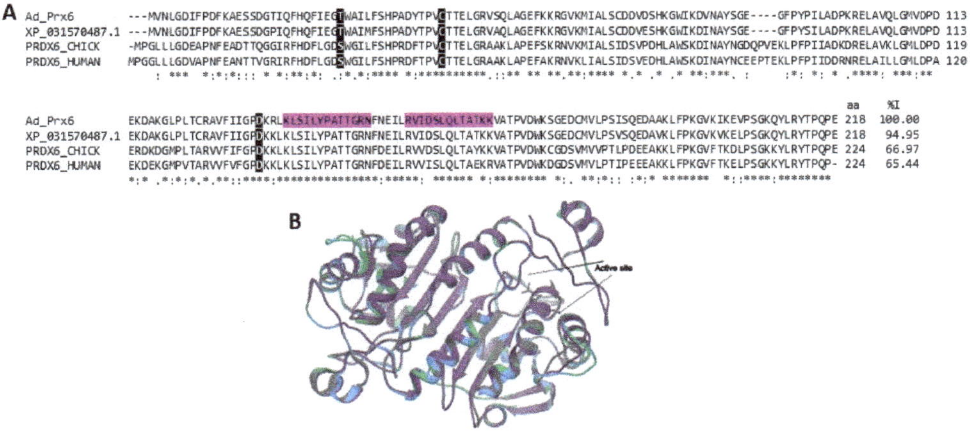

Figure 7. Peroxiredoxin-6 (Prx-6) alignment. (**A**) Alignment of the amino acid sequences of the putative Prx-6 from *A. dowii* (Ad_Prx6 (Transcriptome: c26896_g1_i1)) with Prx-6 from chickens (UniProtKB: PRDX6_CHICK), humans (UniProtKB: PRDX6_HUMAN), and from the sea anemone *Actinia tenebrosa* (REFSEQ: XP_031570487.1). The region covered by the tryptic peptides obtained in the *A. dowii* proteome is presented on a magenta background. On a black background with white letters, the residues related to Phospholipase A2 activity and peroxidative cysteine are highlighted. (**B**) Structure models of PRDX: Ad_Prx6 is gray, XP_031570487.1 is purple, PRDX6_CHICK is green, and PRDX6_HUMAN is blue. The comparison of the Cα chains presented RMSD values that oscillate between 0.032 and 0.342.

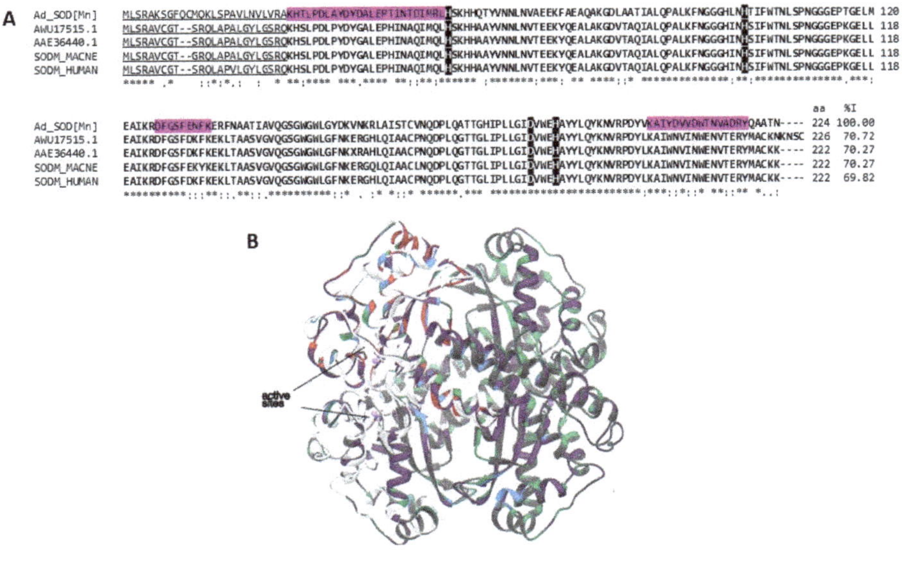

Figure 8. Superoxide dismutase [Mn] alignment. (**A**) Multiple alignment of the amino acid sequence of the putative mitochondrial superoxide dismutase from *A. dowii* (Ad_SOD [Mn] (Transcriptome:

c27625_g1_i1)) with other human SOD2s (UniProtKB: SODM_HUMAN, GenBank: AWU17515.1 and AAE36440.1) and from pig-tailed macaques (UniProtKB: SODM_MACNE). The region covered by the tryptic peptides obtained from the proteomes of *A. dowii* and *L. neglecta* is presented on a magenta background. On a black background with white letters, the residues related to the binding to manganese are highlighted. The region corresponding to the transit peptide is underlined and the region of the mature protein is indicated in bold. (**B**) Structural models of SOD. Ad_SOD [Mn] is red, AWU17515.1 is blue, AAE36440.1 is gray, SODM_MACNE 551 is purple, and SODM_HUMAN is green. The comparison of the Cα chains presented RMSD values that oscillate between 0.014 and 0.536.

2.3.4. Heat Shock Proteins (HSPs)

Heat shock proteins (HSPs) are highly conserved proteins that are inducible by a variety of stressful stimuli and by physiological processes, including cell differentiation and development [70]. At the cellular level, Hsp70 proteins are essential for proteostasis, aiding in the folding of newly synthesized and denatured proteins and, if necessary, in the degradation of aggregated proteins [73]. HSPs are classified into different classes based on their molecular weight, and of these, Hsp70 (70 kDa) and Hsp90 (90 kDa) are the most used classes for stress studies.

Hsp70 has been used as an early stress response biomarker for several marine organisms [74], including some cnidarian species such as *Nematostella vectensis* and *Aurelia coerulea* [40,49]. HSPs with a molecular weight of 70 and 90 kDa have also been shown to function as tumor-derived immunogenic peptide carrier proteins that induce a T cell-mediated immune response against cancer [75]. Furthermore, several studies have reported a mammalian neuroprotective effect of Hsp70, particularly after induction or administration [76]. It has also been reported that Hsp60 and Hsp70 can prevent the oligomerization of amyloidogenic proteins [77]. In the proteomes of *A. dowii* and *L. neglecta*, sequences have been identified that code for probable chaperones Hsp60, Hsp90, and Hsp70. The amino acid sequence of the Hsp70 precursor was identified in the *A. dowii* transcriptome and showed 82.1 and 85.7% identity with several Hsp70s that are part of inventions with potential medical and biotechnological uses (Figure 9A).

The secretion of over-expressed proteins in yeast is increased by the design of gene constructs that express Hsp70 (AAC89780.1) [78], whereas Hsp70 (QTV59613.1) stimulates cholesterol efflux from macrophage foam cells and thus has potential for use in controlling cholesterol homeostasis [79]. Furthermore, Hsp70 (ACQ40951.1) can stimulate the proliferation and cytolytic activity of Natural Killer (NK) cells against tumor cells [80]. In another application, an immunogenic peptide of between eight and fifteen amino acid residues was generated from Hsp70 (QKO41990.1); this peptide induces cytotoxic T lymphocytes against tumor cells via antigen presentation [81]. Finally, an invention that involves a mutant of Hsp70 (QRJ96032.1) that lacks the dendritic cell-binding site is useful for the treatment of vitiligo [82].

Our structural alignment (Figure 9B) shows that the studied HSPs maintain the characteristic structure of HSP70, made up of a nucleotide-binding domain and a protein-binding domain [83]. Except for the HSP7D_MANSE protein that presents an insertion between residues 221 and 421, the other HSPs do not show relevant alterations.

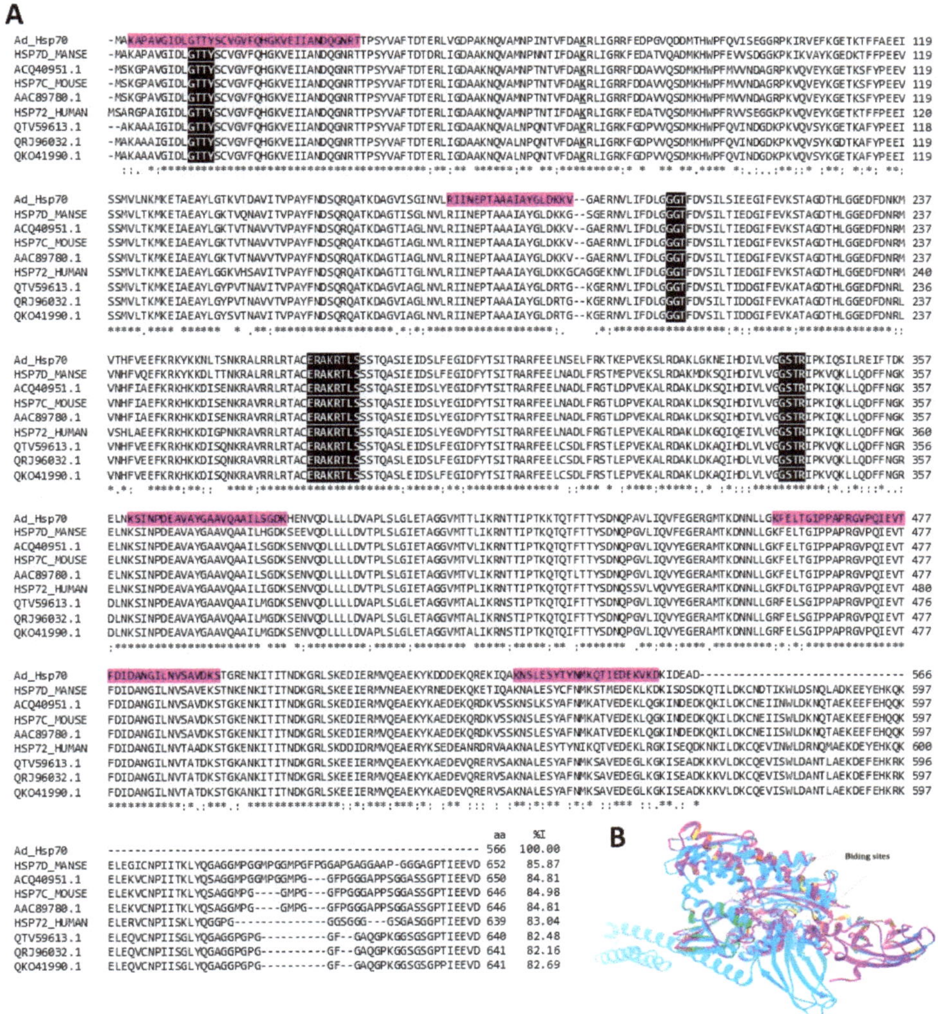

Figure 9. Hsp 70 kDa. (**A**) Multiple alignments of the amino acid sequence of putative *A. dowii* Hsp 70 (Ad_Hsp70 (Transcriptome: c27026_g1_i1)) with other proteins of the same family. In the alignment, the sequences Hsp70 from insects (UniProtKB: HSP7D_MANSE), from humans (UniProtKB: HSP72_HUMAN), from mice (UniProtKB: HSP7C_MOUSE), and several patented Hsp70s for different applications (GenBank: ACQ4089780.1, A QTV59613.1, QRJ96032.1, QKO41990.1). The region covered by the tryptic peptides obtained from the proteomes of *A. dowii* and *L. neglecta* is highlighted by a magenta background. The ATP binding site is underlined and the regions involved in the binding of nucleotide phosphates are highlighted with white letters on a black background. (**B**) Structure models of heat shock proteins: Ad_Hsp70 is green, HSP7D_MANSE is medium blue, ACQ40951.1 is blue, HSP7C_MOUSE is forest green, AAC89780.1 is gray, HSP72_HUMAN is pink, QTV59613.1 is yellow, QRJ96032.1 is red, and QKO41990.1 is purple. The comparison of the Cα chains presented RMSD values that oscillate between 0.007 and 0.77.

2.3.5. Serine Protease Inhibitors (Serpins)

Serine proteases play a fundamental role in blood coagulation, fibrinolysis, complement activation, and inflammation [84]. The excessive activity of these enzymes can trigger pathological states, so it is necessary to regulate their activity. Serpins (serine protease inhibitors) are proteins of ~350–400 amino acid residues, responsible for regulating serine proteases' function through stoichiometric binding to the active site, inactivating their action. Some serpins are secreted and act extracellularly, with some exceptions, such as the Ov-serpin subfamily, which is structurally related to ovalbumin and lacks a signal peptide for its secretion; therefore, members of this subfamily can also act extracellularly. Intracellular serpins have shown the ability to inhibit proteolytic activity during inflammation to act as modulators of the inflammatory response, which makes them candidates for the development of pharmacological formulations for treating inflammatory diseases such as rheumatoid arthritis [85]. We identified the sequence of a likely member of the Ov-serpin subfamily, which we named Ad_Serpin. This probable protein was identified in the CVE of both species and the *A. dowii* tentacle transcriptome. The Ad_Serpin sequence presents 42.05 and 42.40% similarity with human, frog, and mouse serpins (Figure 10A). Figure 10B shows that protease inhibitors maintain the structure of serine protease inhibitors. In addition to general folding, residues that allow for inhibitor interactions, such as Ser195 andCys347, are conserved [86].

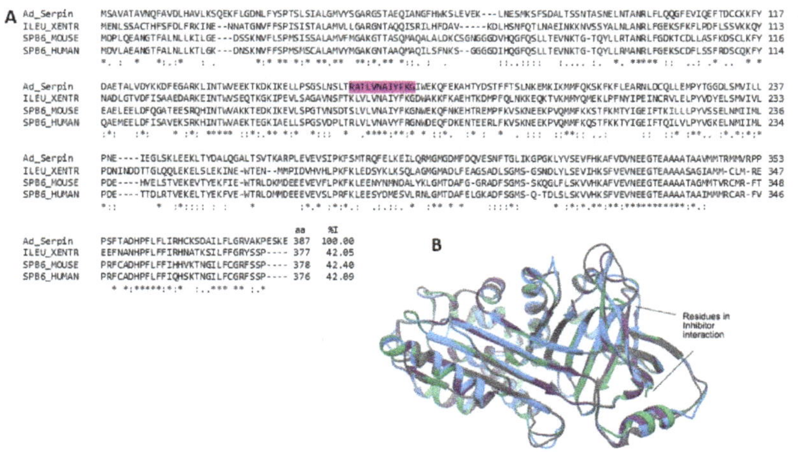

Figure 10. Serpin alignment. (**A**) Multiple alignment of the amino acid sequence of the putative serine protease inhibitor (Ad_Serpin (Transcriptome: c26903_g1_i1)) with the serpins of humans (UniProtKB: SPB6_HUMAN) and mice (UniProtKB: SPB6_MOUSE), and the elastase inhibitor identified in frogs (UniProtKB: ILEU_XENTR). The region covered by the tryptic peptides obtained from the *A. dowii* and *L. neglecta* proteomes is highlighted by a magenta background. (**B**) Structural models of protease inhibitors: Ad_Serpin is blue, ILEU_XENTR is gray, SPB6_MOUSE is green, and SPB6_HUMAN is purple. The comparison of the Cα chains presented RMSD values that oscillate between 0.361 and 0.863.

3. Discussion

The development of new molecules from natural products has grown exponentially in recent decades. Searching for new molecular alternatives to solve health problems and facilitate industrial processes has become attractive. The ocean is an excellent source of natural products due to its biological diversity. Marine bioprospecting has led the way in the exploration of all kinds of molecules with potential applications in various industries and biotechnology, which has allowed us to go beyond the search for "classic" or "more attractive" molecules to molecules that are part of the metabolism and general physiology of organisms of all kingdoms. In this direction, there is a relatively considerable increase in the study of molecules that show a notable difference in their production when organisms inhabit environments whose physical, chemical, and biological parameters can change drastically in one day, which generates immediate adjustments in their metabolism, implying that the production of enzymes and other molecules is affected. During the evolution of these organisms, molecules, such as enzymes that allow for the rapid adaptation of organisms, have modified their biochemical properties to respond to environmental changes. These new biochemical characteristics are attractive for use as molecular tools in biotechnology.

Marine invertebrates are the top group of natural compounds, with attractive applications. Sea anemones are one of the least explored invertebrates from the perspective of marine bioprospecting. These organisms are sessile and have limitations to their ability to move in response to imminent environmental changes that challenge their physiology. However, they can colonize all marine habitats and belong to one of the oldest phyla, with an estimated age of approximately 750 million years [87], suggesting that they have efficiently adapted their molecular machinery to survive environmental changes. The temperatures and UV radiation to which anemones can be exposed and the symbiotic interactions they establish with photosynthetic microalgae generate large amounts of ROS that sea anemones can effectively manage. Sea anemones are still considered unusual species for the bioprospecting of non-toxic molecules to develop biopharmaceutical products. However, proteomic and transcriptomic studies focusing on their venom [31,32,38,87,88], as well as their symbiotic interactions, have revealed the presence of proteins with potential activities that could be evaluated and taken to the second stage to estimate their biotechnological potential. For this reason, we led this marine bioprospecting study in two species of anemones from different geographical areas through proteomic analysis. It was determined that the anemone *A. dowii* that lives in an area with more significant environmental changes, during its physiological adjustments, produces more proteins that respond to stress, while *L. neglecta* produces more antioxidant proteins, possibly due to changes in pH and temperature in its environment. This can be used to determine that metabolic adjustments are evident in environments that are as changing as the interstitial zone. The proteins identified in both proteomes included hydrolase-type enzymes, oxidoreductases, heat shock proteins, adhesion proteins, structural proteins, and protease inhibitors. The presence of these enzymes in *A. dowii* might be explained by the physical–chemical stressors it encounters. However, they contradict the hypothesis that the elevated expression of antioxidant proteins in *L. neglecta* is solely due to radiation and temperature, suggesting that it could be associated with photosynthetic algae living in symbiosis with *L. neglecta*. Such interactions have been shown to promote the production of novel biomolecules of pharmacological interest in holobionts [42], indicating that the environment plays a significant role in shaping the composition of biomolecules in sea anemones.

Several proteins from other organisms have been described in depth, with the possibility of their application in medicine or biotechnology. However, it is essential to carry out experimental studies of the polypeptides identified here to determine the degree of potential for their application.

4. Materials and Methods

4.1. Sea Anemone Collection

A. dowii was collected during low tide in the intertidal zone of the Pacific Ocean coast in Ensenada, Baja California, Mexico, and *L. neglecta* was collected at a 10 m depth in the front reef zone of Puerto Morelos, Quintana Roo, in the Mexican Caribbean. Sea samples were collected and transferred to the laboratory in seawater and then frozen at −80 °C until use.

4.2. Protein Extraction

The *A. dowii* crude venom extract was prepared from three organisms, as described in [89], and *L. neglecta* crude venom extract included five complete individuals that were thawed and deposited in 120 mL of 50 mM sodium phosphate at pH 7.4 with protease inhibitors and mixed with constant stirring at 4 °C for 12 h to stimulate the discharge of the nematocysts. Then, the crude venom extract was subjected to three rounds of freezing and thawing to enhance nematocyst discharge. The suspension was centrifuged at 25,400× *g* for 40 min to remove cell debris. The crude venom extract was filtered through sterile 0.22 µm membranes, and the protein was quantified by the Bradford assay [90]. The analysis of the crude venom extracts electrophoretic profiles used SDS–PAGE in 15% acrylamide [91].

4.3. Proteomic Samples

Protein samples from the crude venom extracts of *A. dowii* and *L. neglecta* were prepared in triplicate for analysis by Shotgun proteomics. All replicates contained 40 µg of total protein in 100 µL TE buffer (10 mM Tris–HCl pH 7.6, 1 mM EDTA pH 8) and 100 µL of 0.3% sodium deoxycholate [*w*/*v*], totaling 300 µL with sterile tetra distilled water. The samples were precipitated with 72% (*w*/*v*) trichloroacetic acid (TCA) for 3 h on ice and subsequently centrifuged at 8436× *g* for 20 min at 4 °C. The pellets were recovered and subjected to additional precipitation with 90% (*v*/*v*) acetone overnight at −30 °C. The samples in acetone were centrifuged at 8436× *g* for 20 min at 4 °C, the pellet was dried for 20 min in a Savant integrated SpeedVac system (Thermo Fisher Scientific, Waltham, MA, USA), and the supernatant was discarded. The samples were sent to the proteomics unit of the Institute de Recherches Cliniques (Montreal, QC, Canada) for processing.

4.4. Protein Digestion

The protein samples were resuspended with 10 µL of 6 M urea and reduced with 45 mM DTT for 30 min at 37 °C. Subsequently, the samples were alkylated with 55 mM iodoacetamide for 20 min at 24 °C in the dark. Each of the samples was diluted with 18.2 Mohm purified water (Merck Millipore, Darmstadt, Germany) H_2O until the urea concentration was 2 M. Then, the samples were digested with 10 µL trypsin solution (5 ng/µL Promega sequencing-grade trypsin, 50 mM ammonium bicarbonate); after 18 h, the reaction was stopped with 5% formic acid. The samples were centrifuged in a vacuum system until drying, then stored at −20 °C until use. Tryptic peptides were resolubilized by shaking in 10 µL of a solution of 0.2% formic acid for 15 min and desalting using C18 ZipTips pipette tips (Millipore, Billerica, MA, USA). The eluted samples were vacuum-dried and resolubilized by stirring for 15 min in 10 µL of a 2% acetonitrile/1% formic acid solution.

LC–MS/MS

Trypsinized samples were analyzed by LC–MS/MS using an Easy-nLC II liquid chromatography system (Proxeon Biosystems, Odense, Denmark) coupled with an Orbitrap Velos LTQ mass spectrometer (ThermoFisher Scientific, Bremen, Germany) using a nano-electrospray (Proxeion) ion source. Chromatography was performed using a 15 cm long, 75 µm internal diameter Self-Pack PicoFrit column, fused to a silica capillary (New Objective, Woburn, MA, USA) and packed with C18 Jupiter 5 µm 300 materials and ÅRP (Phenomenex, Torrance, CA, USA). The buffers used for chromatography were 18.2 Mohm purified water (Merck Millipore) H_2O at 0.2% formic acid (buffer A) and 100% acetonitrile/0.2% formic acid (buffer B). Samples (5 µL) were loaded onto the column at a 600 nL/min flow rate and eluted with a 2-step linear gradient at a 250 nL/min flow rate. Buffer B increased from 2 to 35% in 80 min and then from 35 to 80% in 12 min. LC–MS/MS data acquisition was achieved using a parallel acquisition mode, where the full scan MS data were acquired in the orbitrap (R = 60,000), and the top 15 precursors were isolated and fragmented in the linear ion trap. MS spectra were acquired over m/z 360 to 1700 and used an AGC of 1,000,000. MS/MS spectra were subject to the 1/3 (of the precursor m/z) rule for ion traps and used an AGC of 1,000,000 loadings. Data-dependent scan events used a maximum ion fill time of 100 ms and 1 Microscan. Target ions selected for MS/MS were excluded for 31 s after 2 counts. The nanospray and S-lens voltages were set to 1.3–1.8 kV and 50 V, respectively. The capillary temperature was adjusted to 250 °C. MS/MS conditions were normalized; the collision energy occurred at 35 V; activation q was 0.25; activation time was 10 ms.

4.5. Identification and Quantification of Proteins

Tandem mass spectra were extracted using Mascot Daemon v2.1 (Matrix Science, London, UK). Deconvolution and deisotope of the state of charge were not performed. The protein search was performed against the Refseq_Cnidaria database. The precursor ion mass width was set at ±10 ppm and the maximum fragment mass error was set to 0.6 Da. Trypsin was assumed to be the digesting enzyme, and up to 1 cleavage was allowed to be skipped. Cysteine carbamidomethylation was chosen as a constant modification, and methionine oxidation was determined as a variable modification. MS/MS-based polypeptide identifications were validated with Scaffold software v4.11.1 (Proteome Software Inc., Portland, OR, USA). Scaffold identification parameters were set to at least two unique peptides per protein, with a minimum probability of 99% for proteins [92] and 95% for corresponding peptides [93]. Proteins containing similar peptides that could not be differentiated by MS/MS analysis were only pooled to satisfy parsimony principles.

To compare the proteomic information between both species of sea anemones, quantitative analyses of the peptides without labeling were carried out using the spectral counting method and analyzed using the exponentially modified protein abundance index (emPAI) using the Scaffold software (see Supplementary Material S1). Protein expression was considered significantly different only if the protein ratios differed two-fold, and the volcano plot shows the changes in protein expression. Student's *t*-test was applied to each set of samples with three different biological replicates, and a threshold level of 0.05 was considered to select proteins (Supplementary Material S1). All data were corrected by applying the Benjamini-Hochberg correction. The distribution of the spectra and total unique peptides, as well as the number of proteins identified exclusively in each species and those identified in both, were obtained, and are presented graphically in Venn diagrams. The proteomic indices of both species, such as the emPAI, amino acid sequence coverage, and molecular weight distribution of the identified proteins, were extracted from the Scaffold software and represented in bar graphs. All proteomic analysis graphs were performed with R (www.r-project.org (accessed on 16 July 2020)) and Rstudio (www.rstudio.com (accessed on 17 July 2020)).

4.6. Bioinformatic Analysis

The total proteins identified in each extract were annotated if they met the scaffold parameters mentioned above (Section 4.5) and were identified as having two or more unique peptides in at least two of three biological replicates. For the functional annotation of previously identified proteins in the Refseq_Cnidaria database, each protein was manually curated using BLASTP (https://blast.ncbi.nlm.nih.gov/Blast.cgi, accessed on 4 September 2021) and the UniProtKB data database (https://www.uniprot.org/; accessed on 4 September 2021). In addition, a homemade database generated from the information obtained from the *A. dowii* tentacle transcriptome, previously reported by our group [31,44], was used; this allowed us to identify the complete sequence and improve the annotation process using the UniProtKB database. The functional assignment, cellular location, and biological processes related to the total proteins identified in *A. dowii*, *L.neglecta*, or both samples were obtained from the QuickGO Gene Ontology database (https://www.ebi.ac.uk/QuickGO/; GO version: 4 September 2021) and were presented with Excel 15.28. For some proteins identified in the crude venom extract of *A. dowii*, the restriction to a single peptide present in at least two biological replicates was modified, maintaining the percentage probability of identification of peptides and proteins established for Scaffold, if the protein was also identified in the transcriptome of *A. dowii* [31,44].

Protein–protein interactions between all proteins identified in *A. dowii* and *L. neglecta* proteomes were analyzed for the proteins identified in each crude venom extract using the String v11.5 software (https://string-db.org/, accessed on 4 September 2021). The *N. vectensis* genome database was used as a template, and these data contributed to the identification of each protein, along with its metabolic pathway. The interactions were generated according to the functional and physical relationship between the proteins. The clustering of interactions in *A. dowii* and *L. neglecta* was associated using the Markov Clustering Algorithm (MCL) clustering method using an inflation parameter of 2 [94]. The network edges were set confidentially (the thick line represents the strength of the data support), and the sources of interaction were text mining, databases, experiments, co-expression, neighborhood, gene fusion, and co-occurrence.

Proteins with differential downward or upward expression, identified in both extracts (Section 4.5), were also analyzed and grouped according to their expression and molecular function. The protein–protein interactions obtained with the String v11.5 software identified the metabolic pathways or processes in which the proteins participate that alter their expression; the minimum interaction score required for all analyses in String v11.5 was set at 0.7 (high confidence) [95]. Multiple alignments of some selected sequences from the sea anemone proteomes were analyzed against functionally and structurally characterized orthologs. Alignments were performed with Clustal Omega (1.2.4) (https://www.ebi.ac.uk/Tools/msa/clustalo/, accessed on 4 September 2021). The number of amino acids (aa) and the identity percentages (I%) were calculated considering the total sequence. In each alignment, the characters (*), (:), and (.) indicate positions with fully conserved residues, positions with conservative substitutions, and positions with less conservative substitutions, respectively. The same proteins in the alignment were modeled with SWISS-MODEL [96]. Each model was minimized with YASARA [32]. Finally, the structures were refined using 3D refinement [97] and analyzed using UCSF Chimera [98].

Supplementary Materials: The following supporting information can be downloaded at https://www.mdpi.com/article/10.3390/md22010012/s1, S1, Putative components identified in the crude venom extract in the proteome from *Anthopleura dowii* Verrill, 1869, and *Lebrunia neglecta* Duchassaing & Michelotti, 1860. S2–S4, Interactomes. S5, Differentially expressed proteins. S6, Interactome plot.

Author Contributions: C.R.-A. and S.R.-C. conceived and designed the experiments. S.R.-C., B.M.-Z. and C.R.-A. performed the experiments. S.R.-C., B.M.-Z., C.R.-A., N.S. and R.G.-M. wrote and critically reviewed the paper. All authors have read and agreed to the published version of the manuscript.

Funding: This work was supported by the Programa de Apoyo a Proyectos de Investigación e Innovación Tecnológica (PAPIIT), Universidad Nacional Autónoma de México (grant number IT200819), and Instituto de Ciencias Aplicadas y Tecnología, UNAM (Project 203-2023). Field and lab work was funded by grants to N.S. by the Harte Charitable Foundation through the Harte Research Institute for the Gulf of Mexico Studies, Texas A&M University at Corpus Christi, and Comisión Nacional para el Conocimiento y Uso de la Biodiversidad (CONABIO-NE018).

Institutional Review Board Statement: Specimens were collected according to the guidelines of the National Commission of Aquaculture, Fishing, and Feeding of the Mexican Federal Government (permit number PPF/DGOPA 07332.250810.4060).

Data Availability Statement: The data generated is Supplementary Material.

Acknowledgments: The authors acknowledge the technical assistance of Juan Manuel Barbosa Castillo (IFC, UNAM), David S. Castañeda Carreón, and Roberto P. Rodríguez Bahena (IBT, UNAM). We are also grateful to Denis Faubert at Institut de Recherches Cliniques de Montréal—Proteomics Discovery Platform for MS analysis. S.R.-C. is the receptor of a postdoctoral fellowship from CONAHCyT (CVU 253995, period: October 2022–September 2023) and B.M.-Z. is a Ph.D. student of Programa de Maestría y Doctorado en Ciencias Bioquímicas (UNAM) and is the receptor of a fellowship from CONAHCyT (2019-000002-01NACF-12652).

Conflicts of Interest: The authors declare that they have no competing interest in relation to the publication of this paper.

References

1. Birolli, W.G.; Lima, R.N.; Porto, A.L.M. Applications of Marine-Derived Microorganisms and Their Enzymes in Biocatalysis and Biotransformation, the Underexplored Potentials. *Front. Microbiol.* **2019**, *10*, 1453. [CrossRef]
2. Debashish, G.; Malay, S.; Barindra, S.; Joydeep, M. Marine Enzymes. In *Marine Biotechnology I.*; Ulber, R., Le Gal, Y., Eds.; Advances in Biochemical Engineering/Biotechnology; Springer: Berlin/Heidelberg, Germany, 2005; Volume 96, pp. 189–218, ISBN 978-3-540-25659-5.
3. Bekiari, M. Marine Bioprospecting: Understanding the Activity and Some Challenges Related to Environmental Protection, Scientific Research, Ethics, and the Law. In *Blue Planet Law: The Ecology of Our Economic and Technological World*; da Gloria Garcia, M., Cortês, A., Eds.; Sustainable Development Goals Series; Springer International Publishing: Cham, Switzerland, 2023; pp. 237–252, ISBN 978-3-031-24888-7.
4. Leal, M.C.; Madeira, C.; Brandão, C.A.; Puga, J.; Calado, R. Bioprospecting of Marine Invertebrates for New Natural Products-a Chemical and Zoogeographical Perspective. *Molecules* **2012**, *17*, 9842–9854. [CrossRef] [PubMed]
5. Kvetkina, A.; Kostina, E.; Gladkikh, I.; Chausova, V.; Yurchenko, E.; Bakunina, I.; Pivkin, M.; Anastyuk, S.; Popov, R.; Monastyrnaya, M.; et al. Deep-Sea Anemones Are Prospective Source of New Antimicrobial and Cytotoxic Compounds. *Mar. Drugs* **2021**, *19*, 654. [CrossRef] [PubMed]
6. Aertsen, A.; Meersman, F.; Hendrickx, M.E.G.; Vogel, R.F.; Michiels, C.W. Biotechnology under High Pressure: Applications and Implications. *Trends Biotechnol.* **2009**, *27*, 434–441. [CrossRef] [PubMed]
7. Avila, C.; Angulo-Preckler, C. A Minireview on Biodiscovery in Antarctic Marine Benthic Invertebrates. *Front. Mar. Sci.* **2021**, *8*, 686477. [CrossRef]
8. Bellantuono, A.J.; Granados-Cifuentes, C.; Miller, D.J.; Hoegh-Guldberg, O.; Rodriguez-Lanetty, M. Coral Thermal Tolerance: Tuning Gene Expression to Resist Thermal Stress. *PLoS ONE* **2012**, *7*, e50685. [CrossRef] [PubMed]
9. Ding, D.-S.; Patel, A.K.; Singhania, R.R.; Chen, C.-W.; Dong, C.-D. Effects of Temperature and Salinity on Growth, Metabolism and Digestive Enzymes Synthesis of *Goniopora columna*. *Biololgy* **2022**, *11*, 436. [CrossRef] [PubMed]
10. Paul, S.I.; Majumdar, B.C.; Ehsan, R.; Hasan, M.; Baidya, A.; Bakky, M.A.H. Bioprospecting Potential of Marine Microbial Natural Bioactive Compounds. *J. Appl. Biotechnol. Rep.* **2021**, *8*, 96–108. [CrossRef]
11. Li, Y.; Chen, Z.; Zhang, P.; Gao, F.; Wang, J.; Lin, L.; Zhang, H. Characterization of a Novel Superoxide Dismutase from a Deep-Sea Sea Cucumber (*Psychoropotes verruciaudatus*). *Antioxidants* **2023**, *12*, 1227. [CrossRef]
12. Fontanini, A.; Steckbauer, A.; Dupont, S.; Duarte, C.M. Variable metabolic responses of Skagerrak invertebrates to low O_2 and high CO_2 scenarios. *Biogeosciences* **2018**, *15*, 3717–3729. [CrossRef]
13. Rodrigo, A.P.; Costa, P.M. The Hidden Biotechnological Potential of Marine Invertebrates: The Polychaeta Case Study. *Environ. Res.* **2019**, *173*, 270–280. [CrossRef] [PubMed]
14. Popkova, D.; Otstavnykh, N.; Sintsova, O.; Baldaev, S.; Kalina, R.; Gladkikh, I.; Isaeva, M.; Leychenko, E. Bioprospecting of Sea Anemones (Cnidaria, Anthozoa, Actiniaria) for β-Defensin-like α-Amylase Inhibitors. *Biomedicines* **2023**, *11*, 2682. [CrossRef] [PubMed]
15. Lee, P.G.; Turk, P.E.; Yang, W.T.; Hanlon, R.T. Biological Characteristics and Biomedical Applications of the Squid *Sepioteuthis Lessoniana* Cultured Through Multiple Generations. *Biol. Bull.* **1994**, *186*, 328–341. [CrossRef] [PubMed]
16. Linsmayer, L.B.; Deheyn, D.D.; Tomanek, L.; Tresguerres, M. Dynamic Regulation of Coral Energy Metabolism throughout the Diel Cycle. *Sci. Rep.* **2020**, *10*, 19881. [CrossRef] [PubMed]
17. Dahlhoff, E.P.; Stillman, J.H.; Menge, B.A. Physiological Community Ecology: Variation in Metabolic Activity of Ecologically Important Rocky Intertidal Invertebrates Along Environmental Gradients. *Integr. Comp. Biol.* **2002**, *42*, 862–871. [CrossRef] [PubMed]
18. Lewis, L.S.; Smith, J.E.; Eynaud, Y. Comparative Metabolic Ecology of Tropical Herbivorous Echinoids on a Coral Reef. *PLoS ONE* **2018**, *13*, e0190470. [CrossRef] [PubMed]
19. Sorte, C.J.B.; Jones, S.J.; Miller, L.P. Geographic Variation in Temperature Tolerance as an Indicator of Potential Population Responses to Climate Change. *J. Exp. Mar. Biol. Ecol.* **2011**, *400*, 209–217. [CrossRef]
20. Kuan, P.-L.; You, J.-Y.; Wu, G.-C.; Tseng, Y.-C. Temperature Increases Induce Metabolic Adjustments in the Early Developmental Stages of Bigfin Reef Squid (*Sepioteuthis Lessoniana*). *Sci. Total Environ.* **2022**, *844*, 156962. [CrossRef]
21. Sumi, M.S.; Thazeem, B.; Sunish, K.S. Bioprospective Studies of Pigmented Ink from *Sepioteuthis lessoniana* and Its Molecular Identification Using CO1 Gene. *J. Basic Appl. Zool.* **2023**, *84*, 5. [CrossRef]
22. Doering, T.; Maire, J.; Chan, W.Y.; Perez-Gonzalez, A.; Meyers, L.; Sakamoto, R.; Buthgamuwa, I.; Blackall, L.L.; van Oppen, M.J.H. Comparing the Role of ROS and RNS in the Thermal Stress Response of Two Cnidarian Models, *Exaiptasia diaphana* and *Galaxea fascicularis*. *Antioxidants* **2023**, *12*, 1057. [CrossRef]
23. Lee, Y.; Byeon, E.; Kim, D.-H.; Maszczyk, P.; Wang, M.; Wu, R.S.S.; Jeung, H.-D.; Hwang, U.-K.; Lee, J.-S. Hypoxia in Aquatic Invertebrates: Occurrence and Phenotypic and Molecular Responses. *Aquat. Toxicol.* **2023**, *263*, 106685. [CrossRef]
24. Song, L.; Lv, J.; Wang, L.; Sun, D.; Gao, B.; Liu, P. Characterization of a Chitinase-1 Gene (PtCht-1) from a Marine Crab *Portunus trituberculatus* and Its Response to Immune Stress. *Gene* **2020**, *741*, 144523. [CrossRef]
25. Shock, B.C.; Foran, C.M.; Stueckle, T.A. Effects of Salinity Stress on Survival, Metabolism, Limb Regeneration, and Ecdysis in *Uca pugnax*. *J. Crustac. Biol.* **2009**, *29*, 293–301. [CrossRef]

26. Labrou, N. (Ed.) *Therapeutic Enzymes: Function and Clinical Implications*; Advances in Experimental Medicine and Biology; Springer Singapore, 2019; Volume 1148, ISBN 9789811377082.
27. Bruhn, T.; Schaller, C.; Schulze, C.; Sanchez-Rodriguez, J.; Dannmeier, C.; Ravens, U.; Heubach, J.F.; Eckhardt, K.; Schmidtmayer, J.; Schmidt, H.; et al. Isolation and Characterisation of Five Neurotoxic and Cardiotoxic Polypeptides from the Sea Anemone *Anthopleura elegantissima*. *Toxicon* **2001**, *39*, 693–702. [CrossRef]
28. Logashina, Y.A.; Mosharova, I.V.; Korolkova, Y.V.; Shelukhina, I.V.; Dyachenko, I.A.; Palikov, V.A.; Palikova, Y.A.; Murashev, A.N.; Kozlov, S.A.; Stensvåg, K.; et al. Peptide from Sea Anemone *Metridium senile* Affects Transient Receptor Potential Ankyrin-Repeat 1 (TRPA1) Function and Produces Analgesic Effect. *J. Biol. Chem.* **2017**, *292*, 2992–3004. [CrossRef]
29. Madio, B.; King, G.F.; Undheim, E.A.B. Sea Anemone Toxins: A Structural Overview. *Mar. Drugs* **2019**, *17*, 325. [CrossRef]
30. Xiao, M.; Brugler, M.R.; Broe, M.B.; Gusmão, L.C.; Daly, M.; Rodríguez, E. Mitogenomics suggests a sister relationship of Relicanthus daphneae (Cnidaria: Anthozoa: Hexacorallia: Incerti ordinis) with Actinaria. *Sci. Rep.* **2019**, *9*, 18182. [CrossRef]
31. Ramírez-Carreto, S.; Vera-Estrella, R.; Portillo-Bobadilla, T.; Licea-Navarro, A.; Bernaldez-Sarabia, J.; Rudiño-Piñera, E.; Verleyen, J.J.; Rodríguez, E.; Rodríguez-Almazán, C. Transcriptomic and Proteomic Analysis of the Tentacles and Mucus of *Anthopleura dowii Verrill*, 1869. *Mar. Drugs* **2019**, *17*, E436. [CrossRef]
32. Mitchell, M.L.; Tonkin-Hill, G.Q.; Morales, R.A.V.; Purcell, A.W.; Papenfuss, A.T.; Norton, R.S. Tentacle Transcriptomes of the Speckled Anemone (*Actiniaria*: *Actiniidae*: *Oulactis* Sp.): Venom-Related Components and Their Domain Structure. *Mar. Biotechnol.* **2020**, *22*, 207–219. [CrossRef]
33. Monastyrnaya, M.M.; Kalina, R.S.; Kozlovskaya, E.P. The Sea Anemone Neurotoxins Modulating Sodium Channels: An Insight at Structure and Functional Activity after Four Decades of Investigation. *Toxins* **2022**, *15*, 8. [CrossRef]
34. Prentis, P.J.; Pavasovic, A.; Norton, R.S. Sea Anemones: Quiet Achievers in the Field of Peptide Toxins. *Toxins* **2018**, *10*, 36. Available online: https://www.mdpi.com/2072-6651/10/1/36 (accessed on 19 November 2023).
35. Marino, A.; Valveri, V.; Muià, C.; Crupi, R.; Rizzo, G.; Musci, G.; La Spada, G. Cytotoxicity of the Nematocyst Venom from the Sea Anemone *Aiptasia mutabilis*. *Comp. Biochem. Physiol. Part C Toxicol. Pharmacol.* **2004**, *139*, 295–301. [CrossRef]
36. Sarian, F.D.; Janeček, Š.; Pijning, T.; Ihsanawati; Nurachman, Z.; Radjasa, O.K.; Dijkhuizen, L.; Natalia, D.; van der Maarel, M.J.E.C. A New Group of *Glycoside hydrolase* Family 13 α-Amylases with an Aberrant Catalytic Triad. *Sci. Rep.* **2017**, *7*, 44230. [CrossRef]
37. Vandepas, L.E.; Tassia, M.G.; Halanych, K.M.; Amemiya, C.T. Unexpected Distribution of Chitin and Chitin Synthase across Soft-Bodied Cnidarians. *Biomolecules* **2023**, *13*, 777. [CrossRef]
38. Macrander, J.; Broe, M.; Daly, M. Tissue-Specific Venom Composition and Differential Gene Expression in Sea Anemones. *Genome Biol. Evol.* **2016**, *8*, 2358–2375. [CrossRef]
39. Richier, S.; Sabourault, C.; Courtiade, J.; Zucchini, N.; Allemand, D.; Furla, P. Oxidative Stress and Apoptotic Events during Thermal Stress in the Symbiotic Sea Anemone, *Anemonia viridis*. *FEBS J.* **2006**, *273*, 4186–4198. [CrossRef]
40. Tang, P.-C.; Watson, G.M. Proteomic Identification of Hair Cell Repair Proteins in the Model Sea Anemone *Nematostella vectensis*. *Hear. Res.* **2015**, *327*, 245–256. [CrossRef]
41. Yang, Y.J.; Kim, C.S.; Choi, B.-H.; Cha, H.J. Mechanically Durable and Biologically Favorable Protein Hydrogel Based on Elastic Silklike Protein Derived from Sea Anemone. *Biomacromolecules* **2015**, *16*, 3819–3826. [CrossRef]
42. Morlighem, J.R.L.; Huang, C.; Liao, Q.; Gomes, P.B.; Pérez, C.D.; Prieto-Da-Silva, R.D.B.; Lee, S.M.-Y.; Rádis-Baptista, G. The Holo-Transcriptome of the Zoantharian *Protopalythoa variabilis* (Cnidaria: Anthozoa): A Plentiful Source of Enzymes for Potential Application in Green Chemistry, Industrial and Pharmaceutical Biotechnology. *Mar. Drugs* **2018**, *16*, 207. [CrossRef]
43. Merquiol, L.; Romano, G.; Ianora, A.; D'Ambra, I. Biotechnological Applications of *Scyphomedusae*. *Mar. Drugs* **2019**, *17*, 604. [CrossRef]
44. Ayala-Sumuano, J.-T.; Licea-Navarro, A.; Rudiño-Piñera, E.; Rodríguez, E.; Rodríguez-Almazán, C. Sequencing and de Novo Transcriptome Assembly of *Anthopleura dowii* Verrill (1869), from Mexico. *Genom. Data* **2017**, *11*, 92–94. [CrossRef] [PubMed]
45. Domínguez-Pérez, D.; Campos, A.; Alexei Rodríguez, A.; Turkina, M.V.; Ribeiro, T.; Osorio, H.; Vasconcelos, V.; Antunes, A. Proteomic Analyses of the Unexplored Sea Anemone *Bunodactis verrucosa*. *Mar. Drugs* **2018**, *16*, 42. [CrossRef]
46. Cassoli, J.S.; Verano-Braga, T.; Oliveira, J.S.; Montandon, G.G.; Cologna, C.T.; Peigneur, S.; de Castro Pimenta, A.M.; Kjeldsen, F.; Roepstorff, P.; Tytgat, J.; et al. The Proteomic Profile of *Stichodactyla duerdeni* Secretion Reveals the Presence of a Novel O-Linked Glycopeptide. *J. Proteom.* **2013**, *87*, 89–102. [CrossRef]
47. Wu, C.C.; MacCoss, M.J.; Howell, K.E.; Yates, J.R. A Method for the Comprehensive Proteomic Analysis of Membrane Proteins. *Nat. Biotechnol.* **2003**, *21*, 532–538. [CrossRef]
48. Zheng, Y.Z.; DeMarco, M.L. Manipulating Trypsin Digestion Conditions to Accelerate Proteolysis and Simplify Digestion Workflows in Development of Protein Mass Spectrometric Assays for the Clinical Laboratory. *Clin. Mass Spectrom.* **2017**, *6*, 1–12. [CrossRef]
49. Liu, W.; Mo, F.; Jiang, G.; Liang, H.; Ma, C.; Li, T.; Zhang, L.; Xiong, L.; Mariottini, G.; Zhang, J.; et al. Stress-Induced Mucus Secretion and Its Composition by a Combination of Proteomics and Metabolomics of the Jellyfish *Aurelia coerulea*. *Mar. Drugs* **2018**, *16*, 341. [CrossRef]
50. Sproles, A.E.; Oakley, C.A.; Matthews, J.L.; Peng, L.; Owen, J.G.; Grossman, A.R.; Weis, V.M.; Davy, S.K. Proteomics Quantifies Protein Expression Changes in a Model Cnidarian Colonised by a Thermally Tolerant but Suboptimal Symbiont—PubMed. Available online: https://pubmed.ncbi.nlm.nih.gov/31118473/ (accessed on 19 November 2023).

51. Chen, C.; Hou, J.; Tanner, J.J.; Cheng, J. Bioinformatics Methods for Mass Spectrometry-Based Proteomics Data Analysis. *Int. J. Mol. Sci.* **2020**, *21*, 2873. [CrossRef]
52. Filippova, I.Y.; Dvoryakova, E.A.; Sokolenko, N.I.; Simonyan, T.R.; Tereshchenkova, V.F.; Zhiganov, N.I.; Dunaevsky, Y.E.; Belozersky, M.A.; Oppert, B.; Elpidina, E.N. New Glutamine-Containing Substrates for the Assay of Cysteine Peptidases from the C1 Papain Family. *Front. Mol. Biosci.* **2020**, *7*, 578758. [CrossRef]
53. Recombinant Cathepsin L of Tribolium Castaneum and Its Potential in the Hydrolysis of Immunogenic Gliadin Peptides—PMC. Available online: https://www.ncbi.nlm.nih.gov/pmc/articles/PMC9266932/ (accessed on 19 November 2023).
54. Dunaevsky, Y.E.; Tereshchenkova, V.F.; Belozersky, M.A.; Filippova, I.Y.; Oppert, B.; Elpidina, E.N. Effective Degradation of Gluten and Its Fragments by Gluten-Specific Peptidases: A Review on Application for the Treatment of Patients with Gluten Sensitivity. *Pharmaceutics* **2021**, *13*, 1603. [CrossRef] [PubMed]
55. Kominami, E.; Ishido, K.; Muno, D.; Sato, N. The Primary Structure and Tissue Distribution of Cathepsin C. *Biol. Chem. Hoppe Seyler* **1992**, *373*, 367–373. [CrossRef] [PubMed]
56. Dahiya, N.; Tewari, R.; Hoondal, G.S. Biotechnological Aspects of Chitinolytic Enzymes: A Review. *Appl. Microbiol. Biotechnol.* **2006**, *71*, 773–782. [CrossRef] [PubMed]
57. Stoykov, Y.M.; Pavlov, A.I.; Krastanov, A.I. Chitinase Biotechnology: Production, Purification, and Application. *Eng. Life Sci.* **2015**, *15*, 30–38. [CrossRef]
58. Oyeleye, A.; Normi, Y.M. Chitinase: Diversity, Limitations, and Trends in Engineering for Suitable Applications. *Biosci. Rep.* **2018**, *38*, BSR2018032300. [CrossRef] [PubMed]
59. Hamid, R.; Khan, M.A.; Ahmad, M.; Ahmad, M.M.; Abdin, M.Z.; Musarrat, J.; Javed, S. Chitinases: An Update. *J. Pharm. Bioallied. Sci.* **2013**, *5*, 21–29. [CrossRef] [PubMed]
60. Casadidio, C.; Peregrina, D.V.; Gigliobianco, M.R.; Deng, S.; Censi, R.; Di Martino, P. Chitin and Chitosans: Characteristics, Eco-Friendly Processes, and Applications in Cosmetic Science. *Mar. Drugs* **2019**, *17*, 369. [CrossRef] [PubMed]
61. Rhee, S.G.; Woo, H.A.; Kil, I.S.; Bae, S.H. Peroxiredoxin Functions as a Peroxidase and a Regulator and Sensor of Local Peroxides. *J. Biol. Chem.* **2012**, *287*, 4403–4410. [CrossRef] [PubMed]
62. Burnside, S.W.; Hardingham, G.E. Transcriptional Regulators of Redox Balance and Other Homeostatic Processes with the Potential to Alter Neurodegenerative Disease Trajectory. *Biochem. Soc. Trans.* **2017**, *45*, 1295–1303. [CrossRef] [PubMed]
63. Szeliga, M. Peroxiredoxins in Neurodegenerative Diseases. *Antioxidants* **2020**, *9*, 1203. [CrossRef] [PubMed]
64. Mesika, R.; Reichmann, D. When Safeguarding Goes Wrong: Impact of Oxidative Stress on Protein Homeostasis in Health and Neurodegenerative Disorders. *Adv. Protein Chem. Struct. Biol.* **2019**, *114*, 221–264. [CrossRef]
65. Rhee, S.G. Overview on Peroxiredoxin. *Mol. Cells* **2016**, *39*, 1–5. [CrossRef]
66. Sharapov, M.G.; Novoselov, V.I.; Gudkov, S.V. Radioprotective Role of Peroxiredoxin 6. *Antioxidants* **2019**, *8*, 15. [CrossRef]
67. Kim, K.H.; Lee, W.; Kim, E.E. Crystal Structures of Human Peroxiredoxin 6 in Different Oxidation States. *Biochem. Biophys. Res. Commun.* **2016**, *477*, 717–722. [CrossRef]
68. Sharapov, M.G.; Gudkov, S.V.; Gordeeva, A.E.; Karp, O.E.; Ivanov, V.E.; Shelkovskaya, O.V.; Bruskov, V.I.; Novoselov, V.I.; Fesenko, E.E. Peroxiredoxin 6 Is a Natural Radioprotector. *Dokl. Biochem. Biophys.* **2016**, *467*, 110–112. [CrossRef]
69. Sharapov, M.G.; Novoselov, V.I.; Fesenko, E.E.; Bruskov, V.I.; Gudkov, S.V. The Role of Peroxiredoxin 6 in Neutralization of X-Ray Mediated Oxidative Stress: Effects on Gene Expression, Preservation of Radiosensitive Tissues and Postradiation Survival of Animals. *Free Radic. Res.* **2017**, *51*, 148–166. [CrossRef]
70. Mancini, A. Manganese Superoxide Dismutase Variants and Uses Thereof. U.S. Patent 9976127B2, 22 May 2018.
71. Rosa, A.C.; Corsi, D.; Cavi, N.; Bruni, N.; Dosio, F. Superoxide Dismutase Administration: A Review of Proposed Human Uses. *Molecules* **2021**, *26*, 1844. [CrossRef]
72. Guan, Y.; Hickey, M.J.; Borgstahl, G.E.; Hallewell, R.A.; Lepock, J.R.; O'Connor, D.; Hsieh, Y.; Nick, H.S.; Silverman, D.N.; Tainer, J.A. Crystal Structure of Y34F Mutant Human Mitochondrial Manganese Superoxide Dismutase and the Functional Role of Tyrosine 34. *Biochemistry* **1998**, *37*, 4722–4730. [CrossRef]
73. Hartl, F.U.; Bracher, A.; Hayer-Hartl, M. Molecular Chaperones in Protein Folding and Proteostasis. *Nature* **2011**, *475*, 324–332. [CrossRef]
74. Feder, M.E.; Hofmann, G.E. Heat-Shock Proteins, Molecular Chaperones, and the Stress Response: Evolutionary and Ecological Physiology. *Annu. Rev. Physiol.* **1999**, *61*, 243–282. [CrossRef]
75. Schild, H.; Arnold-Schild, D.; Lammert, E.; Rammensee, H.G. Stress Proteins and Immunity Mediated by Cytotoxic T Lymphocytes. *Curr. Opin. Immunol.* **1999**, *11*, 109–113. [CrossRef]
76. Kim, J.Y.; Barua, S.; Huang, M.Y.; Park, J.; Yenari, M.A.; Lee, J.E. Heat Shock Protein 70 (HSP70) Induction: Chaperonotherapy for Neuroprotection after Brain Injury. *Cells* **2020**, *9*, 2020. [CrossRef]
77. Alberti, G.; Paladino, L.; Vitale, A.M.; Caruso Bavisotto, C.; Conway de Macario, E.; Campanella, C.; Macario, A.J.L.; Marino Gammazza, A. Functions and Therapeutic Potential of Extracellular Hsp60, Hsp70, and Hsp90 in Neuroinflammatory Disorders. *Appl. Sci.* **2021**, *11*, 736. [CrossRef]
78. Weitzel, G.; Li, G.C. Thermal Response of Yeast Cells Overexpressing Hsp70 Genes. *Int. J. Hyperth.* **1993**, *9*, 783–797. [CrossRef]
79. Jensen, T.K.; Ikonen, E.; Gungor, B. Heat Shock Proteins and Cholesterol Homeostasis. WO2017178029A1, 19 October 2017.
80. Multhoff, G. Hsp70 Peptide Stimulating Natural Killer (NK) Cell Activity and Uses Thereof 2009. U.S. Patent 7517948B2, 1 April 2004.

81. Miyakawa, T.; Oka, M.; Hazama, S.; Tamada, K.; Udaka, K. Hsp70-Derived Peptide, and Method of Manufacturing Pharmaceutical Composition, Immunity Inducer, and Antigen-Presenting Cell for Cancer Treatment or Prevention Using the Same. EP 15849707 A 20151007. 2016. Available online: https://patentscope.wipo.int/search/en/detail.jsf?docId=WO2016056596 (accessed on 19 November 2023).
82. Le Poole, I.C.; Guevara-Patino, J.A.; Zloza, A. Mutant HSP70i to Prevent Autoimmune Disease 2013. U.S. Patent 2014 US 2018/0066030 A1, 20 November 2014. Available online: https://patentimages.storage.googleapis.com/e6/c1/ff/8aee26fe13c8bf/US20180066030A1.pdf (accessed on 19 November 2023).
83. Rosenzweig, R.; Nillegoda, N.B.; Mayer, M.P.; Bukau, B. The Hsp70 Chaperone Network. *Nat. Rev. Mol. Cell Biol.* **2019**, *20*, 665–680. [CrossRef]
84. López-Otín, C.; Bond, J.S. Proteases: Multifunctional Enzymes in Life and Disease. *J. Biol. Chem.* **2008**, *283*, 30433–30437. [CrossRef]
85. Wilkinson, D.J. Serpins in Cartilage and Osteoarthritis: What Do We Know? *Biochem. Soc. Trans.* **2021**, *49*, 1013–1026. [CrossRef]
86. Wang, L.; Li, Q.; Wu, L.; Liu, S.; Zhang, Y.; Yang, X.; Zhu, P.; Zhang, H.; Zhang, K.; Lou, J.; et al. Identification of SERPINB1 as a Physiological Inhibitor of Human Granzyme H. *J. Immunol.* **2013**, *190*, 1319–1330. [CrossRef]
87. Madio, B.; Undheim, E.A.B.; King, G.F. Revisiting Venom of the Sea Anemone *Stichodactyla haddoni*: Omics Techniques Reveal the Complete Toxin Arsenal of a Well-Studied Sea Anemone Genus. *J. Proteom.* **2017**, *166*, 83–92. [CrossRef]
88. Ramírez-Carreto, S.; Miranda-Zaragoza, B.; Rodríguez-Almazán, C. Actinoporins: From the Structure and Function to the Generation of Biotechnological and Therapeutic Tools. *Biomolecules* **2020**, *10*, E539. [CrossRef]
89. Ramírez-Carreto, S.; Pérez-García, E.I.; Salazar-García, S.I.; Bernáldez-Sarabia, J.; Licea-Navarro, A.; Rudiño-Piñera, E.; Pérez-Martínez, L.; Pedraza-Alva, G.; Rodríguez-Almazán, C. Identification of a Pore-Forming Protein from Sea Anemone *Anthopleura dowii* Verrill (1869) Venom by Mass Spectrometry. *J. Venom Anim. Toxins. Incl. Trop. Dis.* **2019**, *25*, e147418. [CrossRef]
90. Bradford, M.M. A Rapid and Sensitive Method for the Quantitation of Microgram Quantities of Protein Utilizing the Principle of Protein-Dye Binding. *Anal. Biochem.* **1976**, *72*, 248–254. [CrossRef]
91. Laemmli, U.K. Cleavage of Structural Proteins during the Assembly of the Head of Bacteriophage T4. *Nature* **1970**, *227*, 680–685. [CrossRef]
92. Nesvizhskii, A.I.; Keller, A.; Kolker, E.; Aebersold, R. A Statistical Model for Identifying Proteins by Tandem Mass Spectrometry. *Anal. Chem.* **2003**, *75*, 4646–4658. [CrossRef]
93. Keller, A.; Nesvizhskii, A.I.; Kolker, E.; Aebersold, R. Empirical Statistical Model to Estimate the Accuracy of Peptide Identifications Made by MS/MS and Database Search. *Anal. Chem.* **2002**, *74*, 5383–5392. [CrossRef]
94. Van Dongen, S. Graph Clustering Via a Discrete Uncoupling Process. *SIAM J. Matrix Anal. Appl.* **2008**, *30*, 121–141. [CrossRef]
95. Szklarczyk, D.; Gable, A.L.; Nastou, K.C.; Lyon, D.; Kirsch, R.; Pyysalo, S.; Doncheva, N.T.; Legeay, M.; Fang, T.; Bork, P.; et al. The STRING Database in 2021: Customizable Protein-Protein Networks, and Functional Characterization of User-Uploaded Gene/Measurement Sets. *Nucleic Acids Res.* **2021**, *49*, D605–D612. [CrossRef]
96. Waterhouse, A.; Bertoni, M.; Bienert, S.; Studer, G.; Tauriello, G.; Gumienny, R.; Heer, F.T.; de Beer, T.A.P.; Rempfer, C.; Bordoli, L.; et al. SWISS-MODEL: Homology Modelling of Protein Structures and Complexes. *Nucleic Acids Res.* **2018**, *46*, W296–W303. [CrossRef]
97. Bhattacharya, D.; Nowotny, J.; Cao, R.; Cheng, J. 3Drefine: An Interactive Web Server for Efficient Protein Structure Refinement. *Nucleic Acids Res.* **2016**, *44*, W406–W409. [CrossRef]
98. Pettersen, E.F.; Goddard, T.D.; Huang, C.C.; Couch, G.S.; Greenblatt, D.M.; Meng, E.C.; Ferrin, T.E. UCSF Chimera—A Visualization System for Exploratory Research and Analysis. *J. Comput. Chem.* **2004**, *25*, 1605–1612. [CrossRef]

marine drugs

MDPI

Article

Proteomic Identification of Plasma Components in *Tachypleus tridentatus* and Their Effects on the Longitudinal Bone Growth Rate in Rats

Shu Jiang [1,2,†], Xinjian Qu [1,2,†], Siping Liu [1,2], Jun Wei [1,2], Xiangxi Yi [1,2,*], Yonghong Liu [1,2,*] and Chenghai Gao [1,2,*]

1 Institute of Marine Drugs, Guangxi University of Chinese Medicine, Nanning 530200, China
2 Guangxi Key Laboratory of Marine Drugs, Nanning 530200, China
* Correspondence: yixiangxi2017@163.com (X.Y.); yonghongliu@scsio.ac.cn (Y.L.); gaoch@gxtcmu.edu.cn (C.G.)
† These authors contributed equally to this work.

Abstract: *Tachypleus tridentatus* (*T. tridentatus*) is a marine animal and traditional Chinese medicine. *T. tridentatus* plasma is a valuable resource for important medical and health-based functions. In this experiment, in order to evaluate the effect and mechanism of *T. tridentatus* plasma with respect to the promotion of bone tissue growth in rats, the processes of ultrafiltration and mass spectrometry were first used to separate and identify the components of *T. tridentatus* plasma. Then, a comparison of the effects of the *T. tridentatus* plasma samples, which each possessed different molecular weights, regarding the growth of the long bones of rats was conducted. Finally, transcriptomics, proteomics, and bioinformatics were all used to analyze the biological functions and related signaling pathways of the *T. tridentatus* plasma in order to promote rat bone growth. The results showed that the contents of amino acid residues in peptides are related to the growth promotion that was contained in the 10–30 kDa plasma group. Moreover, the *T. tridentatus* plasma samples were found to be higher in this respect than those in the whole plasma group. In addition, the 10–30 kDa plasma group could significantly promote bone growth activity in rats. The proteomic analysis showed that the proteins that were differentially expressed in the 10–30 kDa plasma group were mainly enriched in the PI3K-AKT signal pathway. Our study suggested that the *T. tridentatus* plasma possesses promising potential for the purposes of clinical use, whereby it can serve the role of a growth-promoting agent.

Keywords: *Tachypleus tridentatus*; longitudinal bone growth; transcriptomics; proteomics

1. Introduction

Tachypleus tridentatus (*T. tridentatus*) is a marine animal and traditional Chinese medicine [1]. *T. tridentatus* possesses anti-inflammatory, analgesic, hemostatic, and antidysentery medicinal effects [2–4]. It is also worth noting that *T. tridentatus* blood that contains hemocyanin possesses important medical application value [5]. The blood extract can be made into a Limulus-reagent (LR), which is widely used for the purposes of endotoxin detection and quantification in injections, as well as in radiopharmaceuticals, vaccines, and other biological products [2,3,6]. Indeed, Tachypleus amebocyte lysate (TAL) can be used for the purpose of accurate and rapid quantitative determination regarding bacterial infections or mycotoxins [7]. Most commonly in recent years, the developing COVID-19 vaccines have increased the demand for LR. Therefore, with such facts in mind, *T. tridentatus* blood is an important and irreplaceable resource.

Short stature refers to a child's height that is lower by more than two standard deviations in comparison to the normal height of individuals of the same race and in a similar environment [8]. Furthermore, the overall body size of mammals is mainly determined by longitudinal bone growth [9]. However, having said this, longitudinal bone growth is also the result of chondrocyte proliferation, as well as the ensuing hypertrophy in growth plates, which are called endochondral ossification and is mainly controlled by a system of the growth hormone-(GH-) insulin-like growth factor-1(IGF-1) axis [10,11]. The most important treatment for patients who possess a short stature is GH injection. However, daily injections and the abuse of GH in order to increase height, as well as the resulting pain that occurs, have drawn increasing attention in regard to medical ethics [12–14]. Shorter people have been reported to possess a lower health-related quality of life in adulthood, even though they no longer fall within the definition of short stature [15]. Therefore, a health product that can promote growth and development has become the top priority of many patients who possess a short stature.

T. tridentatus is one of the main sources of LR. However, less than 1% of *T. tridentatus* blood components have been used as fungal diagnostic reagents. In addition, more than 99% of the remaining plasma is discarded, thereby resulting in a great waste of resources [16]. In order to make full use of the remaining plasma, in this paper, the components remaining in the plasma are first separated and identified when compared to the effects of each molecular weight fragment of *T. tridentatus* plasma sample on the growth of long bones in rats, as well as then explored with respect to the mechanism by which the *T. tridentatus* plasma promotes bone growth in rats. This research not only helps to lay a theoretical foundation for the development of *T. tridentatus* plasma products but also helps to promote the high-value scientific utilization of plasma resources.

2. Results

2.1. *T. tridentatus Plasma Ultrafiltration and HPLC Analysis*

A total of 256.52 g of solid powder samples were obtained as whole plasma (WP) samples after the lyophilization from 20 L of *T. tridentatus* plasma.

Five samples were, in turn, obtained by separating 100 L of *T. tridentatus* plasma with 30 kDa, 10 kDa, 5 kDa, and 2 kDa ultrafiltration membrane modules. The samples were lyophilized in order to yield samples with molecular weights of 0–2 kDa, 2–5 kDa, 5–10 kDa, 10–30 kDa, and >30 kDa—which weighed 13.59 g, 16.50 g, 21.31 g, 10.64 g, and 155.56 g, respectively. The sample of *T. tridentatus* weighing >30 kDa plasma possessed the highest yield at 60.39%. Furthermore, the yields of the 0–2 kDa, 2–5 kDa, and 5–10 kDa samples were 20.80%, 6.41%, and 8.27%, respectively. Additionally, the lowest yield was obtained for the 10–30 kDa sample, which accounted for only 4.13% (Table 1).

Table 1. Ultrafiltration separation of *Tachypleus tridentatus* plasma.

Ultrafiltration-Separated Samples	Molecular Weight (kDa)	Weight (g)	Proportion (%)
1	0–2	53.59	20.80
2	2–5	16.50	6.41
3	5–10	21.31	8.27
4	10–30	10.64	4.13
5	>30	155.56	60.39

Each sample of *T. tridentatus* was collected and analyzed via high-performance liquid chromatography. With respect to this, the results showed that all six samples possessed peaks at t_R = 14.6 min, thereby indicating that the samples contained the same component (Figure 1a).

The candidate proteins of each sample were identified by protein mass spectrometry (MS). Further, the results showed that the >30 kDa sample and WP sample were mainly composed of hemocyanin subunits, while the <30 kDa samples were mainly composed of α-2 macroglobulin and complement-3-like proteins. The contents of arginine, valine, methionine, and hydrophobic amino acid residues in the peptides, which are related to growth-promoting activity contained in the 10–30 kDa sample, were found to be higher than those found in the WP sample (Table S1).

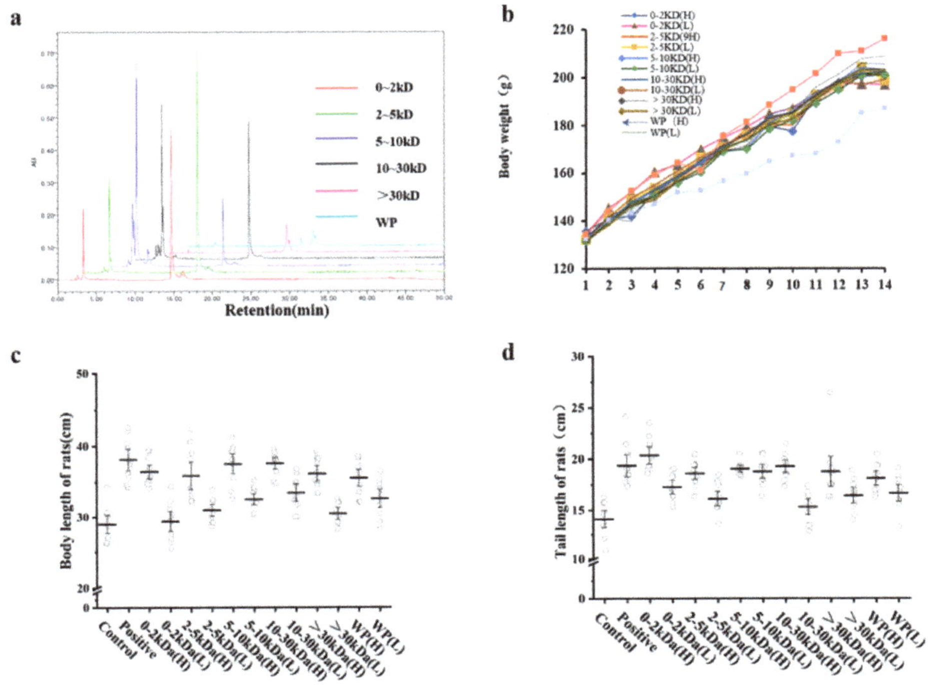

Figure 1. The effects of the *T. tridentatus* plasma on the growth performance of Wistar rats. (**a**) Chromatogram of the different components in the *T. tridentatus* plasma; (**b**) weight gain curve; (**c**) body lengths of the rats during the 14 d of sample administration; and (**d**) tail length of the rats during the 14d of sample administration. ($n = 10$ for each group, as determined by conducting one-way ANOVA; Control: saline group; and positive: recombinant human growth hormone group).

2.2. The Effects of T. tridentatus Plasma on the Growth Activity Performance of Wistar Rats

In order to detect the effect of each *T. tridentatus* plasma sample on the growth and activity performances, Wistar rats were injected subcutaneously with rhGH 200 μg/kg in the positive control group (positive), and an equal volume of saline was injected into the rats in the negative control group (control). The high-dose experimental group was given each sample solution at a dosage of 200 mg/kg/d, and the low-dose experimental group was given each sample solution at a dosage of 100 mg/kg/d.

Changes in body weight, body length, and tail length are intuitive indicators of rat growth and development. On the 12th day, when compared with the control group, there was a significant difference in the body weight of the rats in each group ($p < 0.05$) (Figure 1b).

Furthermore, the body length results showed that the body length of each animal in the high-dose group was significantly different from that of the control group (Figure 1c). Moreover, among the experimental groups, the rats in the 10–30 kDa group possessed the longest body length ($p < 0.05$). When compared with the control group, the tail length results showed that there were significant differences in the tail lengths of the rats in each high-dose group, except in regard to the >30 kDa group. Furthermore, the tails of the rats in each low-dose group were shorter than those in the corresponding high-dose group (Figure 1d). In conclusion, the body weight, body length, and tail length of the rats were positively correlated with the concentration of each sample.

2.3. Effects of T. tridentatus Plasma on Bone Growth in Wistar Rats

Generally, changes in the length of the femur and tibia are indicators that directly reflect the growth of long bones in rats. As such, in order to avoid the influence of water and excess muscle tissue in regard to bone weight, the right femur of each rat was weighed. According to the results, the femur and tibia lengths of the rats in the 10~30 kDa (H) group were the longest ($p > 0.05$) (Figure 2a,b). Furthermore, the growth of long bones in rats is mainly accomplished by endochondral ossification, which is the development of growth plates in the metaphysis of the long diaphysis.

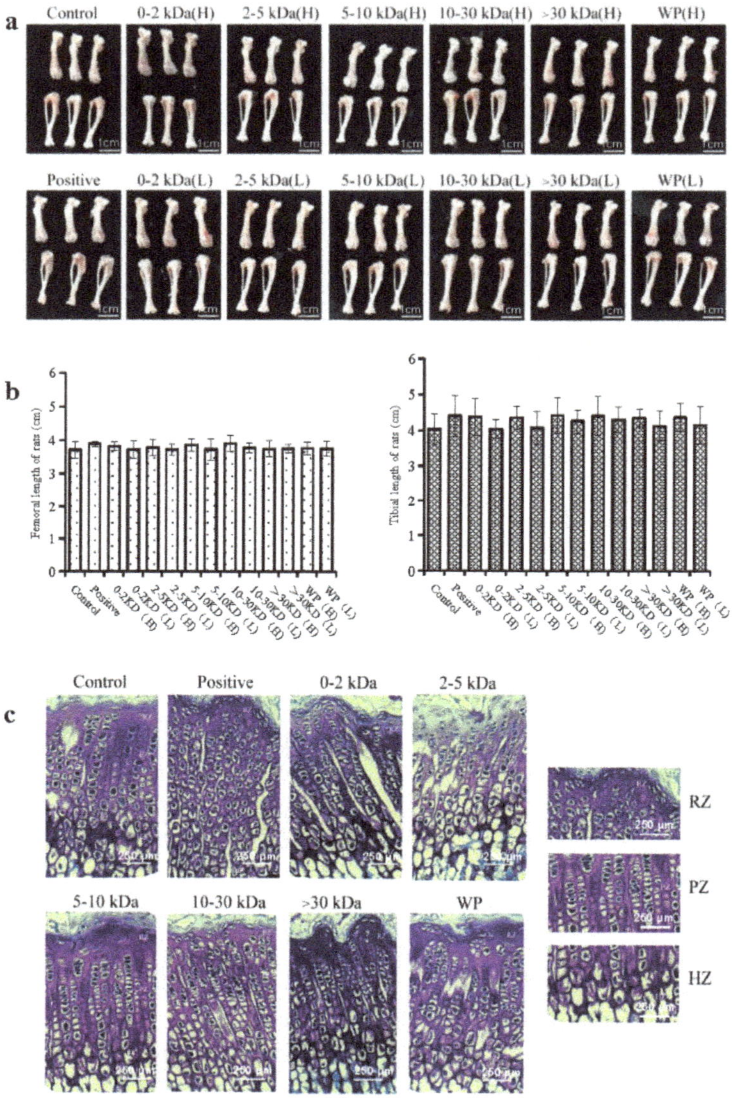

Figure 2. *T. tridentatus* plasma samples promoted normal skeletal growth in the rats after 14 d of subcutaneous injection. (**a**) Representative images of the tibial and femurs of rats, thereby demonstrating their length after 14 d of treatment. The scale bar is 1 cm; (**b**) the length of the femur and tibia in rats; and (**c**) photographs of cresyl violet-stained chondrocytes in the proximal growth plates of the tibias that were obtained from the control, positive (200 µg/kg), and *T. tridentatus* plasma-treated (200 mg/kg/d) rats. (Nonparametric test; RZ: resting zone; PZ: proliferative zone; HZ: hypertrophic zone; and scale bar = 250 µm). Each data point represents one individual rat. All data are shown as the mean ± standard deviation (s.d.). The scale bar is 250 µm.

The cresyl violet staining results with respect to the growth plate of the distal femur in the different groups of rats are shown in Figure 2c and Table 2. In the control group, the boundaries of the growth plates were clear; further, the cells were arranged in regular columns. However, as shown in Table 2, the height of the growth plate and the proliferative zone in the WP (H) and 30 kDa (H) groups were significantly higher than those in the control group ($p < 0.05$). Moreover, the animals in the 2–5 kDa, 5–10 kDa, and 10–30 kDa groups showed significantly increased growth plate heights than those found in the control group ($p < 0.01$).

Table 2. Height of overall growth plate, as well as the proliferative and hypertrophic zone in each group ($n = 10$).

Group	Overall Growth Plate (μm)	Proliferative Zone (μm)	Hypertrophic Zone (μm)
Control	95.31 ± 5.38	62.76 ± 2.18	29.56 ± 3.14
Positive	144.51 ± 2.92 **	96.32 ± 1.47 **	46.12 ± 2.81 **
0–2 kDa (H)	108.10 ± 7.01	73.44 ± 4.03	33.61 ± 1.76
2–5 kDa (H)	121.74 ± 3.89 **	83.07 ± 3.62 **	40.36 ± 3.71 *
5–10 kDa (H)	122.72 ± 6.73 **	83.76 ± 3.19 **	39.88 ± 6.66 *
10–30 kDa (H)	128.74 ± 4.02 **	87.05 ± 0.77 **	42.48 ± 3.62 *
>30 kDa (H)	113.99 ± 8.81 *	76.15 ± 1.04 *	35.84 ± 1.99
WP (H)	110.96 ± 4.62 *	75.12 ± 2.30 *	35.05 ± 3.21

Results represented as mean ± SD. * $p < 0.05$, and, ** $p < 0.01$ vs. control group.

2.4. *Effects of the T. tridentatus Plasma on the Expression Levels of Serum ALP, BMP-2, and IGF-1 in Rats*

The alkaline phosphatase (ALP) protein content detection in serum showed that, when compared with the control group, the positive group, 0–2 kDa (H) group, 2–5 kDa (H) group, 5–10 kDa (H) group, 10–30 kDa (H) group, >30 kDa (H) group, and the WP (H) group possessed significantly increased ALP contents in serum ($p < 0.05$).

When compared with the control group, the 0–2 kDa (H) group, 2–5 kDa (H) group, 5–10 kDa (H) group, and the 10–30 kDa (H) group, a significantly increased serum with bone morphogenetic protein-2 (BMP-2) concentrations ($p < 0.01$)was shown. In addition, the 0–2 kDa (L) group, 2–5 kDa (L) group, 5–10 kDa (L) group, >30 kDa (H) group, WP(H) group, and WP (L) group also showed increased BMP2 expression levels in serum ($p < 0.05$). The serum level of BMP-2 with respect to the rats in the 10–30 kDa (H) group was found to be the highest.

IGF-1 is a standard indicator for detecting bone growth. When compared with the control group, the 0–2 kDa (H) group, 5–10 kDa (H) group, 10–30 kDa (H) group, 10–30 kDa (L) group, >30 kDa (H) group, WP (H) group, and the positive group, significantly increased concentrations of IGF-1 in serum ($p < 0.01$) were observed. Moreover, the 2–5 kDa (H) group, 2–5 kDa (L) group, >30 kDa (L) group, 5–10 kDa (L) group, and WP (L) group showed increased concentrations of IGF-1 in serum ($p < 0.05$). Lastly, with respect to the aforementioned, there was no significant difference between the 0–2 kDa (L) group and the control group (Figure 3a).

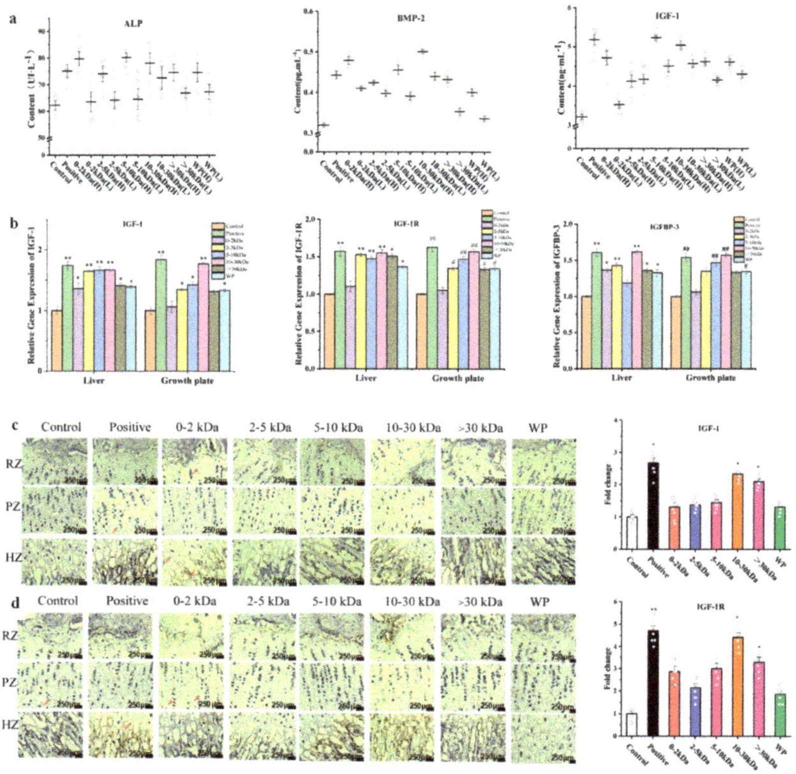

Figure 3. Different plasma components increased the factors related to growth and development in adolescent female rats. (**a**) Effects of different plasma components on total serum ALP, BMP-2, and IGF-1 levels in adolescent rats; and (**b**) quantification of the IGF-1, IGF-1R, and IGFBP-3 mRNA levels in the livers and growth plates that were obtained from the control, positive (200 μg/kg), and *T. tridentatus* plasma-treated (200 mg/kg/d, 100 mg/kg/d) rats (as determined via one-way ANOVA, $n = 5$ for each group). Each data point represents one individual rat. All data are shown as the mean ± standard deviation (s.d.). # $p < 0.05$ and ## $p < 0.01$. (**c**) Representative immunohistochemical staining images of IGF-1 in the rat growth plates after 14 d of treatment; and (**d**) representative immunohistochemical staining images of the IGF-1R in the rat growth plates after 14 d of treatment ($n = 10$ for each group; nonparametric test; RZ: resting zone; PZ: proliferative zone; HZ: hypertrophic zone; and scale bar = 250 μm). * $p < 0.05$ and ** $p < 0.01$.

2.5. *Effects of T. tridentatus Plasma on the Expression of Bone Growth-Related Genes in Rats*

Quantitative RT-PCR was used in order to detect the mRNA expression levels of IGF-1, insulin-like growth factor-1 receptor (IGF-1R), and insulin-like growth factor binding protein 3 (IGFBP-3) in the Wistar rat tibial growth plates and livers (Figure 3b). When compared with the control group, the expression levels of IGF-1 mRNA in the liver of the rhGH group, 2–5 kDa group, and 10–30 kDa group were significantly increased ($p < 0.01$); moreover, the 5–10 kDa group, >30 kDa group, and the WP group also showed higher expressions than the control group ($p < 0.05$). In regard to the rat epiphyseal growth plate, the 2–5 kDa group, the 5–10 kDa group, and the WP group, an upregulated expression of IGF-1 mRNA ($p < 0.05$) was shown.

When compared with the control group, the expression level of IGF-1 mRNA in the liver of the rhGH group, 2–5 kDa group, and 10–30 kDa group was significantly increased ($p < 0.01$). In addition, the 5–10 kDa group, >30 kDa group, and WP group also showed higher expressions than the control group ($p < 0.05$). In regard to the rat epiphyseal growth plate, the 2–5 kDa group, the 5–10 kDa group, and the WP group upregulated expression of IGF-1 mRNA ($p < 0.05$) was shown.

In the rat livers, when compared with the control group, the rhGH group, 2–5 kDa group, and the 10–30 kD group showed a significantly upregulated mRNA expression of IGFBP-3 ($p < 0.01$). Regarding the 0–2 kDa group, >30 kDa group, and the WP group, increased expression levels ($p < 0.05$) were also shown. With respect to the epiphyseal growth plate-when compared with the control group- the rhGH group, 5–10 kDa group and the 10–30 kDa group displayed a significantly increased expression level of IGFBP-3 mRNA ($p < 0.01$). Lastly, the 2–5 kDa group and the WP group also showed increased IGFBP-3 mRNA expression levels ($p < 0.05$).

2.6. Effects of the T. tridentatus Plasma on the Expression of Cartilage-Related Proteins in Rat Growth Plates

An immunohistochemical method was used to detect the expression of cartilage-related proteins in rat growth plates after 14 days of plasma treatment. Moreover, the IGF-1 protein is mainly expressed in chondrocytes in the hypertrophic zone, which are present at the base of the growth plate cartilage and around the trabecular bone (Figure 3c). When compared with the rats in the control group, the rats in the positive group, >30 kDa (H) group, and the 10–30 kDa (H) group showed a significantly upregulated expression of IGF-1 in the proximal tibia tissue ($p < 0.05$).

The IGF-1R protein is expressed in the growth plate matrix, growth plate proliferation zone, and hypertrophy zone of the proximal tibia. When conducting an immunohistochemical analysis, the IGF-1R protein appeared brown (Figure 3d). The results demonstrated that compared with the control group, the rats in the rhGH group possessed a significantly increased expression level of IGF-1R protein ($p < 0.01$), while the animals in the 10–30 kDa and >30 kDa group displayed an upregulated expression of the IGF-1R protein ($p < 0.01$). Having said this there was no significant difference in the expression of this protein between the 0–2 kDa, 2–5 kDa, and 5–10 kDa samples when compared with the control group ($p > 0.05$).

2.7. The Femur Transcriptome of the Rats in the WP and 10–30 kDa Groups

When compared with the control group, the rats in the WP group had 130 differentially expressed genes (DEGs), of which 53 genes were upregulated, and 77 genes were downregulated, as shown in Figure 4a (left). The genes expression level in the WP group are shown in Figure 4b (left). When compared with the control group, 145 DEGs were identified in the 10–30 kDa group, including 47 upregulated genes and 98 downregulated genes, as shown in Figure 4a (right). The relative levels of gene expression in the 10–30 kDa group are shown in Figure 4b (right). As shown by the Gene Ontology (GO) analysis, the differential expression of the genes in the femur transcriptome of rats in the WP group, as well as the 10–30 kDa group, aggregated into biological processes and cellular components after 14 days of treatment (Figure 4c,d).

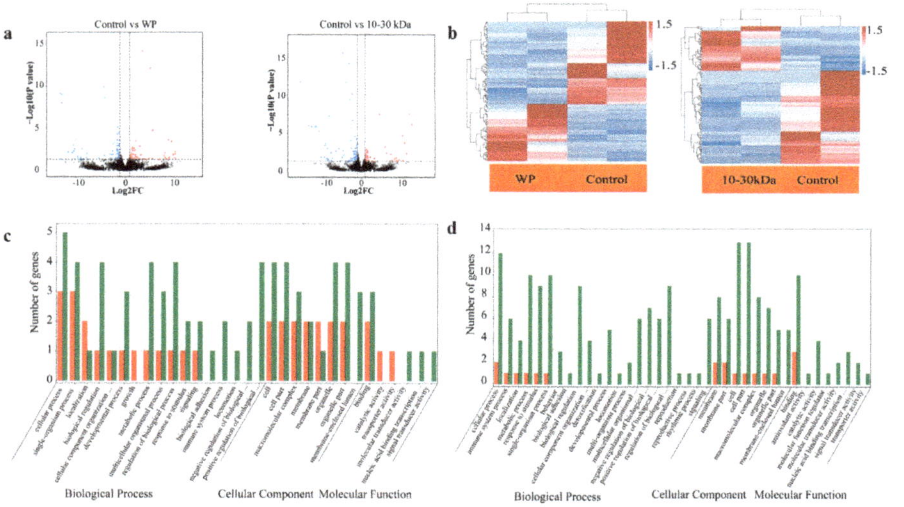

Figure 4. Femur transcriptomes, as determined by RNA sequencing. (**a**) RNA from the rats in the purified whole plasma and the 10–30 kDa group were assessed by RNA-seq; further, the transcript abundances were compared with DEseq2. Transcripts are plotted based upon fold change (between whole plasma, control, and between the 10–30 kDa and control) and adjusted p value (significance of differential expression), with mRNAs shown in lighter (±1 to 2.5 log2-fold change) or darker shades (>2.5 log2-fold change); (**b**) heatmap depicting the relative levels of genes in the whole plasma (left) and 10–30 kDa groups (right) (red represents relatively high expression, and blue represents relatively low expression); (**c**) GO analysis of the differentially expressed genes in the femur transcriptome of rats treated with WP for 14 d; and (**d**) GO analysis of the differentially expressed genes in the femur transcriptome of rats treated with the 10–30 kDa sample for 14 d. Red indicates the upregulated genes and green indicates the downregulated genes in the Go enrichment and classification histograms. Ontology covers cellular component, molecular function, and biological process domains.

2.8. *Quantitative Proteomics Analysis of Rat Femur Tissue in the WP and 10–30 kDa Groups*

The high-throughput identification and screening of differentially expressed proteins (DEPs) in the rat femur tissues of the rats in the WP and 10–30 kDa groups, was based on isotope labeling (iTRAQ) and proteomic technology. After filtering the data in order to exclude low-scoring spectra, 933,431 unique spectra-matching specific peptides were obtained. A total of 29,007 unique peptides were identified, thereby resulting in the identification of 5,860 proteins. A total of 388 DEPs were identified in the 10–30 kDa group when compared to the control group; in addition 926 DEPs were identified in the WP group when compared to the control group (Figure 5a,b). The numbers of DEPs and DEGs for these two comparisons are shown in Table 3.

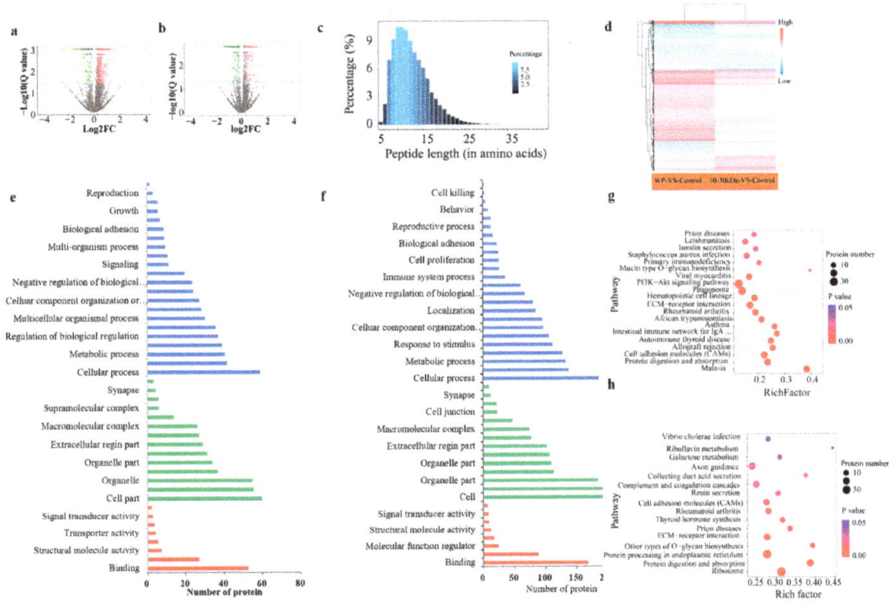

Figure 5. Comparative total proteomics of rats in the WP group and the 10–30 kDa group when compared with the control group. (**a**) Diagram of the total proteins from rat femur tissue in the purified WP group; (**b**) diagram of total protein from rat femur tissue in the purified 10–30 kDa group; (**c**) the percentage of each different peptide length; (**d**) clustering analysis of the DEPs in the comparisons of control group vs. 10–30 kDa group, as well as control group vs. the whole plasma group. Blue indicates downregulation, red indicates upregulation, and white indicates no detectable change in expression; (**e**) GO analysis of the DEPs for femur proteomics of the rats in the WP group; (**f**) GO analysis of the DEGs for femur proteomics of the rats in the 10–30 kDa group (blue indicates molecular function, green indicates cellular component, and red indicates biological process domains); (**g**) statistics of the pathway enrichment of the differentially expressed proteins in the control group vs. whole plasma group comparison; and (**h**) statistics of pathway enrichment of the differentially expressed proteins in the control group vs. the 10–30 kDa group comparison. RichFactor is the ratio of the number of differentially expressed proteins annotated in this pathway term with respect to the number of all proteins annotated in this pathway term.

Table 3. Numbers of DEPs and DEGs in the two comparisons.

Comparisons	Up-Regulated Expression Genes/Protein Number	Down-Regulated Expression Genes/Protein Number	Total Different ExPression Gene/Protein Number
10–30 kDa-vs-Control	47/214	98/174	145/388
Whole plasma-vs-Control	53/574	77/352	130/926

2.9. The Femur Transcriptome of the Rats in the WP and 10–30 kDa Groups

All identified proteins and matching peptide lengths were analyzed and counted. Further, the results showed that most of the peptides were between seven and fifteen amino acid residues in length, which conformed to the rule of trypsin digestion with respect to peptides (Figure 5c). In addition, it must be noted that PRM is a new development in targeted mass spectrometry that is more specific and sensitive than selective response monitoring; it has been widely used for the quantification and detection of target proteins. Moreover, 15 candidate proteins related to long bone growth were screened for the purposes of PRM analysis, which converged with the protein expression levels measured by iTRAQ, thereby confirming that our iTRAQ results were reliable and reproducible (Figure S1).

The protein expression patterns were grouped for the purposes of cluster analysis whilst using Euclidean distance and hierarchical algorithms for the comparisons concerning the control group vs. 10–30 kDa group and control group vs. whole plasma group (Figure 5d).

The functional analysis of the 220 DEPs in the WP group showed that most proteins acted on biological processes, whereby they accounted for 48.07% of the total proteins (Figure 5e). The GO analysis results of the 59 DEPs in the femur tissues of the rats in the 10–30 kDa plasma group were mainly enriched in molecular function, cellular component, and biological regulation. Among them, 14 proteins were enriched in the closed cavity of the cell membrane, six proteins were enriched in cell junctions, and six proteins were enriched in supramolecular complexes (Figure 5f).

The functional classification of the DEPs in each comparison group was subjected to Clusters of Orthologous Groups of proteins (COG/KOG) analysis. The results showed that the functional types of DEPs in the WP group, as well as the 10–30 kDa, group were enriched in cell physiological processes, signal transduction mechanisms, post-translational modifications, protein conformational transitions, chaperones, etc. (Figure 5g,h). In the 10–30 kDa group, the enrichment of proteins related to intracellular transport, secretion transport, vesicular transport, extracellular structure-function, and information storage and processing was higher than those found in the WP group (Figure 5h).

2.10. Kyoto Encyclopedia of Genes and Genomes (KEGG) Pathway Enrichment Analysis of DEGs and DEPs

Correlations between proteomes and transcriptomes are shown in the comparative proteomes and transcriptomes with respect to the comparisons of the control group vs. 10–30 kDa group, as well as the control group vs. the whole plasma group.

The Venn diagram analysis found 5567 genes coexpressed in the control group vs. the whole plasma group, as well as 5569 coexpressed genes in the control group vs. the 10–30 kDa group (Figure 6a). Following a comparison of all the: genes; proteins; DEGs, and DEPs; P02688; D3ZC54; Q4V7F2; D3ZMS3; M0R9U2; Q5M7V3 and D4ACR1 showed the same trend with respect to the comparison of the transcriptomic and proteomic levels in the control group vs. the 10–30 kDa group (Figure 6b). In the comparison of the whole plasma group vs. the control group, there were only two DEGs/DEPs with the same expression trend at the transcriptomic and proteomic levels (Figure 6c).

In order to determine the mode of action with respect to the samples in rats, the screened differential proteins were subjected to KEGG pathway analysis in order to determine the metabolic and signaling pathways involved. The KEGG signaling pathway analysis was performed on 220 differentially expressed proteins in the whole plasma group. In addition, a total of 44 related signaling pathways were retrieved; further, 16 signaling pathways were also significantly enriched. Among these, the main enrichment pathways were ribosome, endoplasmic reticulum protein processing, protein digestion, absorption, etc. (Figure 6d).

The KEGG signaling pathway analysis was performed on 59 DEGs in the 10–30 kDa group. Following this, a total of 43 related signaling pathways were enriched. The phosphatidylinositol 3-hydroxykinase/protein kinase B (PI3K-AKT) signaling pathway (ko04151), as well as the mitogen-activated protein kinase (MAPK) signaling pathway (ko04010) are most closely associated with the growth of long bones. Six pathways were enriched in both the comparisons of the control group vs. the 10–30 kDa group, as well as the control group vs. the whole plasma group (Figure 6e).

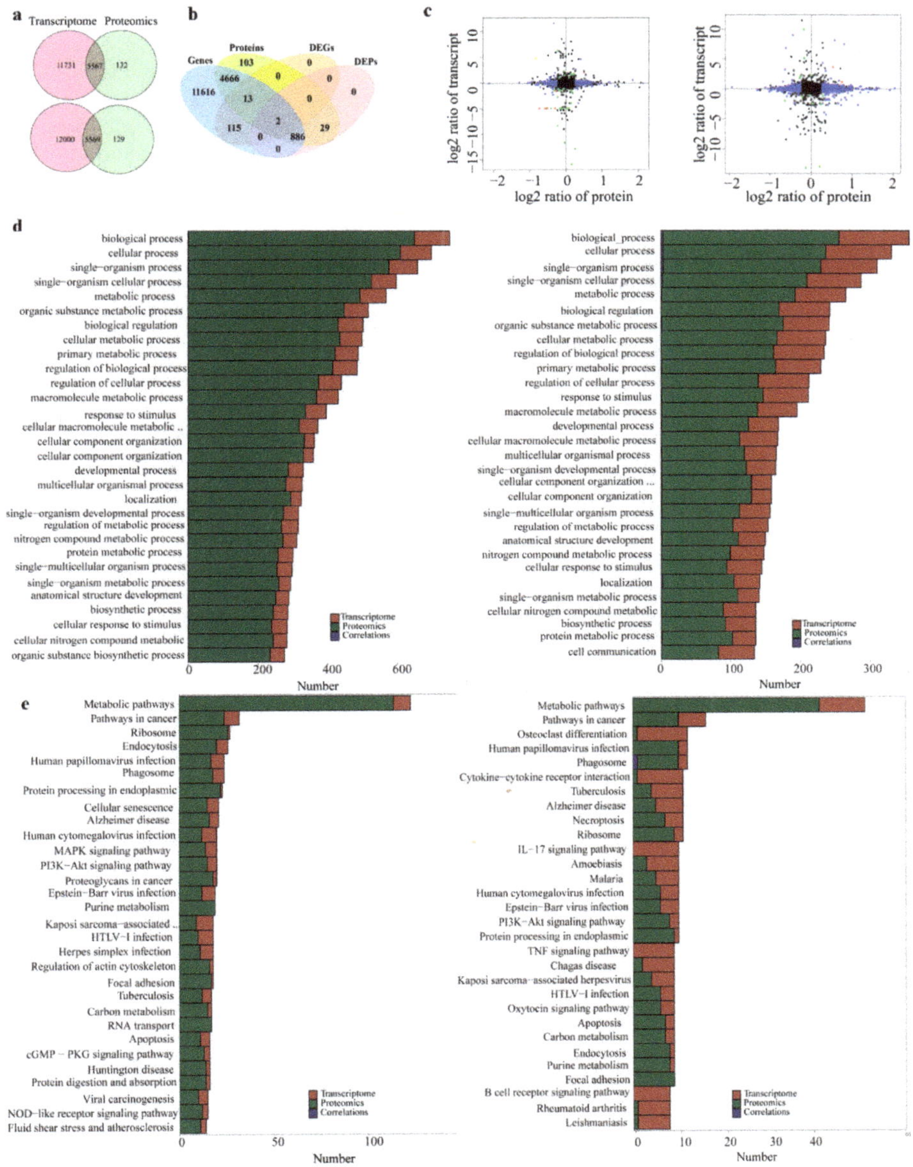

Figure 6. Combined proteomics and transcriptomics analyses. (**a**) Venn diagram of the expressed genes and proteins in the control group vs. the whole plasma group, as well as control group vs. the 10–30 kDa group comparisons; (**b**) comparison of all genes, proteins, DEGs, and DEPs in the control group vs. the whole plasma group (**left**), as well as control group vs. the 10–30 kDa group samples (**right**); (**c**) four-quadrant diagram of control group vs. the whole plasma group (**left**), as well as control group vs. the 10–30 kDa group (**right**) comparisons. (Each dot represents one gene/protein; black dots represent nondifferentiated proteins and genes; green dots represent DEGs but not the non-DEPs; blue dots represent DEPs but not the non-DEGs; and red dots represent both DEGs and DEPs with the same difference trend); (**d**) pathway analysis of the DEGs (**left**) and DEPs (**right**) from rat femurs; (**e**) Pathway correlation analysis in the control group vs. the whole plasma group (**left**), as well as control group vs. the 10–30 kDa group (**right**) comparisons.

3. Discussion

T. tridentatus is an ancient marine arthropod whose evolutionary history can be traced back to approximately 450 million years ago; further, it is known as the famous living fossil [16]. Tachyplesin and Tatritin were isolated from the acidic extracts of *T. tridentatus.* In addition, they act as antimicrobial peptides with respect to the self-defense process in horseshoe crabs against invading microorganisms [4,17,18]. The *T. tridentatus* amoebocyte lysate is crucial in regard to medical testing [6], but its involvement in the industries of China in recent years has been limited due to the decline in the *T. tridentatus* population [5].

Ultrafiltration can separate polypeptides of different molecular weights while maintaining their physiological activities. Furthermore, the plasma proteomics of *T. tridentatus* is an important component of functional genome research, which is currently in its infancy. Most of the proteins identified are high-abundance proteins, such as hemocyanin, actin, α-2 macroglobulin, etc. The protein content in the *T. tridentatus* plasma is as high as 10% and is rich in 12 free amino acids [19]. The results of this study showed that hemocyanin was a highly abundant protein in the whole plasma of *T. tridentatus.* Furthermore, hemocyanin possesses various functions that are related to body energy storage, which is in addition to oxygen transport, the maintenance of osmotic pressure, and immune regulation [20]. Indeed, Alpha-2 macroglobulin is a macromolecular glycoprotein that is widely involved in the regulation and transport of substances in the Limulus plasma. With respect to this, certain studies have shown that α-2 macroglobulin is negatively correlated with the degree of chondrocyte apoptosis and thus can also be used as a carrier of growth factors in order to regulate the growth process of living cells [21].

T. tridentatus plasma can promote the growth and development of Wistar rats by increasing body weight, body length, and tail length in adolescent Wistar rats; in addition, despite this range of growth, the rats do not appear obese. In order to determine the effect of *T. tridentatus* plasma on longitudinal bone growth, we assessed macroscopic and biochemical growth-related factors in an experimental pubertal rat model. The mammalian growth plate consists of three principal zones: resting, proliferative, and hypertrophic zones. In regard to the resting zone, GH begins its actions by direct stimulation of resting stem-like chondrocytes in order to start proliferation [11,22]. The following cell replication in the proliferative zone, as well as with respect to the terminal differentiation and enlargement in the hypertrophic zone, are both primarily caused by circulatory IGF-1 and locally expressed IGF-1. Moreover, IGF-1 can promote bone reconstruction and improve the activity of osteoblasts [23]. In addition, BMP-2 can promote the growth of long bones and the development of epiphyseal cartilage, as well as play an important role in biological bone remodeling and bone formation [24,25]. Moreover, ALP is an important marker of osteoblasts and an important enzyme for the purposes of promoting osteoblast differentiation and mineralization [26–28]. Therefore, IGF-1, BMP2, and ALP are applied to monitor bone growth indicators. With respect to this, the plasma group showed macroscopic changes in total longitudinal bone length after 14 days of treatment. In addition, morphological changes were observed, such as the increased height of the growth plate area and an increased number of proliferating cells. Furthermore, *T. tridentatus* plasma treatment significantly increased the total serum levels of BMP-2, IGF-1R, and IGF-1, as well as induced the upregulated expression of IGF-1 and IGF-1R in the growth plate region. Taken together, these results suggest that *T. tridentatus* plasma extract can promote cell proliferation and longitudinal bone growth in the epiphyseal plate.

In recent years, the use of proteomic techniques has emerged in the field of bone-related research. Based on iTRAQ, differential proteomics technology has been used in order to preliminarily screen the action pathway of the *T. tridentatus* whole plasma group, as well as the 10–30 kDa plasma group samples, in order to promote the growth of long bones in rats. In this study, it was found that the enrichment of proteins related to intracellular transport, secretion transport, vesicle transport, extracellular structure-function, as well as information storage and processing in the 10–30 kDa group was higher than those found in the whole plasma group.

There were examples of main enriched pathways among the differential proteins in the whole plasma group, including ribosome, endoplasmic reticulum protein processing, protein digestion, and protein absorption, as well as 20 related pathways, which were the main pathways involved in the 10–30 kDa plasma group. The differentially expressed proteins in the 10–30 kDa plasma group were mainly enriched in the PI3K-AKT signaling pathway, which were involved in the proliferation, differentiation, growth, and development of the pathway. Indeed, this pathway was speculated to be an important factor in terms of promoting bone tissue growth.

This study shows that there are still many active components in the plasma of *T. tridentatus*, such as hemocyanin and complement-3-like protein. This study establishes a research basis for the removal of high-abundance proteins from the *T. tridentatus* plasma and provides an experimental basis for the development of products that are related to the promotion of long bone growth.

4. Materials and Methods

4.1. Separation of *T. tridentatus* Plasma by Ultrafiltration

The process of *T. tridentatus* plasma ultrafiltration was performed on an ultrafiltration system with different membranes. In short, plasma was pumped with a peristaltic pump (bt100–2j, Lange Constant Flow Pump Co., Ltd. Baoding, China). All experiments were carried out with an ultrafiltration module (Bona Membrane Technology Co., Ltd., Jinan, China) in sequence with different nominal molecular weight cut-off values (i.e., 30, 10, 5, and 2 kDa) for the purposes of microporous polysulfone hollow fiber membranes (where the effective membrane area was 0.15 m^2) [29]. Ultimately, the five fractions (0–2 kDa, 2–5 kDa, 5–10 kDa, 10–30 kDa, and >30 kDa) that were obtained from the *T. tridentatus* plasma were freeze-dried under vacuum at −50 °C, then severally stored at −20 °C.

4.2. HPLC Analysis of the Five Fractions

All sample solutions were filtered through a 0.45 μm membrane filter prior to analysis. The five fractions as described above were analyzed on a Waters 2695 Series HPLC System (Milford, MA, USA), which was equipped with a PDA detector and an Agilent 300 StableBond-C18 column (Palo Alto, CA, USA, 150 × 4.6 mm, 5 μm) at 30 °C with a flow rate of 1.0 mL/min. In addition, the mobile phase was composed of 0.1% trifluoroacetic acid (A) and acetonitrile (B) eluted with the following gradient: 0–10 min (5%), 10–15 min (5%–55%), 15–25 min (55%), 25–30 min (55%–100%), and 30–40 min (100%) solvent B. The injection volume was 10 μL, and the UV spectra were recorded at 260 nm.

4.3. UPLC-Q-Exactive Orbitrap Ms Analysis

UPLC analysis was performed using an Ultimate 3000 system (Dionex, Sunnyvale, CA, USA), which was equipped with an online vacuum degasser, a quaternary pump, and an automatic sampler. Samples were separated with HYPERSIL GOLD C18 column (100 × 2.1 mm, 1.9 μm). Water that possessed 0.1% formic acid (A) and acetonitrile (B) was used as the mobile phase. Elution was carried out at a flow rate of 0.3 mL/min at 35 °C with the following gradient: 0–2 min, 2–5% (B); 2–5 min, 5–8% (B); 5–10 min, 8–15% (B); 10–15 min, 15–18% (B); 15–20 min, 18–40% (B); 20–24 min, 40–80% (B); 24–26 min, 80% (B); 26–27 min, 80–2% (B); and 27–30 min, 2% (B).

4.4. Experimental Rats and Method of Administration

A total of 140 female Wistar rats (weighting 90 ± 10 g) aged 28 days were purchased from SJA Laboratory Animal Co., Ltd (Changsha, China). These 140 Wistar rats were randomly divided into 14 groups according to their body weight, with 10 rats in each group. The rats in the positive group were subcutaneously injected with 200 μg/kg of rhGH, and each treatment was given at high and low dosages of the solutions of the different molecular weight segments of T. tridentatus plasma. The high dosage subcutaneous injection was 200 mg/kg/d, while the low dosage was 100 mg/kg/d. The rats in the control group were subcutaneously injected with an equal volume of normal saline for a period of 14 consecutive days. All the experimental protocols and procedures that were conducted complied with the regulations of the Animal Ethics Committee of the Guangxi University of Traditional Chinese Medicine, as well as followed the Guidelines of Animal Care set by the Chinese Ministry of Health (GB/T 35823-2018).

4.5. Tissue Collection

After 5 days of acclimatization, the rats were randomly divided into the following fourteen groups (n = 10 rats in each group): control, recombinant human GH (rhGH) 200 μg/kg (positive group); 0–2 kDa 200 mg/kg (0–2 kDa (H) group); 0–2 kDa 100 mg/kg (0–2 kDa (L) group); 2–5 kDa 200 mg/kg (2–5 kDa (H) group); 2–5 kDa 100 mg/kg (2–5 kDa (L) group); 5–10 kDa 200 mg/kg (5–10 kDa (H) group); 5–10 kDa 100 mg/kg (5–10 kDa (L) group); 10–30 kDa 200 mg/kg (10–30 kDa (H) group); 10–30 kDa 100 mg/kg (10–30 kDa (L) group); >30 kDa 200 mg/kg (>30 kDa (H) group), >30 kDa 100 mg/kg (>30 kDa (L) group), whole plasma 200 mg/kg (WP (H) group); and the whole plasma 100 mg/kg (WP (L) group). The rats in the plasma treatment and positive groups were subcutaneously injected with different plasma fractions, or rhGH, once daily (8:30 a.m.) for 14 consecutive days. The control group accepted the hypodermic injection with an equivalent volume of normal saline. The body weight and food intake were recorded every day after starting the experiment. Finally, on the 15th day, the rats were sacrificed in order to collect the required tissues for the following experiments. Moreover, both tibias and femurs were removed surgically.

4.6. Rat Growth Plate Height

Cresyl violet (CV) was purchased from Sigma-Aldrich (St. Louis, MO, USA), which can be used for chondrocyte staining to measure the height of growth plate. The heights of the overall growth plate, as well as the resting, proliferative, and hypertrophic zones, were measured at three different locations by ImageJ software. In addition, the proliferative zone (PZ) was measured from the flat chondrocytes aligned with the long axis of the bone that was presumed to be proliferative. The hypertrophic zone (HZ) was measured from chondrocytes that possessed nucleus and cytoplasm, which were easily distinguishable based on their size. The resting zone (RZ) was measured by subtracting the heights of the proliferative and hypertrophic zones from the overall height of the growth plate.

4.7. Enzyme-Linked Immunosorbent Assay (ELISA)

Regarding the quantitative analysis of the serum ALP, BMP-2, and IGF-1 concentrations, a sandwich assay was carried out, in duplicate, in a 96-well plate with an ELISA kit, according to the manufacturer's protocol.

4.8. Immunohistochemistry

In order to detect ALP, BMP-2, and IGF-1 expression in the growth plate, the dehydrated sagittal sections of the tibia were pretreated as per the process described previously [30]. The sections were washed, treated with biotinylated rabbit antibody diluted at 1/200, and then incubated with an avidin-biotin complex reagent diluted at 1/100 for 1 h each. Sections were developed with a 0.05% 3,3-diaminobenzidine solution that contained hydrogen peroxide.

4.9. Real-Time Quantitative Polymerase Chain Reaction (PCR) Analysis

After sacrifice, the livers of the rats were promptly removed, rinsed, and stored at −80 °C. The total RNA from the liver samples was extracted using QIAzol reagent (Qiagen, Hilden, North Westphalia, Germany) and converted into cDNA with a transcription kit (Applied Biosystems, Foster City, CA, USA) according to the instructions. A quantitative PCR was performed on a real-time PCR system. The primers were designed and synthesized by Huiyuan Technology Co., Ltd. (Jinan, China). With respect to this, the primer sequences are shown in Table 4.

Table 4. Real-time primer sequence.

Gene	Primer Sequence
IGF-1 F	5′-GCTATGGCTCCAGCATTCG-3′
IGF-1 R	5′-TCCGGAAGCAACACTCATCC-3′
IGFBP-3 F	5′-GGAAAGACGACGTGCATTG-3′
IGFBP-3 R	5′-GCGTATTTGAGCTCCACGTT-3′
IGF-1R F	5′-GGTCTCTAAGGCCAGAGGTGGA-3′
IGF-1R R	5′-GACGAACTTGTTGGCATTGAGGTA-3′
β-actin F	5′-CGTTGACATCCGTAAAGAC-3′
β-actin R	5′-TAGGAGCCAGGGCAGTA-3′

4.10. Transcriptomic Expression Profiling

Total RNA was extracted using TRIzol reagent kit (Invitrogen, Carlsbad, CA, USA) according to the manufacturer's protocol. RNA quality was assessed on an Agilent 2100 bio-analyzer (Agilent Technologies, Palo Alto, CA, USA) and checked using RNase-free agarose gel electrophoresis. After the total RNA was extracted, the eukaryotic mRNA was enriched with Oligo(dT) beads, while the prokaryotic mRNA was enriched by removing rRNA with a Ribo-ZeroTM Magnetic Kit (Epicentre, Madison, WI, USA). Then, the enriched mRNA was broken up into short fragments by utilizing a fragmentation buffer. Next, it was reverse transcribed into cDNA with random primers. The second-strand cDNA was synthesized by DNA polymerase I, RNase H, dNTPs, and a buffer. Then, the cDNA fragments were purified with a QiaQuick PCR extraction kit (Qiagen, Venlo, The Netherlands). Next, a base was added and then, finally, ligated to Illumina sequencing adapters. The ligation products were size selected via agarose gel electrophoresis, PCR amplified, and then sequenced using an Illumina Novaseq6000, by Gene Denovo Biotechnology Co. (Guangzhou, China).

4.11. Differential Gene Expression Analysis

In R software, the CPM value of the edgeR package was used for the purposes of correction, the genes with low expression in the samples were filtered, and the genes with RPKM values were output via using the RPKM function for the next analysis. The differentially expressed genes (DEGs) were identified through the limma package in R software. Furthermore, the genes meeting the cut-off criteria of $logFC > 2.5$ and $FDR < 0.05$ were regarded as DEGs. Moreover, the differentially expressed genes were visualized in volcano maps with the ggplot2 package (3rd edition) in R software (4.0).

4.12. Functional Annotation and Signaling Pathway Analysis

The GO and the KEGG analyses of the genes were conducted in order to study their main functions and participation with respect to the signaling pathways. In this paper, the clusterProfiler and enrichplot package in R software was used to explore the functions between the core genes and the involved signaling pathways. In addition, the ggplot2 package was used to visualize the output results.

4.13. Quantification of DEPs by PRM

In each group, a total of 15 DEPs were selected for the purposes of validation by PRM. Protein extraction and tryptic digestion were performed following the same protocol used for the label-free quantitative proteomics experiment.

4.14. Statistical Analysis

The statistical analysis and graphical representation of the data were carried out usingthe SPSS software, version 22.0 (SPSS Inc., Chicago, IL, USA). Further, a one-way analysis of variance was used for the purposes of multiple comparisons. The statistical significance was set at $p < 0.05$. All values were presented as mean ± SD.

Supplementary Materials: The following supporting information can be downloaded at: https://www.mdpi.com/article/10.3390/md21020111/s1, Figure S1: The Comparison of protein expression by iTRAQ and PRM; Table S1: List of proteins from MS analysis; Table S2: Significantly different expressed proteins in the control vs. whole plasma comparison; Table S3: Significantly different expressed proteins in control vs. 10–30 kDa comparison.

Author Contributions: Conceptualization, C.G., X.Y. and Y.L.; methodology, S.J.; software, S.L.; validation, J.W.; formal analysis, S.J. and X.Q.; investigation, S.J. and S.L.; resources, C.G.; data curation, S.J. and S.L.; Writing—review and editing, S.J., X.Q. and C.G.; visualization, S.J. and X.Q.; supervision, C.G., X.Y. and Y.L.; project administration, C.G. and Y.L.; funding acquisition, C.G. and Y.L. All authors have read and agreed to the published version of the manuscript.

Funding: This study was supported by the National Natural Science Foundation of China (U20A20101), the Natural Science Foundation of Guangxi Province (2020GXNSFGA297002), Special Fund for Bagui Scholars of Guangxi, Guangxi University of Chinese Medicine "Guipai Traditional Chinese Medicine Inheritance and Innovation Team" Project (2022A007), and High-level Talent of Guangxi Traditional Chinese Medicine (2022C008).

Institutional Review Board Statement: The animal study protocol was approved by the Animal Ethics Committee of the Guangxi University of Traditional Chinese Medicine, as well as followed the Guidelines of Animal Care set by the Chinese Ministry of Health (GB/T 35823-2018).

Informed Consent Statement: Not applicable.

Data Availability Statement: The data presented in this study are available upon request from the corresponding author.

Conflicts of Interest: The authors declare no conflict of interest.

References

1. Ying, Z.; Bao, Y.; Li, Y.; Ye, G.; Zhang, S.; Xu, P.; Zhu, J.; Xie, X. Impact of Different Diets on Adult Tri-Spine Horseshoe Crab, *Tachypleus tridentatus*. *J. Ocean Univ. China* **2022**, *21*, 541–548. [CrossRef]
2. Wang, W.F.; Xie, X.Y.; Chen, K.; Chen, X.L.; Zhu, W.L.; Wang, H.L. Immune Responses to Gram-Negative Bacteria in Hemolymph of the Chinese Horseshoe Crab, *Tachypleus tridentatus*. *Front. Immunol.* **2020**, *11*, 584808. [CrossRef]
3. Kang, X.; Jing, M.; Zhang, G.; He, L.; Hong, P.; Deng, C. The Ameliorating Effect of Plasma Protein from *Tachypleus tridentatus* on Cyclophosphamide-Induced Acute Kidney Injury in Mice. *Mar. Drugs* **2019**, *17*, 227. [CrossRef]
4. Wang, W.F.; Xie, X.Y.; Huang, Y.; Li, Y.K.; Liu, H.; Chen, X.L.; Wang, H.L. Identification of a Novel Antimicrobial Peptide From the Ancient Marine Arthropod Chinese Horseshoe Crab, *Tachypleus tridentatus*. *Front. Immunol.* **2022**, *13*, 794779. [CrossRef] [PubMed]
5. Liu, X.; Xu, Z.; Chang, X.; Fang, J.K.H.; Song, J.; He, J.; Tai, Z.; Zhu, Q.; Hu, M. Enhanced immunity and hemocytes proliferation by three immunostimulants in tri-spine horseshoe crab *Tachypleus tridentatus*. *Fish Shellfish Immunol.* **2021**, *115*, 112–123. [CrossRef]
6. Li, G.; Cheng, J.; Mei, S.; Wu, T.; Ye, T. *Tachypleus tridentatus* Lectin Enhances Oncolytic Vaccinia Virus Replication to Suppress In Vivo Hepatocellular Carcinoma Growth. *Mar. Drugs* **2018**, *16*, 200. [CrossRef] [PubMed]
7. Chen, S.C.; Yen, C.H.; Yeh, M.S.; Huang, C.J.; Liu, T.Y. Biochemical properties and cDNa cloning of two new lectins from the plasma of Tachypleus tridentatus: Tachypleus plasma lectin 1 and 2+. *J. Biol. Chem.* **2001**, *276*, 9631–9639. [CrossRef] [PubMed]
8. Bouvattier, C.; Martinerie, L.; Vautier, V. The Year in Growth and Short Stature. *Arch. Pediatr.* **2022**, *28* (Suppl. S1), 28/8S21–28/8S26. [CrossRef]
9. Frost, H.M.; Schonau, E. On longitudinal bone growth, short stature, and related matters: Insights about cartilage physiology from the Utah paradigm. *J. Pediatr. Endocrinol. Metab.* **2001**, *14*, 481–496. [CrossRef]
10. Ning, B.H.; Londono, I.; Laporte, C.; Villemure, I. Validation of an in vivo micro-CT-based method to quantify longitudinal bone growth of pubertal rats. *Bone* **2022**, *154*, 116207. [CrossRef]
11. Blutke, A.; Schneider, M.R.; Wolf, E.; Wanke, R. Growth hormone (GH)-transgenic insulin-like growth factor 1 (IGF1)-deficient mice allow dissociation of excess GH and IGF1 effects on glomerular and tubular growth. *Physiol. Rep.* **2016**, *4*, e12709. [CrossRef]
12. Choi, H.S.; Kwon, A.; Suh, J.; Song, K.; Chae, H.W.; Kim, H.S. Effect of long-acting growth hormone treatment on endogenous growth hormone secretion in prepubertal patients with idiopathic short stature: A preliminary study. *Growth Horm. IGF Res.* **2022**, *66*, 101486. [CrossRef]
13. Pillai, S.S.; Topor, L.S. 50 Years of Growth in the Assessment of Short Stature. *J. Pediatr.* **2021**, *232*, 94.

14. El-Shafie, A.M.; Kasemy, Z.A.; Omar, Z.A.; Alkalash, S.H.; Salama, A.A.; Mahrous, K.S.; Hewedy, S.M.; Kotb, N.M.; Abd El-Hady, H.S.; Eladawy, E.S.; et al. Prevalence of short stature and malnutrition among Egyptian primary school children and their coexistence with Anemia. *Ital. J. Pediatr.* **2020**, *46*, 91. [CrossRef] [PubMed]
15. Patel, R.; Bajpai, A. Evaluation of Short Stature in Children and Adolescents. *Indian. J. Pediatr.* **2021**, *88*, 1196–1202. [CrossRef] [PubMed]
16. Kwan, K.Y.; Bopp, J.; Huang, S.; Chen, Q.; Wang, C.C.; Wang, X.; Zhen, W.; Zhu, J.; Huang, X. Ontogenetic resource use and trophic dynamics of endangered juvenile Tachypleus tridentatus among diversified nursery habitats in the northern Beibu Gulf, China. *Integr. Zool.* **2021**, *16*, 908–928. [CrossRef]
17. Nakamura, T.; Furunaka, H.; Miyata, T.; Tokunaga, F.; Muta, T.; Iwanaga, S.; Niwa, M.; Takao, T.; Shimonishi, Y. Tachyplesin, a class of antimicrobial peptide from the hemocytes of the horseshoe crab (*Tachypleus tridentatus*). Isolation and chemical structure. *J. Biol. Chem.* **1988**, *263*, 16709–16713. [CrossRef]
18. Kawano, K.; Yoneya, T.; Miyata, T.; Yoshikawa, K.; Tokunaga, F.; Terada, Y.; Iwanaga, S. Antimicrobial peptide, tachyplesin I, isolated from hemocytes of the horseshoe crab (*Tachypleus tridentatus*). NMR determination of the beta-sheet structure. *J. Biol. Chem.* **1990**, *265*, 15365–15367. [CrossRef]
19. Nakamura, S.; Iwanaga, S.; Harada, T.; Niwa, M. A clottable protein (coagulogen) from amoebocyte lysate of Japanese horseshoe crab (Tachypleus tridentatus). Its isolation and biochemical properties. *J. Biochem.* **1976**, *80*, 1011–1021. [CrossRef]
20. Wu, F.; Xie, Z.; Yan, M.; Li, Q.; Song, J.; Hu, M.; Wang, Y. Classification and characterization of hemocytes from two Asian horseshoe crab species *Tachypleus tridentatus* and *Carcinoscorpius rotundicauda*. *Sci. Rep.* **2019**, *9*, 7095. [CrossRef]
21. Kwan, B.K.; Chan, A.K.; Cheung, S.G.; Shin, P.K. Responses of growth and hemolymph quality in juvenile Chinese horseshoe crab *Tachypleus tridentatus* (Xiphosura) to sublethal tributyltin and cadmium. *Ecotoxicology* **2015**, *24*, 1880–1895. [CrossRef] [PubMed]
22. Soliman, A.; Rogol, A.D.; Elsiddig, S.; Khalil, A.; Alaaraj, N.; Alyafie, F.; Ahmed, H.; Elawwa, A. Growth response to growth hormone (GH) treatment in children with GH deficiency (GHD) and those with idiopathic short stature (ISS) based on their pretreatment insulin-like growth factor 1 (IGFI) levels and at diagnosis and IGFI increment on treatment. *J. Pediatr. Endocrinol. Metab.* **2021**, *34*, 1263–1271. [CrossRef] [PubMed]
23. Tripathi, T.; Gupta, P.; Rai, P.; Sharma, J.; Gupta, V.K.; Singh, N.; Verma, M. Longitudinal evaluation of the association between Insulin-like growth factor-1, Bone specific alkaline phosphatase and changes in mandibular length. *Sci. Rep.* **2019**, *9*, 11582. [CrossRef]
24. Sotunde, O.; Brett, N.; Agellon, S.; Vanstone, C.; Weiler, H. Associations of Insulin-like Growth Factor-1, Insulin-like Growth Factor Binding Protein-3, Bone and Body Composition Variables in Children 2 to 8 y. *J. Bone Miner. Res.* **2018**, *33*, 152.
25. Hsu, S.L.; Chou, W.Y.; Hsu, C.C.; Ko, J.Y.; Jhan, S.W.; Wang, C.J.; Lee, M.S.; Hsu, T.C.; Cheng, J.H. Shockwave Therapy Modulates the Expression of BMP2 for Prevention of Bone and Cartilage Loss in the Lower Limbs of Postmenopausal Osteoporosis Rat Model. *Biomedicines* **2020**, *8*, 614. [CrossRef]
26. Richardson, V.; Govil, A.; Choi, B. OsseoGEN((R)): ALP Promoting Osteoinductive Enriched Viable Cellular Bone Matrix. *Tissue Eng. Part A* **2022**, *28*, 251–252.
27. Abdelfattah, H.N.; Lehenkari, P.; Hukkanen, J.; Hakkola, J. Rifampicin induces the bone form of alkaline phosphatase (ALP) in humans. *Toxicol. Lett.* **2019**, *314*, S194.
28. Llorente-Pelayo, S.; Docio, P.; Lavin-Gomez, B.A.; Garcia-Unzueta, M.T.; de las Cuevas, I.; de la Rubia, L.; Cabero-Perez, M.J.; Gonzalez-Lamuno, D. Modified Serum ALP Values and Timing of Apparition of Knee Epiphyseal Ossification Centers in Preterm Infants with Cholestasis and Risk of Concomitant Metabolic Bone Disease of Prematurity. *Nutrients* **2020**, *12*, 3854. [CrossRef]
29. Xie, J.H.; Shen, M.Y.; Nie, S.P.; Zhao, Q.; Li, C.; Xie, M.Y. Separation of water-soluble polysaccharides from Cyclocarya paliurus by ultrafiltration process. *Carbohydr. Polym.* **2014**, *101*, 479–483. [CrossRef]
30. Cui, D.P.; Zhao, D.W.; Huang, S.B. Structural characterization of a safflower polysaccharide and its promotion effect on steroid-induced osteonecrosis in vivo. *Carbohyd. Polym.* **2020**, *233*, 115856. [CrossRef]

marine drugs

MDPI

Review

Unveiling the Anti-Aging Potential of Marine Natural Bioproducts

Nedeljka Rosic [1,2]

[1] Faculty of Health, Southern Cross University, Gold Coast, QLD 4225, Australia; nedeljka.rosic@scu.edu.au
[2] Marine Ecology Research Centre, Southern Cross University, Lismore, NSW 2480, Australia

Abstract: Aging is a natural process resulting in the progressive impairment of multiple functions in the human body, leading to a decline in cellular functionality and the development of aging-related diseases. External stress factors, such as ultraviolet (UV) radiation, pollution, and toxin exposure, increase oxidative stress, damage cellular repair mechanisms, and speed up aging processes. With the rise in the world's aging population, there are enlarged demands for the use of sustainable natural products in food, nutrient supplements and cosmetics that can slow down aging and prolong healthy life and longevity. Algae, including both macroalgae and microalgae, have been recognised as a source of valuable proteins, amino acids, fatty acids, vitamins, and minerals useful for human consumption and medical applications. With increasing demands for nutraceutical and pharmaceutical bioproducts from environmentally friendly resources, the biotechnological industry, over recent decades, has had to provide new, advanced solutions using modern high-throughput omics technologies. The application of proteomics in the area of discoveries of natural products with anti-aging properties has become more popular for wide industry applications. New proteomics profiling provides a better understanding of changes occurring in protein and peptide content, their structure, function and interactions, as well as the regulatory processes and molecular pathways. Mass spectrometry-based proteomics has been used for a wide range of applications including protein identification, characterisation, as well as quantification of proteins within the proteome and sub-proteome. The application of chemical proteomics facilitated the identification of natural products approach and included the synthesis of probes and target fishing, allowing the advanced identification of proteins of interest. This review focuses on marine macro- and microalgal anti-aging compounds and novel proteomics approaches, providing recent experimental evidence of their involvement in anti-aging processes that should facilitate their use in innovative approaches and sustainable biotechnological applications.

Keywords: algae; bioactive compounds; anti-aging; skin care; proteomics

1. Introduction

Natural compounds, including their structural analogues, historically played an important role in drug discovery and in treating various infectious diseases, as well as other malicious conditions [1]. In civilizations worldwide, natural products have been used as a source of food, water, and energy since ancient times [2]. The interest in applying natural bioactive compounds in skin care products has also risen, along with the availability of marine ingredients [3]. Global industry development and an increase in the world population, with more than 1 billion people living in highly nature-dependent tropical regions, has resulted in increasing demands for natural products worldwide [4]. As the world population is estimated to rise to nearly 10 billion people by 2050, higher demands for natural resources, specifically food resources, are anticipated [5].

The demand for anti-aging products has tremendously increased over the past decade, with a global market valued at more than USD 48 billion [6]. The forecast analyses of future trends in the anti-aging product market for 2025–2033 predict further growth, reaching USD 82 billion by 2033 [6]. With increasing awareness of aging processes and the benefit of using anti-aging products, especially for skin care [7], plus the estimated rise in the world's elderly population, reaching 2.1 billion by 2050 [8], the need for anti-aging products is likely to grow exponentially in the future.

The marine environment covers more than 70% of the planet and is the world's largest ecosystem, encompassing extreme conditions, including high salinity, very low and high temperatures, and low and high light conditions [9,10]. With limited land and freshwater availability and further climate challenges, new environmentally sustainable resources are required, and the advanced use of algae presents a green industry of the future that could minimise the use of fossil fuels [11,12]. The biodiversity of marine organisms has been used as a rich source of natural products, extracted from microorganisms and phytoplankton, red, green, and brown macroalgae, and animals such as sponges, cnidarians, molluscs, tunicates, echinoderms, and other marine groups [13]. The application of marine algae and its derivatives has been recognised as cheap, freely available, easily scalable, and capable of meeting the rising needs of natural products [1,14]. The sustainable advantages of using algae for mass production include lowering CO_2 emissions, reducing the use of limited agricultural lands, generating products with high protein content and other nutritious values, and protecting biodiversity [5]. Algae are photosynthetic organisms capable of successfully growing in diverse environmental conditions, demonstrating phenotypic plasticity and the ability to endure and adapt to stressful environmental conditions such as extreme temperatures and UV radiation [9,15–18]. Algae, classified into multicellular macroalgae and unicellular microalgae, are recognised for their various biotechnological applications, such as in human and animal food, as a source of bioactive compounds with rich therapeutic properties important for new drug discoveries, as well in other industry applications and innovations [19,20].

Macroalgae, known as seaweed, are found in marine and freshwater environments and have diverse potential in biotechnology [21]. Based on their pigmentation and chemical signature, marine macroalgae can be divided into three main groups: brown seaweed (Phaeophyceae), red seaweed (Rhodophyceae), and green seaweed (Chlorophyceae). Since ancient times, macroalgae have been used in human and animal foods across the globe, from Asia to America [22]. Recently, due to promising nutritional and pharmaceutical values, with rich content of proteins and other promising bioactive compounds, macroalgae applications extended toward various pharmacologically critical applications in multiple industries, from bioremediation to other ecological roles [23,24]. Therefore, seaweeds have been produced on a larger scale in some Asian countries due to increasing demands for marine natural products [25].

Microalgae, single-cell photosynthetic microorganisms, are genetically diverse and found in marine, freshwater, and soil ecosystems [26]. These unicellular organisms can rapidly and flexibly grow in a wide range of conditions [27]. Consisting of diverse valuable products, microalgae have been used in the human diet [26] as a source of drugs and nutrients [28–30]. From a sustainable perspective, microalgal species present a promising alternative to fossil fuels as a new biofuel, while using lower environmental resources and being capable of mass production in even marginal areas using wastewater [31]. Multiple microalgal species, such as *Chlorella* sp. and *Arthrospira* sp., have a high protein content as a sustainable source of edible proteins for food consumption [5,32], while others have been recognised for biofuel potential [33,34]. The recent development of new technologies and approaches has aimed to increase the industry application of micro-algae, for example, to improve protein digestibility [5] and its use in human food [20]. However, consumer acceptance still presents some challenges [35]. The application of polysaccharides from microalgae has been targeted to be improved by enhancing its production, extraction, and purification conditions, specifically targeting exopolysaccharides due to their pharmacological properties [36]. Other biotechnological applications of microalga include the use of phenolic compounds and carotenoids [37], sources of vitamins [29], as well as for reducing carbon footprints [12] and biofuel production [27,33].

The majority of marine ingredients that are used in anti-aging cosmetics products come from algae, including red, brown, and microalgae [3]. Consequently, macro- and microalgal species have shown great potential as sources of natural products for further applications in cosmetics, anti-aging products, and other biotechnological applications (Figure 1). The natural process of aging leads to the body's functional qualities gradually deteriorating, lowering fitness and increasing the risk of developing chronic illnesses [35]. Degeneration of the main body functions occurs, and aging is sped up by exposure to environmental stressors (e.g., UV radiation, toxin exposure), hereditary factors, and other external factors [36]. Human skin, beyond its role in protection, presents the largest organ that exists, and with aging, it loses its elasticity and moisturising ability [3,37]. Damaging UV radiation, pollution, and exposure to other environmental toxins are examples of external stressors raising oxidative stress, harming cellular repair systems, accelerating cellular death, and promoting aging [15,38,39]. As the world's population ages, there is a greater need for natural solutions that can be applied topically or consumed through diet to slow down the aging process and extend longevity. These contemporary developments are putting more and more pressure on the biotechnology sector to employ cutting-edge, high-throughput omics technologies and offer novel solutions using eco-friendly, sustainable resources.

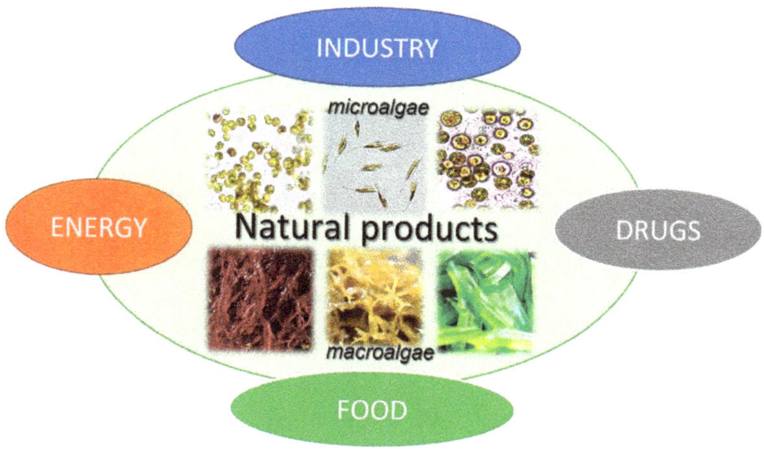

Figure 1. Applications of natural marine products in the 21st century.

In this review, the aims were (1) to discuss recent biotechnological trends in anti-aging natural products from marine algae, (2) to provide an overview of new movements in natural anti-aging products, and (3) to evaluate the use of modern technology, such as proteomics in discoveries of anti-aging natural products.

2. Aging

Aging is a natural and complex process that includes the incremental over-time accumulations of errors in cellular structures, tissues, and organs, leading to a decrease in body fitness and the development of chronic diseases [40]. The number of age-related diseases will continue to rise with the constant increase in the proportion of the aging population (Figure 2). The major factors contributing to the development of aging include internal factors such as genetic predisposition and telomere shortening and also exposure to certain detrimental external factors like toxins and radiation, which have been found to cumulatively impact the progression of aging [41]. Specific signs of aging include molecular signs (e.g., genomic instability, epigenetic changes, mitochondrial dysfunction), as well as cellular (e.g., cellular senescence) and systemic changes [35]. Genetic factors have been proposed to contribute to around 25% of the variability in human longevity [42], with more than 50 genes (e.g., apolipoprotein E gene and gene *FOXO3* [43]) identified via genome-wide studies to have implications for human aging and longevity, with epigenetic changes also playing a key role in regulatory cellular processes of aging [44]. Accumulation of senescent cells and telomere shortening are also recognised intrinsic factors of aging [45]. Telomeres, as repetitive DNA sequences, are found at the end of chromosomes, and they have a protective role in preserving coding parts of the chromosomes during cell replication [46]. As cells proliferate, telomere shortening occurs over time, limiting the number of possible cell divisions and later resulting in mitochondrial dysfunction, genome instability, apoptosis, accumulation of senescent cells, and aging [45,47]. Oxidative stress due to the accumulation of reactive oxygen species (ROS) also leads to genome instability due to damaging impacts on DNA and other macromolecules, contributing to telomere dysfunction and cellular senescence [48]. The immune system declines with aging, resulting in chronic inflammation and a lowering of the body's ability to fight infection [40]. Geriatric syndromes and chronic diseases such as cardiovascular, osteoarthritis, diabetes, and neurodegeneration following dementia, Alzheimer's, and Parkinson's diseases are commonly developed in aging populations [49].

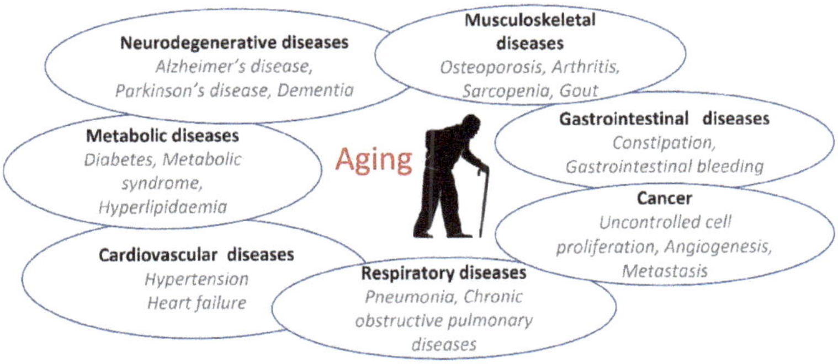

Figure 2. The schematic overview of aging-related diseases.

3. Anti-Aging Compounds

Due to beneficial pharmacological activities, various anti-aging compounds from natural resources have been explored [3,36,50]. A number of algal species have shown a potential to be used in cosmetics, presenting cost-effective and sustainable opportunities to be utilised in biotechnology [3]. Metabolites obtained from marine resources such as seaweeds and microalgae demonstrated significant potential anti-aging properties for use in cosmetics and other applications (Table 1). Bioactive compounds and their derivatives were shown to delay aging and prolong life via maintaining redox homeostasis, reducing oxidative stress, anti-inflammatory actions, and UV-absorbing properties [50,51]. Useful anti-aging compounds found in algae comprise polyunsaturated fatty acids, vitamins, polyphenols, mycosporine-like amino acids, and trace elements and minerals (Table 1). These diverse bioactive molecules are often found as a mixture of beneficial anti-aging compounds in various algal species, such as brown algae kelp, that enhance wound healing and skin regeneration [52] or red macroalgae, like *Porphyra linearis* and *Phycocalidia acanthophora*, that contain antioxidants and anti-inflammatory and UV-absorbing molecules [53]. Another red macroalgae, *Hypnea musciformis*, which is widely distributed around the Mediterranean Sea, contains a range of pharmacologically promising bioactive molecules (i.e., polysaccharides, flavonoids, and phlorotannins) with anticancer effects reported for the liver and intestinal cancer cells [54,55]. Red seaweed *Asparagopsis taxiformis* extracts have antioxidative, antiviral, and anticancer properties [56], as well as anti-methanogenic activity that is found to be useful for the reduction and mitigation of methane-associated climate change [57].

3.1. Polyunsaturated Fatty Acids

Polyunsaturated fatty acids (PUFAs) encompass omega-3 (n-3) and omega-6 (n-6) fatty acid classes and are essential nutrients that are used in human and animal food [58]. Very long PUFAs (20 or 22 carbon chains) from algae are of particular industry interest [59]. PUFAs were reported in microalgal species such as *Chlorella* sp., *Nannochloropsis oculata*, *Pseudochoricystis ellipsoidea*, and *Botryococcus braunii* [60,61], as well as various seaweed species coming from green, red, and brown macroalgae [62,63]. Eicosapentaenoic acid (EPA) and docosahexaenoic acid (DHA) omega-3 fatty acids are beneficial PUFAs that are now successfully sourced from seaweeds instead of fish, which helps in an attempt to protect marine ecosystems from overfishing [64]. In PUFA-rich microalgae *Phaeodactylum tricornutum*, n-3 fatty acids were reported at a concentration of 58 mg/g, which is more than double compared to that in fish (21 mg/g), with beneficial EPA at 53 mg/g, while fish contains 7 mg/g [65]. A combination of n-3 and n-6 PUFAs and their appropriate ratios were determined to be beneficial for reducing body inflammation, as well as cardiovascular and nervous system disorders [62]. PUFAs isolated from various macroalgae (*Ulva lactuca*, *Chondrus crispus*, *Laminaria hyperborea*, *Fucus serratus*, *Undaria pinnatifida*, *Palmaria palmata*, *Ascophyllum nodosum*, *Caulerpa taxifolia*, and *Sargassum natans*) were within the concentration of 2–14 mg/g dry matter (DM) and the n-6: n-3 ratio of around 1, which is within the recommended range of 5 or below [66] for beneficial anti-inflammatory actions and a positive pharmacological impact on nervous and cardiovascular systems [62].

Mechanisms of PUFA actions that contribute to anti-aging processes include maintaining redox homeostasis in vivo, decreasing oxidative stress, reducing telomere shortening and downregulating the antioncogene expression [67]. Multiple studies have shown the benefits of taking PUFAs for prevention and/or treatment such as in the case of age-related musculoskeletal diseases (i.e., sarcopenia, osteoarthritis, and osteoporosis) [68] and neurodegenerative diseases, like Alzheimer's disease (AD) and Parkinson's disease (PD) [58]. In the case of sarcopenia, there is an age-related loss of strength and muscle mass caused by oxidative stress and the accumulation of ROS in the mitochondria of skeletal muscle cells, leading to mitochondrial dysfunction, and speeding up telomere depletion and cellular senescence [69]. In human studies conducted on older adults, improved skeletal muscle growth and functionality were reported, leading to a lower risk of developing sarcopenia due to the consumption of omega-3 fatty acids compared to the control group [70,71]. In studies in vitro, PUFAs were found to have a cytoprotective effect via lowering the level of ROS by increasing the cellular activities of antioxidative enzymes superoxide dismutase and glutathione peroxidase in cells, resulting in their cytoprotective effect and repair of mitochondrial function [72]. Mechanisms of PUFA action in sarcopenia prevention beyond the reduction of oxidative stress also include reduction of inflammation and improvement of mitochondrial health [68]. Animal studies, in the case of age-related neurogenerative diseases, provided evidence linking higher brain DHA concentration with improved cognitive health, highlighting the importance of PUFAs in neuroprotection and possible prevention of AD and PD [73].

3.2. Vitamins

Marine algae accumulate various vitamins, and their precursors with particularly high concentrations of vitamin A (especially precursor β-carotene), but also B1, B2, B6, B12, C, D, and E reported in microalgae in the mature developmental state [28,74], as well as in macroalgal species [21]. Due to rapid growth, microalgae were found to be a potentially very good sustainable source of vitamin D, with additional biotechnological adjustments further enhancing vitamin D production [75]. Exposure to UVB radiation was found to stimulate the synthesis of vitamin D [76], with genetic manipulation and calcium homeostasis also positively impacting vitamin D production under heat stress and nitrogen deprivation [75]. High vitamin quantities were reported per dry weight for *Tetraselmis suecica* (4280 μg/g provitamin A and 6323 μg/g vitamin E), and *Pavlova lutheri* (1162 μg/g vitamin B_{12} and 837 μg/g vitamin C) [77]. Microalgae were also found to be valuable sources of Vitamin E, accumulating tocopherols and tocotrienols forms, useful in cellular protection and chemoprevention [78]. Carnitine is a non-essential amino acid that can be synthesised in the human body from other essential amino acids (e.g., lysine and methionine), and it also needs sufficient levels of vitamins and trace elements as co-factors (e.g., ascorbic acid—vitamin C, ferrous iron, and niacin—vitamin B) [79]. Deficiency in any of these vitamins and trace elements may lead to carnitine deficiencies impacting fatty acid metabolism [80], leading to oxidative stress and aging [81]. There are multiple modes of vitamin actions that contribute to anti-aging processes, including antioxidative action and positive effects on the gut microbiota and gut health [40]. The neuroprotective effect was reported for vitamin E via the reduction of oxidative stress that led to a lower number of p53-positive brain cells [82], while vitamin C combined with some active mixes (containing other bio-peptides, and hyaluronic acid) reduced oxidative stress and stimulated collagen synthesis [83].

3.3. Trace Elements and Minerals

During aging, even in healthy elderly people, changes are occurring in the levels of trace elements (TEs), with a drop in levels of zinc (Zn), selenium (Se), and manganese (Mn) and a rise in iron (Fe), copper (Cu), and iodine (I) concentrations [84]. The lower levels of trace elements and minerals were found to be linked with premature signs of aging, such as premature greying, oxidative stress, inflammation, and development of chronic diseases [36,85]. Significantly lower levels of Zn and vitamin D were reported in the blood serum of participants with diagnosed acne conditions compared to the control group [86]. Skin conditions like inflammation and acne vulgaris were found to benefit from Zn supplements [48], indicating the importance of this TE in skin health. Valuable resources of TEs and minerals were found in edible seaweeds with high amounts of macro minerals such as sodium (Na), potassium (K), calcium (Ca), and magnesium (Mg), and TEs (e.g., Fe, Zn, Mn, Cu) [23,87,88]. In the biomass of five different microalgae (*Porphyridium cruentum, Isochrysis galbana, Phaeodactylum tricornutum, Tetraselmis suecica,* and *Nannochloropsis gaditana*), major minerals and TEs were found in good quantities, including other valuable ingredients [89].

3.4. Polyphenols

Polyphenols (PPs) are micronutrients found in macro- and microalgae that are used in the human diet [90]. These secondary metabolites are highly hydrophilic and are divided into different groups depending on the phenol ring numbers. PPs such as phenolic acids, phlorotannins, and flavonoids possess beneficial pharmacological properties ranging from antioxidant to anti-inflammatory and anticancer that are important in reducing aging and the impact of age-related diseases [91]. During the aging process, oxidative stress and inflammation may lead to DNA epigenetic modifications, impacting the level of DNA methylation and histone content variations and resulting in changes in gene expression [91,92]. Dietary PPs were found to impact these age-related changes via their antioxidative, anti-inflammatory, and anticancer activities [92]. PPs, such as phlorotannins, isolated from marine brown macroalgae *Laminaria hyperborea*, were found to promote wound healing [93] and to act as antioxidants by scavenging harmful ROS produced under UV radiation [94]. Phlorotannins are a heterogenous group of polymerised phenolic compounds (i.e., phloroglucinol, eckol, fucodiphloroethol G, phlorofucofuroeckol A, 7-phloroeckol, dieckol, and 6,6′-bieckol) [95], which are uniquely found in brown algae and are characterised by high antioxidative potential, as well as anti-allergic, anti-inflammatory, anticancer activity and the role in neuroprotection [96]. The antioxidative activity of phlorotannins was found to minimise UV-induced oxidative stress in large kelps [94], while in extracts isolated from Irish macroalgae *Fucus serratus*, antioxidant capacity was enhanced via the application of reverse-phase flash chromatography fractionation [97]. Flavonoids (e.g., flavones, isoflavones, anthocyanidins, and others) also demonstrate beneficial pharmacological properties, including antioxidant, anti-inflammatory, and anticancer [91], with both microalgae and macroalgae being a rich source of these PPs [98]. PPs are recognised for their anti-viral properties, especially effective on enveloped viruses (i.e., targeting spike and membrane viral proteins), with a potential for being in the pharmaceutical industry for the treatment of viral diseases [99]. In addition, anti-bacterial activities were reported for PP extracts isolated from seaweeds *Corallina officinalis*, *Ulva lactuca*, and *Pterocladia capillacea* [100]. Polyphenol extracts isolated from seaweeds were found to have a beneficial neuroprotective role as antioxidants contributing to the prevention of age-related diseases such as cancer, cardiovascular diseases, arthritis, and other challenges of aging [91], as well as in neurodegenerative diseases such as Alzheimer's disease and Parkinson's disease [101]. Ethyl acetate fractions isolated from the brown algae *Padina tetrastromatica* that were tested on human skin cell line (i.e., K-MEL-28) were not toxic at the concentration of up to 100 μg/mL, while demonstrating positive photoprotective and anti-aging potential with a strong sun protection factor (SPF) [102]. Major PPs identified in this ethyl acetate fraction were phlorotannins capable of absorption of UVA and UVB and antioxidative action by mopping up ROS produced under UV exposure [103,104]. PPs isolated from the red macroalgae *Symphyocladia latiuscula* demonstrated a neuroprotective effect when tested on animals with diabetic peripheral neuropathy [105], while edible brown and green macroalgae polyphenol extracts showed high antioxidant activity and moderate anti-bacterial activities [106]. Microalgae are also a promising source of PPs with concentration variation impacted by external conditions [107], indicating the possible biotechnological potential for increasing mass production under controlled cultivation conditions for future sustainable production of PPs.

Table 1. Common anti-aging compounds extracted from algae.

Compound Types	Compounds	Example (Molecular Formula, Structure)	Modes of Anti-Aging Action [REF]	Sources (REF)
Polyunsaturated fatty acids (PUFAs)	Omega-3 PUFA Omega-6 PUFA	Eicosapentaenoic acid ($C_{20}H_{30}O_2$, 20:5) 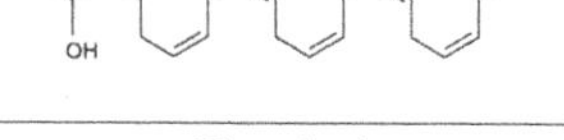	Maintaining the in vivo redox homeostasis; by lowering oxidative stress and reducing telomere shortening; by down-regulating the antioncogene expression [67].	Macroalgae [62] Microalgae [59,108]
Vitamins	Vitamins A, B, C, D, E	Vitamin A ($C_{20}H_{30}O$) 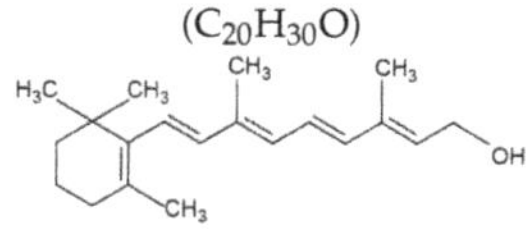	Antioxidative; modulating the gut microbiota; improving inetene function; neuroprotective effect; enhancing fat metabolism [50]	Macroalgae [21] Microalgae [29]

Table 1. *Cont.*

Compound Types	Compounds	Example (Molecular Formula, Structure)	Modes of Anti-Aging Action [REF]	Sources (REF)
Trace elements (TEs) and minerals	Zinc, copper, selenium, sodium, potassium, calcium	Zn, Cu, Se, Na, K, Ca	Maintaining the in vivo cellular homeostasis; metalloenzymes [36]	Macroalgae [87,88] Microalgae [74]
Polyphenols (PPs)	Flavonoids, phlorotannins, phenolic acids, stilbenes, lignans	Phlorotannin: phloroglucinol ($C_6H_6O_3$) 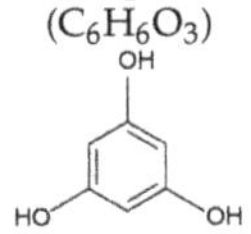	Antioxidant; anti-inflammatory; anticancer properties [91]	Macroalgae [99,100,105,106] Microalgae [107,109]
Mycosporine-like amino acids (MAAs)	Mycosporine glycine, shinorine, porphyra-334, mycosporine-2-glycine, palythine	MAA direct precursor: 4-deoxygadusol ($C_8H_{12}O_5$) 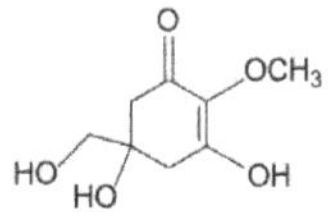	UV-absorbing property (max range 310–360 nm); antioxidative; anti-inflammatory; anti-adipogenic [51,110]	Macroalgae [111–114] Microalgae [115–118]
Marine algal polysaccharides (MAPs)	Alginate, carrageenan, fucoidan, ulvan, and laminarin	Fucoidan ($(C_6H_9O_3SO_3)_n$) 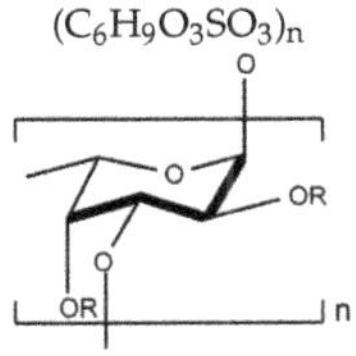	Antioxidants; anti-inflammation; antitumor [119,120]	Macroalgae [53,121] Microalgae [122,123]

3.5. Amino Acids

Amino acids, the building blocks of proteins, are crucial for cellular signalling and could influence the aging process [124]. Mycosporine-like amino acids (MAAs) are small, low-molecular-weight (<400 Da), water-soluble UV-absorbing molecules that are produced via two pathways, shikimate and/or phosphate pathways [125–128]. MAAs are found to be temperature and light-stable compounds that exhibit strong photoprotective properties and the capacity to absorb UVA (320–400 nm) and UVB (280–320 nm) light without the production of harmful free radicals [116,129–131]. MAAs have high molar extinction coefficients with maximum UV absorption between 310 nm and 362 nm [114,132,133].

From MAA direct precursor, 4-deoxygadusol (4-DG), MAA biosynthetic processes lead to further synthesis of more than 30 various molecules that are divided into primary and secondary MAAs [134]. Selected primary MAAs, commonly studied for biotechnological applications, are presented in Table 2. With UV-absorbance capacities within UVA and UVB ranges, MAAs can effectively release absorbed radiation as heat, lacking the production of ROS [135]. These free radicals such as superoxide anion (O_2^-), hydrogen peroxide (H_2O_2), and hydroxyl radicals (OH^-), are also by-products of normal cellular metabolism and, in smaller quantities, they play a role as signalling molecules and in other ordinary physiological processes such as wound healing and repair of tissue [136]. Beyond the role in cellular communication, when present in higher concentration (e.g., in response to UV radiation or other stress factors), elevated levels of ROS may lead to oxidative stress, DNA damage, and cellular aging [137]. Additionally, MAAs have an antioxidative role by mopping up excessive ROS from the body and, therefore, enhancing tolerance to a number of abiotic stress factors [138]. These secondary metabolites are also temperature-

and photostable, acting in the prevention of UV-induced thymine dimer formation from damaging DNA molecules [139–141]. Therefore, as MAAs play an important role in cellular protection from abiotic stress factors, these molecules have been recognised for the potential to be used in cosmetics for UV skin protection as organic sunscreens [51].

The concentration and profile of MAAs vary in different algal species but are influenced by variations in external environments, including UV levels, temperature, salinity, and nutrients [130,138,142]. For example, seasonal variations in temperature and UV levels were found to impact MAA content and profile in red algae *Iridaea tuberculosis*, *Nothogenia fastigiate*, and *Corallina officinalis*, with MAA content reaching from 0.4 to 1 mg/g of dried mass weight (DW) depending on the season and the species [142]. Nitrogen (N) status was also found to impact the MAA profile, with decreased shinorine levels and increased polythene content in red algae *Asparagopsis armata* grown in tanks when N fluxed was raised [143]. Enriched nutrient conditions (i.e., in nitrogen and phosphorous ions) and elevated UV levels also led to an increase in MAA levels in red macroalgae *Hydropuntia cornea*, *Gracilariopsis longissima*, and *Halopithys incurva* [144]. Generally, the highest MAA levels (above 2 mg/g of DW) were reported in red macroalgae (above 10 mg/g DW) [145] with representative species from the orders Gracilariales, Bangiales, Gelidiales, and Gigartinales [111]. In microalgae, MAA concentration also showed seasonal variation with a positive impact on UV radiation on the accumulation of MAAs [142]. In phytoplankton and copepod *Cyclops abyssorum tantric* sampled from an alpine lake, there was a substantial increase of more than threefold in MAA content during summer compared to winter [146]. The conditions used during MAA extraction and final extract storage were also important for the yield of MAAs and could significantly influence MAA content estimation [147]. In red macroalgae *Phycocalidia acanthophora* and *P. linearis*, the levels of desirable bioactivities varied in different types of extracts (i.e., aqueous extracts vs. the hydroethanolic solution), indicating the importance of extraction procedures for optimisation and improving targeted activities [53].

The anti-aging effects of MAAs have been mainly assessed in primary MAAs, including shinorine, porphyra-334, palythine, and mycosporine-glycine (Table 2). MAAs were confirmed to have anti-inflammatory, antioxidative, and UV-absorbing activities, with anti-aging and skin-firming properties that are beneficial for cosmetic products [51,110,148,149]. MAAs are considered environmentally friendly sunscreens found in edible seaweeds and many other aquatic species [111,116,131,150–152]. For assessing anti-aging activities, cellular lines used included human keratinocyte cell line HaCaT and 2,2-diphenyl-1-picryhydrazyl (DPPH) assay [153]. Procollagen levels in human skin fibroblast were evaluated via an ELISA assay [154], while anti-inflammatory and antioxidant effects of MAAs were confirmed on macrophages [155], with immunomodulatory effects [156]. The application of MAAs in cosmetics as more efficient sunscreens with advanced UV-absorbing and anti-aging properties is an exciting area that still needs to be further explored [157,158].

3.6. Polysaccharides

Marine algal polysaccharides (MAPs) are essential molecules building cell walls in marine algae [120]. In macroalgae, these macromolecules are quite abundant, reaching from 15% up to 76% of DW [63], while in microalgae, MAPs were reported to make up to 20% of DW [159]. Pharmaceutical activities reported for MAPs include antioxidative, anti-inflammatory, and anti-tumour, which are beneficial for anti-aging, as well as antiviral and hypolipidemic [120]. Polysaccharides are biopolymers made from monosaccharides (i.e., simple sugar) linked with glycosidic bonds. These molecules, beyond being used as food, have a wide industry application, including roles as emulsifiers, thickeners, and stabilisers [160], with some MAPs produced at massive scale (e.g., agar, alginate, starch, and cellulose) [161]. Alginate, isolated mainly from brown seaweeds, showed similarity to extracellular matrices found in living tissue and, for that reason, has been applied in wound healing dressings with enhanced antioxidative capacity when supplemented with silver [162,163]. Fucoidans, a sulphated polysaccharide, isolated from brown seaweeds like *Turbinaria ornata*, have been recognised for anti-inflammatory, antioxidant, and anticancer activities, with anti-photoaging effects [164]. Anti-aging effects of MAPs were confirmed using animal models with antioxidative activity resulting in upregulation of the gene expression of antioxidant enzymes, resulting in the reduction of oxidative stress, as well as regulation of the expression of other age-linked genes and pathways, reduction of telomere attrition and impact on immune modulation [119].

Table 2. Primary mycosporine-like amino acids (MAAs); algal sources with representative species and maximum UV absorbance.

Compound Name (Molecular Formula; Structure)	Source of the Compounds [Reference]	UV-Absorbing Maximum (λ_{max})
Shinorine ($C_{13}H_{20}N_2O_8$) 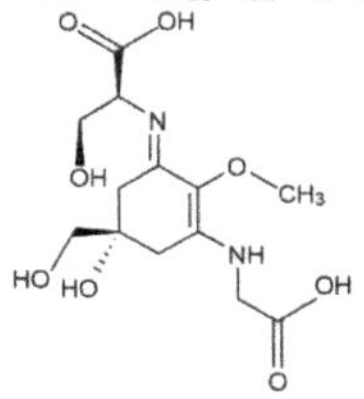	Macroalgae Red seaweeds [111,112,114,130,142,145,147,165–167] *Rhodymenia* spp., *Acanthophora spicifera* *Gelidium corneum, Georgiella confluens; Gelidium amansii, Gracilaria confervoides, Gracilaria* sp., *Bostrychia scorpioides,*	333 nm
Porphyra-334 ($C_{14}H_{22}N_2O_8$) 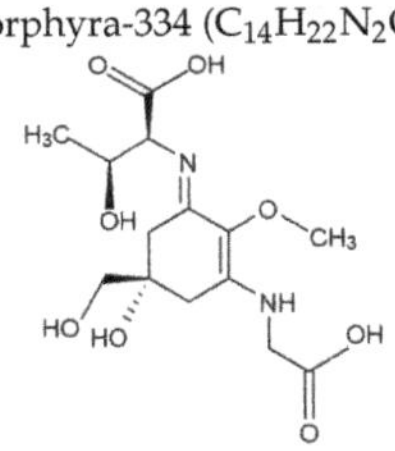	*Porphyra dioica* Brown seaweeds [111,113,168] *Ecklonia radiata, Dictyota bartayresii, Dictyosiphon foeniculaceus, Pilayella littoralis, Ecklonia radiata, Halopteris scoparia, Hydroclathrus clathratus, Sargassum oligocystum* Green seaweeds [111] *Prasiola crispa*	334 nm
Mycosporine-glycine ($C_{10}H_{15}NO_6$) 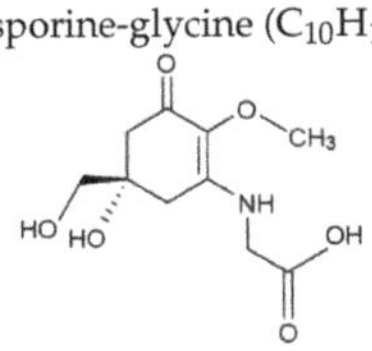	Microalgae [153,169–171] *Chlamydomonas nivalis, Cyclops abyssorum tatricus, Alexandrium* sp., *Chlamydomonas hedleyi, Gloeodinium viscum, Gymnodinium catenatum, Acetabularia mediterranea* Other MAA sources	310 nm
Mycosporine-2-glycine ($C_{12}H_{18}N_2O_7$) 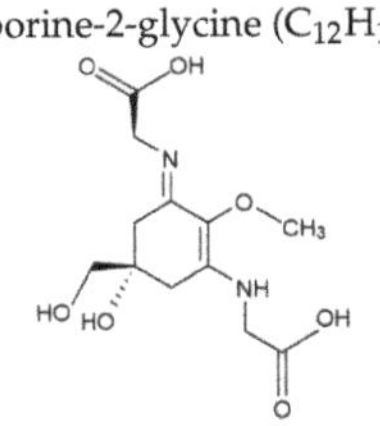	Cyanobacteria [117,172–174] *Nostoc commune, Aphanothece halophytica, Lyngbya* sp. Fungi and Animals [112,116,175] *Ascochyta pisi, Knufia cryptophialidica, Emiliania huxleyi; Gymnodinium linucheae*	332 nm
Palythine ($C_{10}H_{16}N_2O_5$) 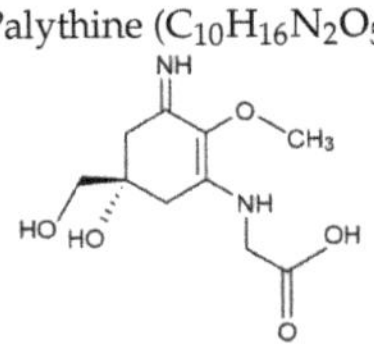	Corals, Sea Anemones, Jellyfish	320 nm

4. In Vitro, Ex Vivo, and In Vivo Experimental Assessments of Natural Products for Anti-Aging Properties

Experimental in vitro and in vivo studies provided specific evidence regarding the anti-aging activities of NPs of interest [3,6,7,36,50,83,110]. Various in vitro assays have been developed to assess the desired anti-aging activities ranging from antioxidant, anti-elastase, anti-collagenase, anti-hyaluronidase, and anti-inflammatory to anti-senescence activity [176]. Some of the studies combined in vitro, ex vivo (using human skin organ culture), and in vivo clinical trials when assessing NP anti-aging activities [177,178]. The advantages of in vitro assays compared to in vivo models include improved reproducibility [176] and a reduction in the number of animals used in research [157].

The excessive production of ROS in the skin, as the result of UV exposure, leads to ROS accumulation, oxidative stress, and skin aging (i.e., loss of skin elasticity, thinner skin, and more wrinkling) (Figure 3). However, ROS are also produced during regular cellular metabolism, mainly in mitochondria and the endoplasmic reticulum, and are important for redox signalling processes [137,179]. Exposure to damaging UV radiation and other stress factors (e.g., pollution) results in excessive ROS levels that cannot be removed by natural mitigation processes (i.e., antioxidants, enzymes, and pigments), causing elevated intracellular ROS levels and oxidative stress, which leads to cellular damage in macromolecules (i.e., lipids, proteins, and DNA) [39,180]. Cultured keratinocytes, human skin cells making up to 90% of all skin epidermal cells, are used for in vitro testing of anti-aging properties of natural products [153,181]. The effect of targeted NPs and their anti-aging capacities have been assessed using a range of spectrophotometric assays such as antioxidative assays (DPPH and ABTS assays), then a tyrosinase inhibition assay for evaluating anti-pigmentation activity, while collagenase and elastase inhibition assays have been used for confirming the anti-wrinkle activity of NPs [176].

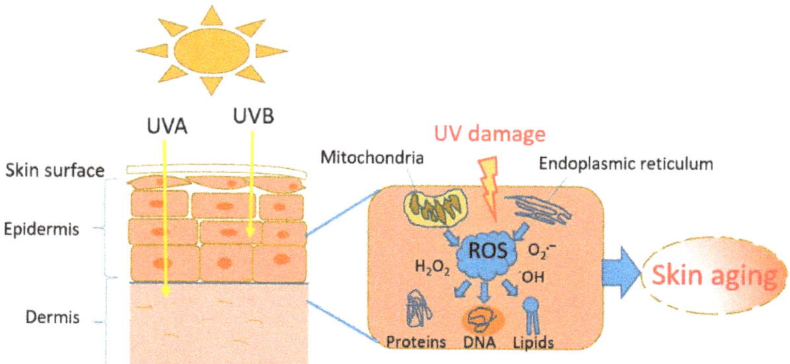

Figure 3. Skin exposure to UV radiation (both, UVA and UVB rays) results in the production and accumulation of ROS such as superoxide ($O_2^{\bullet -}$), hydroxyl radical ($^{\bullet}OH$), and hydrogen peroxide (H_2O_2), leading to oxidative stress and damage to macromolecules (DNA, lipids, proteins), resulting in aging.

In vitro antioxidative assays such as the DPPH have been used to evaluate the ability of NPs to remove free radicals and protect cells from ROS-induced damage. This assay was used to test MAAs' ability to shield human keratinocyte cells from UV radiation (300–400 nm) [153]. These HaCaT cells were irradiated with 15 min of UV (275 kJ/m^2, corresponding to ~90 min of high UV summer sun exposure) and supplemented with various concentrations of MAA (range of: 0, 0.03, 0.15, and 0.3 mM). A strong antioxidative capacity and ability to scavenge free radicals were demonstrated for mycosporine-glycine using a DHHP assay [153]. Similarly, the phosphatidylcholine peroxidation inhibition assay also confirmed strong mycosporine-glycine antioxidative activity [150]. Using a DPPH assay confirmed the ability of polyphenols (i.e., lignans) to scavenge ROS and their antioxidative properties in a concentration-dependent manner [181]. Anti-inflammatory and antioxidant actions of MAP, fucoidans, were confirmed in vitro, but also in vivo using the zebrafish model, indicating strong potential for their addition to cosmetic products [164].

Omega-3 polyunsaturated fatty acids EPA and DHA, using in vitro cultures of HaCaT keratinocyte and CCD922SK fibroblast cell lines, inhibited UVB-induced inflammation by reducing the secretion of IL-8 [182]. In vivo study with healthy human volunteers that used nutritional supplementation of omega-3 over a 10-week period, in the case of EPA, resulted in the reduced production of proinflammatory lipids in UV-exposed skin, while the addition of DHA resulted in lower UV-induced migration of Langerhans cells from the epidermis, which impacts adaptive immune responses [183]. Therefore, more in vivo studies are needed for the improved clinical assessment of the beneficial effects of NPs. However, with ethical limitations and the restricted availability of human volunteers, clinical evidence may take longer than anticipated, and more in vitro models will still be needed to complement pre-clinical trials around the efficiency of NPs in anti-aging processes.

5. Proteomics for Natural Product Discovery

Using omics technologies, such as proteomics, has allowed faster discovery and functional characterisation of novel marine natural products and their use as new drug therapies in wound healing and cosmetics [184,185]. The proteome, representing all proteins within cells or organisms, is explored to give a better understanding of proteins' function and structure, cellular regulatory processes, and protein interactions [186,187]. On the more applied side, the proteomics approaches have been applied to identify specific proteins important in disease detection or development, allocate new drug targets, and follow up on protein expression changes in response to particular conditions/therapies [188]. The more recent interdisciplinary field of chemical proteomics combines proteomics with chemistry to explore and identify drug–target interactions, including the synthesis of probes and target fishing, and better understand the relationship between NPs and drug targets [189].

The application of the proteomic approach in red algae, *Asparagopsis taxiformis*, included the use of different protein extracts from various developmental stages (i.e., sporophyte and gametophyte stages) and provided the most comprehensive overview of the protein profile from the algal development stages [190]. During the process of sample extraction and protein identification, the workflow included the separation of water-insoluble and water-soluble proteins that were separately analysed via high-performance liquid chromatography (HPLC) and mass spectrometry (MS)-based proteomics. In water-soluble and -insoluble fractions and from two protein databases (genome and transcriptome), totals of 741 and 2007 unique non-redundant proteins were identified, respectively. In different extracts, variations in protein profiles were noted, and the set of targeted proteins was identified for further exploration of promising bioactivities. Via this proteomics analysis, 40 secreted proteins were confirmed, including 10 rhodophyte collagen-alpha-like proteins that should be further explored for potential use in anti-aging [190] applications.

In macroalgae, the protein extraction step was identified as a challenging step for the recovery of high-quality MS data due to complex polysaccharides negatively impacting protein purification and downstream proteomic analyses [190,191]. The improved methodology for both high-throughput top-down (i.e., analysis of intact proteins) and bottom-up (i.e., analysis of digested proteins into peptides) proteomic analyses and overcoming these issues has been proposed to include the use of an MS-compatible surfactant, 4-hexylphenylazosulfonate, which was found to facilitate protein extraction and rapid enzymatic digestion, increasing the efficiency of MS-based proteomics analyses [191].

Microalgae have great potential with diverse nutrients that could be applied in the human diet and other industry applications [37]. Peptidomics studies in *Tetradesmus obliquus* identified ~500 bioactive peptides such as angiotensin-converting enzyme (ACE)-inhibitory and antioxidant bioactive peptides with four peptides selected for further analyses [192]. The proteome profile of MAA-producing *Chlamydomonas* species under limited nitrogen conditions contained more proteins with lower N content compared to the control condition, indicating the ability of this microalgae to adapt under stressful starvation conditions [193]. A better understanding of the impact of N status could be used to enhance the biotechnological potential of microalgae for MAA biosynthesis. In the green microalga *Dunaliella parva*, protein expression changes resulted in differential protein levels impacting multiple processes such as photosynthesis, stress response, lipid metabolism, as well as carbohydrate and nitrogen metabolism [194]. Understanding lipid metabolism and its regulation under N-limited conditions is critical in improving ways to enhance the synthesis of valuable NPs such as n-3 PUFA [195]. In the green microalgae *Chlorella vulgaris*, with low copper ions (Cu^{2+}), proteome changes impacted fatty acid biosynthesis and the carbon fixation process, with severely reduced protein numbers (from 581 to 369 proteins) and reduced growth [196]. Therefore, proteomics profiling allowed a better understanding of protein changes occurring in metabolic pathways, especially of lipid metabolism and photosynthesis, which are modified during stress, and these are relevant for other biotechnological applications to improve efficient use and control of growth conditions [197]. Many of the isolated bioactive peptides present tremendous potential not only in medicine but also in industry applications and as nutraceuticals in human health [185]. Consequently, a better understanding of lipid metabolism, stress responses, and other metabolic changes using the proteomics approach provides an innovative way to optimise and enhance algal cultivation conditions for the improved industrial production of valuable anti-aging NPs.

6. Conclusions

Aging is the inevitable life progression that involves multiple stages, with the gradual deterioration of physiological functions and metabolic processes, and disease development. Slowing aging processes and prolonging the lifespan could be positively influenced by nutritional changes and pharmacological supplements. Anti-aging natural products, including polyphenols, polysaccharides, vitamins, and others, are becoming more popular with the increase in demand due to the rapid rise of the elderly world population. Marine natural products have been shown to have promising anti-aging properties due to their anti-inflammatory, antioxidant, photoprotective, anticancer, and anti-pigmentation activities. Interest in using marine ingredients has steadily increased over the past few decades, with the application of anti-aging natural compounds in various cosmetics products. Marine algae are the most prevalent group found in the marine environment, capable of quicker reproduction and mass growth compared to terrestrial counterparts and consequently more economically valuable. Macro- and microalgae contain beneficial NPs with anti-aging properties, and these biomolecules, by manipulation of external conditions, can be accumulated and produced in mass quantities to facilitate their applications in biotechnology. Advanced technologies of bioinformatics, including proteomics, genomics, transcriptomics, and metagenomics, increased understanding of biosynthetic pathways, genetic polymorphisms, gene regulation, and the best ways to enhance biosynthesis and applications of natural products. Furthermore, omics technologies allowed for enhanced cultivation conditions and improved the understanding and application of these anti-aging products in human health. Finally, marine NPs, as organic and environmentally friendly molecules, are meeting high safety standards and allowing for the innovative use of these compounds with anti-aging properties in various biotechnological applications, promising the sustainable economic growth of this sector in the future.

Funding: This research received no external funding.

Institutional Review Board Statement: Not applicable.

Data Availability Statement: Not applicable.

Acknowledgments: The author would like to thank the three anonymous reviewers, as well as Isidora Skrlin, for their critical reviews of this paper.

Conflicts of Interest: The authors declare no conflict of interest.

References

1. Atanasov, A.G.; Zotchev, S.B.; Dirsch, V.M.; Orhan, I.E.; Banach, M.; Rollinger, J.M.; Barreca, D.; Weckwerth, W.; Bauer, R.; Bayer, E.A.; et al. Natural products in drug discovery: Advances and opportunities. *Nat. Rev. Drug Discov.* **2021**, *20*, 200–216. [CrossRef] [PubMed]
2. Mendes, M.C.; Navalho, S.; Ferreira, A.; Paulino, C.; Figueiredo, D.; Silva, D.; Gao, F.; Gama, F.; Bombo, G.; Jacinto, R.; et al. Algae as Food in Europe: An Overview of Species Diversity and Their Application. *Foods* **2022**, *11*, 1871. [CrossRef] [PubMed]
3. Resende, D.I.S.P.; Ferreira, M.; Magalhães, C.; Sousa Lobo, J.M.; Sousa, E.; Almeida, I.F. Trends in the use of marine ingredients in anti-aging cosmetics. *Algal Res.* **2021**, *55*, 102273. [CrossRef]
4. Fedele, G.; Donatti, C.I.; Bornacelly, I.; Hole, D.G. Nature-dependent people: Mapping human direct use of nature for basic needs across the tropics. *Glob. Environ. Change* **2021**, *71*, 102368. [CrossRef]
5. Williamson, E.; Ross, I.L.; Wall, B.T.; Hankamer, B. Microalgae: Potential novel protein for sustainable human nutrition. *Trends Plant Sci.* **2024**, *29*, 370–382. [CrossRef] [PubMed]
6. Vrushali Bothare. Anti-Aging Products Market Size. In *Market Research Report*; Straits Research: Pune, India, 2024. Available online: https://straitsresearch.com/report/anti-aging-products-market (accessed on 19 January 2025).
7. Andrade, L.F.; Hernandez, L.E.; Mashoudy, K.D.; Lalama, M.J.; Saaraswat, M.; Scheinkman, R.J.; Hu, S. A Cost-Based Analysis of Anti-aging Products Across Four Major United States Retailers. *Cureus* **2023**, *15*, e46596. [CrossRef]
8. World Health Organization. *Ageing and Health*; WHO Food Additives Series; WHO: Geneva, Switzerland, 2024.
9. Rosic, N.; Remond, C.; Mello-Athayde, M.A. Differential impact of heat stress on reef-building corals under different light conditions. *Mar. Environ. Res.* **2020**, *158*, 104947. [CrossRef]
10. Ewere, E.E.; Rosic, N.; Bayer, P.E.; Ngangbam, A.; Edwards, D.; Kelaher, B.P.; Mamo, L.T.; Benkendorff, K. Marine heatwaves have minimal influence on the quality of adult Sydney rock oyster flesh. *Sci. Total Environ.* **2021**, *795*, 148846. [CrossRef]
11. Onyeaka, H.; Miri, T.; Obileke, K.; Hart, A.; Anumudu, C.; Al-Sharify, Z.T. Minimizing carbon footprint via microalgae as a biological capture. *Carbon Capture Sci. Technol.* **2021**, *1*, 100007. [CrossRef]
12. Ralph, P.J.; Pernice, M. Save the planet with green industries using algae. *PLoS Biol.* **2023**, *21*, e3002061. [CrossRef]
13. Carroll, A.R.; Copp, B.R.; Davis, R.A.; Keyzers, R.A.; Prinsep, M.R. Marine natural products. *Nat. Prod. Rep.* **2023**, *40*, 275–325. [CrossRef] [PubMed]
14. Carroll, A.R.; Copp, B.R.; Davis, R.A.; Keyzers, R.A.; Prinsep, M.R. Marine natural products. *Nat. Prod. Rep.* **2020**, *37*, 175–223. [CrossRef] [PubMed]
15. Rosic, N.; Delamare-Deboutteville, J.; Dove, S. Heat stress in symbiotic dinoflagellates: Implications on oxidative stress and cellular changes. *Sci. Total Environ.* **2024**, *944*, 173916. [CrossRef] [PubMed]
16. Schaum, C.E.; Collins, S. Plasticity predicts evolution in a marine alga. *Proc. R. Soc. B Biol. Sci.* **2014**, *281*, 20141486. [CrossRef] [PubMed]
17. Brown, K.T.; Bender-Champ, D.; Kubicek, A.; van der Zande, R.; Achlatis, M.; Hoegh-Guldberg, O.; Dove, S.G. The Dynamics of Coral-Algal Interactions in Space and Time on the Southern Great Barrier Reef. *Front. Mar. Sci.* **2018**, *5*, 181. [CrossRef]
18. Ramanan, R.; Kim, B.H.; Cho, D.H.; Oh, H.M.; Kim, H.S. Algae-bacteria interactions: Evolution, ecology and emerging applications. *Biotechnol. Adv.* **2016**, *34*, 14–29. [CrossRef]
19. Patel, A.K.; Singhania, R.R.; Awasthi, M.K.; Varjani, S.; Bhatia, S.K.; Tsai, M.-L.; Hsieh, S.-L.; Chen, C.-W.; Dong, C.-D. Emerging prospects of macro- and microalgae as prebiotic. *Microb. Cell Factories* **2021**, *20*, 112. [CrossRef]
20. Thakur, A.; Sharma, D.; Saini, R.; Suhag, R.; Thakur, D. Cultivating blue food proteins: Innovating next-generation ingredients from macro and microalgae. *Biocatal. Agric. Biotechnol.* **2024**, *60*, 103278. [CrossRef]
21. Adarshan, S.; Sree, V.S.S.; Muthuramalingam, P.; Nambiar, K.S.; Sevanan, M.; Satish, L.; Venkidasamy, B.; Jeelani, P.G.; Shin, H. Understanding Macroalgae: A Comprehensive Exploration of Nutraceutical, Pharmaceutical, and Omics Dimensions. *Plants* **2023**, *13*, 113. [CrossRef]

22. Rosic, N.; Thornber, C. Biotechnological Potential of Macroalgae during Seasonal Blooms for Sustainable Production of UV-Absorbing Compounds. *Mar. Drugs* **2023**, *21*, 633. [CrossRef]
23. Leandro, A.; Pereira, L.; Gonçalves, A.M.M. Diverse Applications of Marine Macroalgae. *Mar. Drugs* **2019**, *18*, 17. [CrossRef] [PubMed]
24. Moreira, A.; Cruz, S.; Marques, R.; Cartaxana, P. The underexplored potential of green macroalgae in aquaculture. *Rev. Aquac.* **2022**, *14*, 5–26. [CrossRef]
25. Buschmann, A.H.; Camus, C.; Infante, J.; Neori, A.; Israel, Á.; Hernández-González, M.C.; Pereda, S.V.; Gomez-Pinchetti, J.L.; Golberg, A.; Tadmor-Shalev, N.; et al. Seaweed production: Overview of the global state of exploitation, farming and emerging research activity. *Eur. J. Phycol.* **2017**, *52*, 391–406. [CrossRef]
26. Dolganyuk, V.; Belova, D.; Babich, O.; Prosekov, A.; Ivanova, S.; Katserov, D.; Patyukov, N.; Sukhikh, S. Microalgae: A Promising Source of Valuable Bioproducts. *Biomolecules* **2020**, *10*, 1153. [CrossRef]
27. Barsanti, L.; Gualtieri, P. Is exploitation of microalgae economically and energetically sustainable? *Algal Res.* **2018**, *31*, 107–115. [CrossRef]
28. Coronado-Reyes, J.A.; González-Hernández, J.C. Vitamins from microalgae. In *Handbook of Food and Feed from Microalgae*; Jacob-Lopes, E., Queiroz, M.I., Maroneze, M.M., Zepka, L.Q., Eds.; Academic Press: New York, NY, USA, 2023; Chapter 10, p. 111.
29. Del Mondo, A.; Smerilli, A.; Sané, E.; Sansone, C.; Brunet, C. Challenging microalgal vitamins for human health. *Microb. Cell Factories* **2020**, *19*, 201. [CrossRef]
30. Maroneze, M.M.; Martinez, A. Minerals and trace elements in microalgal biomass. In *Handbook of Food and Feed from Microalgae*; Jacob-Lopes, E., Queiroz, M.I., Maroneze, M.M., Zepka, L.Q., Eds.; Academic Press: New York, NY, USA, 2023; Chapter 9; p. 103.
31. Kumar, R.; Hegde, A.S.; Sharma, K.; Parmar, P.; Srivatsan, V. Microalgae as a sustainable source of edible proteins and bioactive peptides—Current trends and future prospects. *Food Res. Int.* **2022**, *157*, 111338. [CrossRef]
32. García, J.L.; de Vicente, M.; Galán, B. Microalgae, old sustainable food and fashion nutraceuticals. *Microb. Biotechnol.* **2017**, *10*, 1017–1024. [CrossRef]
33. Larkum, A.W.D.; Ross, I.L.; Kruse, O.; Hankamer, B. Selection, breeding and engineering of microalgae for bioenergy and biofuel production. *Trends Biotechnol.* **2012**, *30*, 198–205. [CrossRef]
34. Spolaore, P.; Joannis-Cassan, C.; Duran, E.; Isambert, A. Commercial applications of microalgae. *J. Biosci. Bioeng.* **2006**, *101*, 87–96. [CrossRef]
35. Guo, J.; Huang, X.; Dou, L.; Yan, M.; Shen, T.; Tang, W.; Li, J. Aging and aging-related diseases: From molecular mechanisms to interventions and treatments. *Signal Transduct. Target. Ther.* **2022**, *7*, 391. [CrossRef]
36. Bjørklund, G.; Shanaida, M.; Lysiuk, R.; Butnariu, M.; Peana, M.; Sarac, I.; Strus, O.; Smetanina, K.; Chirumbolo, S. Natural Compounds and Products from an Anti-Aging Perspective. *Molecules* **2022**, *27*, 7084. [CrossRef] [PubMed]
37. Ding, J.; Wu, B.; Chen, L. Application of Marine Microbial Natural Products in Cosmetics. *Front. Microbiol.* **2022**, *13*, 892505. [CrossRef] [PubMed]
38. Rong, D.; Su, Y.; Jia, D.; Zeng, Z.; Yang, Y.; Wei, D.; Lu, H.; Cao, Y. Experimentally validated oxidative stress-associated prognostic signatures describe the immune landscape and predict the drug response and prognosis of SKCM. *Front. Immunol.* **2024**, *15*, 1387316. [CrossRef] [PubMed]
39. Bertram, C.; Hass, R. Cellular responses to reactive oxygen species-induced DNA damage and aging. *Biol. Chem.* **2008**, *389*, 211–220. [CrossRef]
40. Jaul, E.; Barron, J. Age-Related Diseases and Clinical and Public Health Implications for the 85 Years Old and Over Population. *Front. Public Health* **2017**, *5*, 335. [CrossRef]
41. Luo, J.; Ganesan, K.; Xu, B. Unlocking the Power: New Insights into the Anti-Aging Properties of Mushrooms. *J. Fungi* **2024**, *10*, 215. [CrossRef]
42. Passarino, G.; De Rango, F.; Montesanto, A. Human longevity: Genetics or Lifestyle? It takes two to tango. *Immun. Ageing* **2016**, *13*, 12. [CrossRef]
43. Willcox, B.J.; Donlon, T.A.; He, Q.; Chen, R.; Grove, J.S.; Yano, K.; Masaki, K.H.; Willcox, D.C.; Rodriguez, B.; Curb, J.D. FOXO3A genotype is strongly associated with human longevity. *Proc. Natl. Acad. Sci. USA* **2008**, *105*, 13987–13992. [CrossRef]
44. Morris, B.J.; Willcox, B.J.; Donlon, T.A. Genetic and epigenetic regulation of human aging and longevity. *Biochim. Biophys. Acta Mol. Basis Dis.* **2019**, *1865*, 1718–1744. [CrossRef]
45. Navarro, C.; Salazar, J.; Díaz, M.P.; Chacin, M.; Santeliz, R.; Vera, I.; D'Marco, L.; Parra, H.; Bernal, M.C.; Castro, A.; et al. Intrinsic and environmental basis of aging: A narrative review. *Heliyon* **2023**, *9*, e18239. [CrossRef] [PubMed]
46. Zhu, Y.; Liu, X.; Ding, X.; Wang, F.; Geng, X. Telomere and its role in the aging pathways: Telomere shortening, cell senescence and mitochondria dysfunction. *Biogerontology* **2019**, *20*, 1–16. [CrossRef] [PubMed]
47. Jurk, D.; Wilson, C.; Passos, J.F.; Oakley, F.; Correia-Melo, C.; Greaves, L.; Saretzki, G.; Fox, C.; Lawless, C.; Anderson, R.; et al. Chronic inflammation induces telomere dysfunction and accelerates ageing in mice. *Nat. Commun.* **2014**, *2*, 4172. [CrossRef]

48. Coluzzi, E.; Leone, S.; Sgura, A. Oxidative Stress Induces Telomere Dysfunction and Senescence by Replication Fork Arrest. *Cells* **2019**, *8*, 19. [CrossRef] [PubMed]
49. Franceschi, C.; Garagnani, P.; Morsiani, C.; Conte, M.; Santoro, A.; Grignolio, A.; Monti, D.; Capri, M.; Salvioli, S. The Continuum of Aging and Age-Related Diseases: Common Mechanisms but Different Rates. *Front. Med.* **2018**, *5*, 61. [CrossRef]
50. Song, L.; Zhang, S. Anti-Aging Activity and Modes of Action of Compounds from Natural Food Sources. *Biomolecules* **2023**, *13*, 1600. [CrossRef]
51. Rosic, N.N. Mycosporine-Like Amino Acids: Making the Foundation for Organic Personalised Sunscreens. *Mar. Drugs* **2019**, *17*, 638. [CrossRef]
52. Chandika, P.; Ko, S.C.; Jung, W.K. Marine-derived biological macromolecule-based biomaterials for wound healing and skin tissue regeneration. *Int. J. Biol. Macromol.* **2015**, *77*, 24–35. [CrossRef]
53. Moreira, B.R.; Vega, J.; Sisa, A.D.A.; Bernal, J.S.B.; Abdala-Díaz, R.T.; Maraschin, M.; Figueroa, F.L.; Bonomi-Barufi, J. Antioxidant and anti-photoaging properties of red marine macroalgae: Screening of bioactive molecules for cosmeceutical applications. *Algal Res.* **2022**, *68*, 102893. [CrossRef]
54. Elhady, S.S.; Habib, E.S.; Abdelhameed, R.F.A.; Goda, M.S.; Hazem, R.M.; Mehanna, E.T.; Helal, M.A.; Hosny, K.M.; Diri, R.M.; Hassanean, H.A.; et al. Anticancer Effects of New Ceramides Isolated from the Red Sea Red Algae *Hypnea musciformis* in a Model of Ehrlich Ascites Carcinoma: LC-HRMS Analysis Profile and Molecular Modeling. *Mar. Drugs* **2022**, *20*, 63. [CrossRef]
55. Begolli, R.; Chatziangelou, M.; Samiotaki, M.; Goutas, A.; Barda, S.; Goutzourelas, N.; Kevrekidis, D.P.; Malea, P.; Trachana, V.; Liu, M.; et al. Transcriptome and proteome analysis reveals the anti-cancer properties of *Hypnea musciformis* marine macroalga extract in liver and intestinal cancer cells. *Hum. Genom.* **2023**, *17*, 71. [CrossRef] [PubMed]
56. Ponte, J.M.S.; Seca, A.M.L.; Barreto, M.C. Asparagopsis Genus: What We Really Know About Its Biological Activities and Chemical Composition. *Molecules* **2022**, *27*, 1787. [CrossRef] [PubMed]
57. Ridoutt, B.; Lehnert, S.A.; Denman, S.; Charmley, E.; Kinley, R.; Dominik, S. Potential GHG emission benefits of *Asparagopsis taxiformis* feed supplement in Australian beef cattle feedlots. *J. Clean. Prod.* **2022**, *337*, 130499. [CrossRef]
58. Şimşek, H.; Uçar, A. Polyunsaturated fatty acids as a nutraceutical for age-related neurodegenerative diseases: Current knowledge and future directions. *Clin. Nutr. Open Sci.* **2024**, *56*, 65–73. [CrossRef]
59. Harwood, J.L. Algae: Critical Sources of Very Long-Chain Polyunsaturated Fatty Acids. *Biomolecules* **2019**, *9*, 708. [CrossRef]
60. Liang, K.; Zhang, Q.; Cong, W. Enzyme-Assisted Aqueous Extraction of Lipid from Microalgae. *J. Agric. Food Chem.* **2012**, *60*, 11771–11776. [CrossRef]
61. Ramesh Kumar, B.; Deviram, G.; Mathimani, T.; Duc, P.A.; Pugazhendhi, A. Microalgae as rich source of polyunsaturated fatty acids. *Biocatal. Agric. Biotechnol.* **2019**, *17*, 583–588. [CrossRef]
62. van Ginneken, V.J.; Helsper, J.P.; de Visser, W.; van Keulen, H.; Brandenburg, W.A. Polyunsaturated fatty acids in various macroalgal species from North Atlantic and tropical seas. *Lipids Health Dis.* **2011**, *10*, 104. [CrossRef]
63. Lourenço-Lopes, C.; Fraga-Corral, M.; Jimenez-Lopez, C.; Pereira, A.G.; Garcia-Oliveira, P.; Carpena, M.; Prieto, M.A.; Simal-Gandara, J. Metabolites from Macroalgae and Its Applications in the Cosmetic Industry: A Circular Economy Approach. *Resources* **2020**, *9*, 101. [CrossRef]
64. Lenihan-Geels, G.; Bishop, K.S.; Ferguson, L.R. Alternative sources of omega-3 fats: Can we find a sustainable substitute for fish? *Nutrients* **2013**, *5*, 1301–1315. [CrossRef]
65. Stiefvatter, L.; Lehnert, K.; Frick, K.; Montoya-Arroyo, A.; Frank, J.; Vetter, W.; Schmid-Staiger, U.; Bischoff, S.C. Oral Bioavailability of Omega-3 Fatty Acids and Carotenoids from the Microalgae *Phaeodactylum tricornutum* in Healthy Young Adults. *Mar. Drugs* **2021**, *19*, 700. [CrossRef]
66. Gonzalez-Becerra, K.; Barron-Cabrera, E.; Muñoz-Valle, J.F.; Torres-Castillo, N.; Rivera-Valdes, J.J.; Rodriguez-Echevarria, R.; Martinez-Lopez, E. A Balanced Dietary Ratio of n-6:n-3 Polyunsaturated Fatty Acids Exerts an Effect on Total Fatty Acid Profile in RBCs and Inflammatory Markers in Subjects with Obesity. *Healthcare* **2023**, *11*, 2333. [CrossRef]
67. Chen, J.; Wei, Y.; Chen, X.; Jiao, J.; Zhang, Y. Polyunsaturated fatty acids ameliorate aging via redox-telomere-antioncogene axis. *Oncotarget* **2017**, *8*, 7301–7314. [CrossRef] [PubMed]
68. Chen, H.; Xiong, R.; Cheng, J.; Ye, J.; Qiu, Y.; Huang, S.; Li, M.; Liu, Z.; Pang, J.; Zhang, X.; et al. Effects and Mechanisms of Polyunsaturated Fatty Acids on Age-Related Musculoskeletal Diseases: Sarcopenia, Osteoporosis, and Osteoarthritis—A Narrative Review. *Nutrients* **2024**, *16*, 3130. [CrossRef] [PubMed]
69. Picca, A.; Calvani, R. Molecular Mechanism and Pathogenesis of Sarcopenia: An Overview. *Int. J. Mol. Sci.* **2021**, *22*, 3032. [CrossRef] [PubMed]
70. Bae, Y.-J.; Cui, X.-S.; Shin, S.-H. Increased Omega-3 Fatty Acid Intake Is Associated with Low Grip Strength in Elderly Korean Females. *Nutrients* **2022**, *14*, 2374. [CrossRef]
71. Dupont, J.; Wauters, E.; Dedeyne, L.; Vercauteren, L.; Amini, N.; Lapauw, L.; Matthys, C.; Verschueren, S.; Tournoy, J.; Koppo, K.; et al. Are dietary intake and nutritional status of specific polyunsaturated fatty acids correlated with sarcopenia outcomes in community-dwelling older adults with sarcopenia?—Exploratory results from ENHANce. *BMC Geriatr.* **2023**, *23*, 272. [CrossRef]

72. Ghzaiel, I.; Zarrouk, A.; Nury, T.; Libergoli, M.; Florio, F.; Hammouda, S.; Ménétrier, F.; Avoscan, L.; Yammine, A.; Samadi, M., et al. Antioxidant Properties and Cytoprotective Effect of *Pistacia lentiscus* L. Seed Oil against 7β-Hydroxycholesterol-Induced Toxicity in C2C12 Myoblasts: Reduction in Oxidative Stress, Mitochondrial and Peroxisomal Dysfunctions and Attenuation of Cell Death. *Antioxidants* **2021**, *10*, 1772. [CrossRef]
73. Kerdiles, O.; Layé, S.; Calon, F. Omega-3 polyunsaturated fatty acids and brain health: Preclinical evidence for the prevention of neurodegenerative diseases. *Trends Food Sci. Technol.* **2017**, *69*, 203–213. [CrossRef]
74. Fabregas, J.; Herrero, C. Vitamin content of four marine microalgae. Potential use as source of vitamins in nutrition. *J. Ind. Microbiol.* **1990**, *5*, 259–263. [CrossRef]
75. Ouedrhiri, W.; Bennis, I.; El Arroussi, H. Recent advances in microalgae-based vitamin D metabolome: Biosynthesis, and production. *Bioresour. Technol.* **2024**, *407*, 131078. [CrossRef] [PubMed]
76. Araújo, R.G.; Alcantar-Rivera, B.; Meléndez-Sánchez, E.R.; Martínez-Prado, M.A.; Sosa-Hernández, J.E.; Iqbal, H.M.N.; Parra-Saldivar, R.; Martínez-Ruiz, M. Effects of UV and UV-vis Irradiation on the Production of Microalgae and Macroalgae: New Alternatives to Produce Photobioprotectors and Biomedical Compounds. *Molecules* **2022**, *27*, 5334. [CrossRef]
77. De Roeck-Holtzhauer, Y.; Quere, I.; Claire, C. Vitamin analysis of five planktonic microalgae and one macroalga. *J. Appl. Phycol.* **1991**, *3*, 259–264. [CrossRef]
78. Galasso, C.; Gentile, A.; Orefice, I.; Ianora, A.; Bruno, A.; Noonan, D.M.; Sansone, C.; Albini, A.; Brunet, C. Microalgal Derivatives as Potential Nutraceutical and Food Supplements for Human Health: A Focus on Cancer Prevention and Interception. *Nutrients* **2019**, *11*, 1226. [CrossRef] [PubMed]
79. Flanagan, J.L.; Simmons, P.A.; Vehige, J.; Willcox, M.D.P.; Garrett, Q. Role of carnitine in disease. *Nutr. Metab.* **2010**, *7*, 30. [CrossRef]
80. Kendler, B.S. Carnitine: An overview of its role in preventive medicine. *Prev. Med.* **1986**, *15*, 373–390. [CrossRef]
81. Liu, D.; Zeng, X.; Li, L.; Ou, Z.L. Carnitine promotes recovery from oxidative stress and extends lifespan in *C. elegans*. *Aging* **2020**, *13*, 813–830. [CrossRef]
82. La Fata, G.; van Vliet, N.; Barnhoorn, S.; Brandt, R.M.C.; Etheve, S.; Chenal, E.; Grunenwald, C.; Seifert, N.; Weber, P.; Hoeijmakers, J.H.J.; et al. Vitamin E Supplementation Reduces Cellular Loss in the Brain of a Premature Aging Mouse Model. *J. Prev. Alzheimers Dis.* **2017**, *4*, 226–235.
83. Martín-Martínez, A.; Sánchez-Marzo, N.; Martínez-Casanova, D.; Abarquero-Cerezo, M.; Herranz-López, M.; Barrajón-Catalán, E.; Matabuena-Yzaguirre, M. High global antioxidant protection and stimulation of the collagen synthesis of new anti-aging product containing an optimized active mix. *J. Cosmet. Dermatol.* **2022**, *21*, 3993–4000. [CrossRef]
84. Baudry, J.; Kopp, J.F.; Boeing, H.; Kipp, A.P.; Schwerdtle, T.; Schulze, M.B. Changes of trace element status during aging: Results of the EPIC-Potsdam cohort study. *Eur. J. Nutr.* **2020**, *59*, 3045–3058. [CrossRef]
85. Ahuja, K.; Lio, P. The role of trace elements in dermatology: A systematic review. *J. Integr. Dermatol.* **2023**. Available online: https://www.jintegrativederm.org/article/73228-the-role-of-trace-elements-in-dermatology-a-systematic-review (accessed on 19 January 2025).
86. Kazeminejad, A.; Hajheydari, Z.; Taghian, S.S.; Gholizadeh, N. Serum zinc, selenium, and vitamin D levels in patients with acne vulgaris: A case–control study. *J. Cosmet. Dermatol.* **2024**, *23*, 4249–4254. [CrossRef] [PubMed]
87. Rupérez, P. Mineral content of edible marine seaweeds. *Food Chem.* **2002**, *79*, 23–26. [CrossRef]
88. Circuncisão, A.R.; Catarino, M.D.; Cardoso, S.M.; Silva, A.M.S. Minerals from Macroalgae Origin: Health Benefits and Risks for Consumers. *Mar. Drugs* **2018**, *16*, 400. [CrossRef] [PubMed]
89. Di Lena, G.; Casini, I.; Lucarini, M.; Sánchez del Pulgar, J.; Aguzzi, A.; Caproni, R.; Gabrielli, P.; Lombardi-Boccia, G. Chemical characterization and nutritional evaluation of microalgal biomass from large-scale production: A comparative study of five species. *Eur. Food Res. Technol.* **2020**, *246*, 323–332. [CrossRef]
90. Manach, C.; Scalbert, A.; Morand, C.; Rémésy, C.; Jiménez, L. Polyphenols: Food sources and bioavailability. *Am. J. Clin. Nutr.* **2004**, *79*, 727–747. [CrossRef] [PubMed]
91. Arora, I.; Sharma, M.; Sun, L.Y.; Tollefsbol, T.O. The Epigenetic Link between Polyphenols, Aging and Age-Related Diseases. *Genes* **2020**, *11*, 1094. [CrossRef]
92. Zhor, C.; Wafaa, L.; Ghzaiel, I.; Kessas, K.; Zarrouk, A.; Ksila, M.; Ghrairi, T.; Latruffe, N.; Masmoudi-Kouki, O.; El Midaoui, A.; et al. Effects of polyphenols and their metabolites on age-related diseases. *Biochem. Pharmacol.* **2023**, *214*, 115674. [CrossRef]
93. Halm, H.; Lüder, U.H.; Wiencke, C. Induction of phlorotannins through mechanical wounding and radiation conditions in the brown macroalga Laminaria hyperborea. *Eur. J. Phycol.* **2011**, *46*, 16–26. [CrossRef]
94. Cruces, E.; Huovinen, P.; Gómez, I. Phlorotannin and Antioxidant Responses Upon Short-term Exposure to UV Radiation and Elevated Temperature in Three South Pacific Kelps. *Photochem. Photobiol.* **2012**, *88*, 58–66. [CrossRef]
95. Kim, S.-K.; Wijesekara, I. Role of Marine Nutraceuticals in Cardiovascular Health. In *Sustained Energy for Enhanced Human Functions and Activity*; Bagchi, D., Ed.; Academic Press: New York, NY, USA, 2017; p. 273.

96. Kumar, L.R.G.; Paul, P.T.; Anas, K.K.; Tejpal, C.S.; Chatterjee, N.S.; Anupama, T.K.; Mathew, S.; Ravishankar, C.N. Phlorotannins-bioactivity and extraction perspectives. *J. Appl. Phycol.* **2022**, *34*, 2173–2185. [CrossRef] [PubMed]
97. Heffernan, N.; Smyth, T.J.; Soler-Villa, A.; Fitzgerald, R.J.; Brunton, N.P. Phenolic content and antioxidant activity of fractions obtained from selected Irish macroalgae species (*Laminaria digitata, Fucus serratus, Gracilaria gracilis* and *Codium fragile*). *J. Appl. Phycol.* **2015**, *27*, 519–530. [CrossRef]
98. Goiris, K.; Muylaert, K.; Voorspoels, S.; Noten, B.; De Paepe, D.; Baart, G.J.E.; De Cooman, L. Detection of flavonoids in microalgae from different evolutionary lineages. *J. Phycol.* **2014**, *50*, 483–492. [CrossRef]
99. Besednova, N.N.; Andryukov, B.G.; Zaporozhets, T.S.; Kryzhanovsky, S.P.; Fedyanina, L.N.; Kuznetsova, T.A.; Zvyagintseva, T.N.; Shchelkanov, M.Y. Antiviral Effects of Polyphenols from Marine Algae. *Biomedicines* **2021**, *9*, 200. [CrossRef] [PubMed]
100. El-Gammal, M.I.; Abou-Dobara, M.I.; Ibrahim, H.A.H.; Abdulhafith, S.A.; Okbah, M.A. Polyphenols in selected marine algae and aromatic herbs with antimicrobial properties: A comparative study. *Egypt. J. Aquat. Res.* **2024**, *50*, 71–77. [CrossRef]
101. Lomartire, S.; Gonçalves, A.M.M. Marine Macroalgae Polyphenols as Potential Neuroprotective Antioxidants in Neurodegenerative Diseases. *Mar. Drugs* **2023**, *21*, 261. [CrossRef]
102. Sharma, G.; John, J. Identification of polyphenols using UPLC-QTOF MS/MS, in-vitro photoprotective and antiaging activities of brown macroalga *Padina tetrastromatica*. *Algal Res.* **2023**, *75*, 103255. [CrossRef]
103. Sanjeewa, K.K.A.; Kim, E.-A.; Son, K.-T.; Jeon, Y.-J. Bioactive properties and potentials cosmeceutical applications of phlorotannins isolated from brown seaweeds: A review. *J. Photochem. Photobiol. B Biol.* **2016**, *162*, 100–105. [CrossRef]
104. Bedoux, G.; Hardouin, K.; Burlot, A.S.; Bourgougnon, N. Chapter Twelve—Bioactive Components from Seaweeds: Cosmetic Applications and Future Development. In *Advances in Botanical Research*; Bourgougnon, N., Ed.; Academic Press: New York, NY, USA, 2014; p. 345.
105. Samaddar, S.; Koneri, R. Polyphenols of marine red macroalga *Symphyocladia latiuscula* ameliorate diabetic peripheral neuropathy in experimental animals. *Heliyon* **2019**, *5*, e01781. [CrossRef]
106. Kumar, Y.; Tarafdar, A.; Kumar, D.; Saravanan, C.; Badgujar, P.C.; Pharande, A.; Pareek, S.; Fawole, O.A. Polyphenols of Edible Macroalgae: Estimation of In Vitro Bio-Accessibility and Cytotoxicity, Quantification by LC-MS/MS and Potential Utilization as an Antimicrobial and Functional Food Ingredient. *Antioxidants* **2022**, *11*, 993. [CrossRef]
107. Cichoński, J.; Chrzanowski, G. Microalgae as a Source of Valuable Phenolic Compounds and Carotenoids. *Molecules* **2022**, *27*, 8852. [CrossRef] [PubMed]
108. Chen, W.; Li, T.; Du, S.; Chen, H.; Wang, Q. Microalgal polyunsaturated fatty acids: Hotspots and production techniques. *Front. Bioeng. Biotechnol.* **2023**, *11*, 1146881. [CrossRef] [PubMed]
109. Vieira, M.V.; Turkiewicz, I.P.; Tkacz, K.; Fuentes-Grünewald, C.; Pastrana, L.M.; Fuciños, P.; Wojdyło, A.; Nowicka, P. Microalgae as a Potential Functional Ingredient: Evaluation of the Phytochemical Profile, Antioxidant Activity and In-Vitro Enzymatic Inhibitory Effect of Different Species. *Molecules* **2021**, *26*, 7593. [CrossRef] [PubMed]
110. Kageyama, H.; Waditee-Sirisattha, R. Antioxidative, anti-inflammatory, and anti-aging properties of mycosporine-like amino acids: Molecular and cellular mechanisms in the protection of skin-aging. *Mar. Drugs* **2019**, *17*, 222. [CrossRef]
111. Sun, Y.; Zhang, N.; Zhou, J.; Dong, S.; Zhang, X.; Guo, L.; Guo, G. Distribution, Contents, and Types of Mycosporine-Like Amino Acids (MAAs) in Marine Macroalgae and a Database for MAAs Based on These Characteristics. *Mar. Drugs* **2020**, *18*, 43. [CrossRef]
112. Rosic, N.N.; Braun, C.; Kvaskoff, D. Extraction and Analysis of Mycosporine-Like Amino Acids in Marine Algae. *Methods Mol. Biol.* **2015**, *1308*, 119–129.
113. Karsten, U.; Sawall, T.; Wiencke, C. A survey of the distribution of UV-absorbing substances in tropical macroalgae. *Phycol. Res.* **1998**, *46*, 271–279.
114. Orfanoudaki, M.; Hartmann, A.; Karsten, U.; Ganzera, M. Chemical profiling of mycosporine-like amino acids in twenty-three red algal species. *J. Phycol.* **2019**, *55*, 393–403. [CrossRef]
115. Hartmann, A.; Becker, K.; Karsten, U.; Remias, D.; Ganzera, M. Analysis of mycosporine-like amino acids in selected algae and cyanobacteria by hydrophilic interaction liquid chromatography and a novel MAA from the red alga *Catenella repens*. *Mar. Drugs* **2015**, *13*, 6291–6305. [CrossRef]
116. Sinha, R.P.; Singh, S.P.; Häder, D.P. Database on mycosporines and mycosporine-like amino acids (MAAs) in fungi, cyanobacteria, macroalgae, phytoplankton and animals. *J. Photochem. Photobiol. B Biol.* **2007**, *89*, 29–35. [CrossRef]
117. Pathak, J.; Rajneesh; Maurya, P.K.; Singh, S.P.; Häder, D.-P.; Sinha, R.P. Cyanobacterial Farming for Environment Friendly Sustainable Agriculture Practices: Innovations and Perspectives. *Front. Environ. Sci.* **2018**, *6*, 7. [CrossRef]
118. Singh, S.P.; Kumari, S.; Rastogi, R.P.; Singh, K.L.; Sinha, R.P. Mycosporine-like amino acids (MAAs): Chemical structure, biosynthesis and significance as UV-absorbing/screening compounds. *Indian J. Exp. Biol.* **2008**, *46*, 7–17. [PubMed]
119. Guo, X.; Luo, J.; Qi, J.; Zhao, X.; An, P.; Luo, Y.; Wang, G. The Role and Mechanism of Polysaccharides in Anti-Aging. *Nutrients* **2022**, *14*, 5330. [CrossRef] [PubMed]

120. Xu, S.Y.; Huang, X.; Cheong, K.L. Recent Advances in Marine Algae Polysaccharides: Isolation, Structure, and Activities. *Mar. Drugs* **2017**, *15*, 388. [CrossRef]
121. Miguel, S.P.; D'Angelo, C.; Ribeiro, M.P.; Simões, R.; Coutinho, P. Chemical Composition of Macroalgae Polysaccharides from Galician and Portugal Coasts: Seasonal Variations and Biological Properties. *Mar. Drugs* **2023**, *21*, 589. [CrossRef]
122. Li, S.; Guo, W.; Zhang, M.; Zeng, M.; Wu, H. Microalgae polysaccharides exert antioxidant and anti-inflammatory protective effects on human intestinal epithelial cells in vitro and dextran sodium sulfate-induced mouse colitis in vivo. *Int. J. Biol. Macromol.* **2024**, *254*, 127811. [CrossRef]
123. Severo, I.A.; Dias, R.R.; do Nascimento, T.C.; Deprá, M.C.; Maroneze, M.M.; Zepka, L.Q.; Jacob-Lopes, E. Microalgae-derived polysaccharides: Potential building blocks for biomedical applications. *World J. Microbiol. Biotechnol.* **2022**, *38*, 150. [CrossRef]
124. Canfield, C.-A.; Bradshaw, P.C. Amino acids in the regulation of aging and aging-related diseases. *Transl. Med. Aging* **2019**, *3*, 70–89. [CrossRef]
125. Shick, J.M.; Dunlap, W.C. Mycosporine-like amino acids and related gadusols: Biosynthesis, accumulation, and UV-protective functions in aquatic organisms. *Annu. Rev. Physiol.* **2002**, *64*, 223–262. [CrossRef]
126. Balskus, E.P.; Walsh, C.T. The genetic and molecular basis for sunscreen biosynthesis in cyanobacteria. *Science* **2010**, *329*, 1653–1656. [CrossRef]
127. Rosic, N.N. Phylogenetic analysis of genes involved in mycosporine-like amino acid biosynthesis in symbiotic dinoflagellates. *Appl. Microbiol. Biotechnol.* **2012**, *94*, 29–37. [CrossRef]
128. Garcia-Pichel, F.; Wingard, C.E.; Castenholz, R.W. Evidence Regarding the UV Sunscreen Role of a Mycosporine-Like Compound in the Cyanobacterium *Gloeocapsa* sp. *Appl. Environ. Microbiol.* **1993**, *59*, 170–176. [CrossRef] [PubMed]
129. Torres, P.; Santos, J.P.; Chow, F.; Pena Ferreira, M.J.; dos Santos, D.Y.A.C. Comparative analysis of in vitro antioxidant capacities of mycosporine-like amino acids (MAAs). *Algal Res.* **2018**, *34*, 57–67. [CrossRef]
130. Vega, J.; Schneider, G.; Moreira, B.R.; Herrera, C.; Bonomi-Barufi, J.; Figueroa, F.L. Mycosporine-Like Amino Acids from Red Macroalgae: UV-Photoprotectors with Potential Cosmeceutical Applications. *Appl. Sci.* **2021**, *11*, 5112. [CrossRef]
131. Figueroa, F.L. Mycosporine-Like Amino Acids from Marine Resource. *Mar. Drugs* **2021**, *19*, 18. [CrossRef]
132. Rosic, N.N.; Dove, S. Mycosporine-like amino acids from coral dinoflagellates. *Appl. Environ. Microbiol.* **2011**, *77*, 8478–8486. [CrossRef]
133. D'Agostino, P.M.; Woodhouse, J.N.; Liew, H.T.; Sehnal, L.; Pickford, R.; Wong, H.L.; Burns, B.P.; Neilan, B.A. Bioinformatic, phylogenetic and chemical analysis of the UV-absorbing compounds scytonemin and mycosporine-like amino acids from the microbial mat communities of Shark Bay, Australia. *Environ. Microbiol.* **2019**, *21*, 702–715. [CrossRef] [PubMed]
134. Portwich, A.; Garcia-Pichel, F. Biosynthetic pathway of mycosporines (mycosporine-like amino acids) in the cyanobacterium *Chlorogloeopsis* sp. strain PCC 6912. *Phycologia* **2003**, *42*, 384–392. [CrossRef]
135. Rastogi, R.P.; Incharoensakdi, A. UV radiation-induced biosynthesis, stability and antioxidant activity of mycosporine-like amino acids (MAAs) in a unicellular cyanobacterium *Gloeocapsa* sp. CU2556. *J. Photochem. Photobiol. B Biol.* **2014**, *130*, 287–292. [CrossRef]
136. Bhattacharyya, A.; Chattopadhyay, R.; Mitra, S.; Crowe, S.E. Oxidative stress: An essential factor in the pathogenesis of gastrointestinal mucosal diseases. *Physiol. Rev.* **2014**, *94*, 329–354. [CrossRef]
137. Schieber, M.; Chandel, N.S. ROS function in redox signaling and oxidative stress. *Curr. Biol.* **2014**, *24*, R453–R462. [CrossRef] [PubMed]
138. Oren, A.; Gunde-Cimerman, N. Mycosporines and mycosporine-like amino acids: UV protectants or multipurpose secondary metabolites? *FEMS Microbiol. Lett.* **2007**, *269*, 1–10. [CrossRef] [PubMed]
139. Ikehata, H.; Ono, T. The mechanisms of UV mutagenesis. *J. Radiat. Res.* **2011**, *52*, 115–125. [CrossRef] [PubMed]
140. Carreto, J.I.; Carignan, M.O. Mycosporine-like amino acids: Relevant secondary metabolites. chemical and ecological aspects. *Mar. Drugs* **2011**, *9*, 387–446. [CrossRef]
141. Conde, F.R.; Churio, M.S.; Previtali, C.M. Experimental study of the excited-state properties and photostability of the mycosporine-like amino acid palythine in aqueous solution. *Photochem. Photobiol. Sci.* **2007**, *6*, 669–674. [CrossRef]
142. Jofre, J.; Celis-Plá, P.S.M.; Figueroa, F.L.; Navarro, N.P. Seasonal Variation of Mycosporine-Like Amino Acids in Three Subantarctic Red Seaweeds. *Mar. Drugs* **2020**, *18*, 75. [CrossRef]
143. Lopez Figueroa, F.; Bueno, A.; Korbee, N.; Santos, R.; Mata, L.; Schuenhoff, A. Accumulation of Mycosporine-like Amino Acids in *Asparagopsis armata* Grown in Tanks with Fishpond Effluents of Gilthead Sea Bream, *Sparus aurata*. *J. World Aquac. Soc.* **2008**, *39*, 692–699. [CrossRef]
144. Álvarez-Gómez, F.; Korbee, N.; Figueroa, F.L. Effects of UV Radiation on Photosynthesis, Antioxidant Capacity and the Accumulation of Bioactive Compounds in *Gracilariopsis longissima*, *Hydropuntia cornea* and *Halopithys incurva* (Rhodophyta). *J. Phycol.* **2019**, *55*, 1258–1273. [CrossRef]
145. Karsten, U.; Sawall, T.; West, J.; Wiencke, C. Ultraviolet sunscreen compounds in epiphytic red algae from mangroves. *Hydrobiologia* **2000**, *432*, 159–171. [CrossRef]

146. Tartarotti, B.; Sommaruga, R. Seasonal and ontogenetic changes of mycosporine-like amino acids in planktonic organisms from an alpine lake. *Limnol. Oceanogr.* **2006**, *51*, 1530–1541. [CrossRef]
147. Sun, Y.; Han, X.; Hu, Z.; Cheng, T.; Tang, Q.; Wang, H.; Deng, X.; Han, X. Extraction, Isolation and Characterization of Mycosporine-like Amino Acids from Four Species of Red Macroalgae. *Mar. Drugs* **2021**, *19*, 615. [CrossRef] [PubMed]
148. Wada, N.; Sakamoto, T.; Matsugo, S. Mycosporine-Like Amino Acids and Their Derivatives as Natural Antioxidants. *Antioxidants* **2015**, *4*, 603–646. [CrossRef] [PubMed]
149. Lee, J.H.; Kim, H.-S.; Seo, H.H.; Song, M.Y.; Kulkarni, A.; Choi, Y.-H.; Kim, K.W.; Moh, S.H. Antiaging Effects of Algae-Derived Mycosporine-Like Amino Acids (MAAs) on Skin. In *Textbook of Aging Skin*; Farage, M.A., Miller, K.W., Maibach, H.I., Eds.; Springer: Berlin/Heidelberg, Germany, 2014; p. 1.
150. Dunlap, W.C.; Yamamoto, Y. Small-molecule antioxidants in marine organisms: Antioxidant activity of mycosporine-glycine. *Comp. Biochem. Physiol. Part B Biochem. Mol. Biol.* **1995**, *112*, 105–114. [CrossRef]
151. Rosic, N.N. Recent advances in the discovery of novel marine natural products and mycosporine-like amino acid UV-absorbing compounds. *Appl. Microbiol. Biotechnol.* **2021**, *105*, 7053–7067. [CrossRef]
152. Geraldes, V.; Pinto, E. Mycosporine-Like Amino Acids (MAAs): Biology, Chemistry and Identification Features. *Pharmaceuticals* **2021**, *14*, 63. [CrossRef]
153. Suh, S.S.; Hwang, J.; Park, M.; Seo, H.H.; Kim, H.S.; Lee, J.H.; Moh, S.H.; Lee, T.K. Anti-inflammation activities of mycosporine-like amino acids (MAAs) in response to UV radiation suggest potential anti-skin aging activity. *Mar. Drugs* **2014**, *12*, 5174–5187. [CrossRef]
154. Ryu, J.; Park, S.J.; Kim, I.H.; Choi, Y.H.; Nam, T.J. Protective effect of porphyra-334 on UVA-induced photoaging in human skin fibroblasts. *Int. J. Mol. Med.* **2014**, *34*, 796–803. [CrossRef]
155. Tarasuntisuk, S.; Palaga, T.; Kageyama, H.; Waditee-Sirisattha, R. Mycosporine-2-glycine exerts anti-inflammatory and antioxidant effects in lipopolysaccharide (LPS)-stimulated RAW 264.7 macrophages. *Arch. Biochem. Biophys.* **2019**, *662*, 33–39. [CrossRef]
156. Becker, K.; Hartmann, A.; Ganzera, M.; Fuchs, D.; Gostner, J.M. Immunomodulatory Effects of the Mycosporine-Like Amino Acids Shinorine and Porphyra-334. *Mar. Drugs* **2016**, *14*, 119. [CrossRef]
157. Rosic, N.; Climstein, M.; Boyle, G.M.; Thanh Nguyen, D.; Feng, Y. Exploring Mycosporine-like Amino Acid UV-Absorbing Natural Products for a New Generation of Environmentally Friendly Sunscreens. *Mar. Drugs* **2023**, *21*, 253. [CrossRef]
158. Singh, A.; Čížková, M.; Bišová, K.; Vítová, M. Exploring Mycosporine-Like Amino Acids (MAAs) as Safe and Natural Protective Agents against UV-Induced Skin Damage. *Antioxidants* **2021**, *10*, 683. [CrossRef] [PubMed]
159. Caetano, P.A.; do Nascimento, T.C.; Fernandes, A.S.; Nass, P.P.; Vieira, K.R.; Maróstica Junior, M.R.; Jacob-Lopes, E.; Zepka, L.Q. Microalgae-based polysaccharides: Insights on production, applications, analysis, and future challenges. *Biocatal. Agric. Biotechnol.* **2022**, *45*, 102491. [CrossRef]
160. Kraan, S. Algal Polysaccharides, Novel Applications and Outlook. In *Carbohydrates—Comprehensive Studies on Glycobiology and Glycotechnology*; Chang, C.-F., Ed.; IntechOpen: Rijeka, Croatia, 2012.
161. Lin, J.; Jiao, G.; Kermanshahi-Pour, A. Algal Polysaccharides-Based Hydrogels: Extraction, Synthesis, Characterization, and Applications. *Mar. Drugs* **2022**, *20*, 306. [CrossRef] [PubMed]
162. Wiegand, C.; Heinze, T.; Hipler, U.C. Comparative in vitro study on cytotoxicity, antimicrobial activity, and binding capacity for pathophysiological factors in chronic wounds of alginate and silver-containing alginate. *Wound Repair Regen.* **2009**, *17*, 511–521. [CrossRef] [PubMed]
163. Lee, K.Y.; Mooney, D.J. Alginate: Properties and biomedical applications. *Prog. Polym. Sci.* **2012**, *37*, 106–126. [CrossRef]
164. Wang, L.; Jayawardena, T.U.; Hyun, J.; Wang, K.; Fu, X.; Xu, J.; Gao, X.; Park, Y.; Jeon, Y.-J. Antioxidant and anti-photoaging effects of a fucoidan isolated from *Turbinaria ornata*. *Int. J. Biol. Macromol.* **2023**, *225*, 1021–1027. [CrossRef]
165. Hoyer, K.; Karsten, U.; Wiencke, C. Induction of sunscreen compounds in Antarctic macroalgae by different radiation conditions. *Mar. Biol.* **2002**, *141*, 619–627.
166. Choi, S.Y.; Lee, S.Y.; Kim, H.G.; Jeong, J.C.; Batara, D.C.; Kim, S.H.; Cho, J.Y. Shinorine and porphyra-334 isolated from laver (*Porphyra dentata*) inhibit adipogenesis in 3T3-L1 cells. *Food Sci. Biotechnol.* **2022**, *31*, 617–625. [CrossRef]
167. Lalegerie, F.; Lajili, S.; Bedoux, G.; Taupin, L.; Stiger-Pouvreau, V.; Connan, S. Photo-protective compounds in red macroalgae from Brittany: Considerable diversity in mycosporine-like amino acids (MAAs). *Mar. Environ. Res.* **2019**, *147*, 37–48. [CrossRef]
168. Figueroa, F.L.; Domínguez-González, B.; Korbee, N. Vulnerability and acclimation to increased UVB radiation in three intertidal macroalgae of different morpho-functional groups. *Mar. Environ. Res.* **2014**, *97*, 30–38. [CrossRef]
169. Sommaruga, R.; Garcia-Pichel, F. UV-absorbing mycosporine-like compounds in planktonic and benthic organisms from a high-mountain lake. *Arch. Hydrobiol.* **1999**, *144*, 255–269. [CrossRef]
170. Görünmek, M.; Ballık, B.; Çakmak, Z.E.; Çakmak, T. Mycosporine-like amino acids in microalgae and cyanobacteria: Biosynthesis, diversity, and applications in biotechnology. *Algal Res.* **2024**, *80*, 103507. [CrossRef]

171. Raj, S.; Kuniyil, A.M.; Sreenikethanam, A.; Gugulothu, P.; Jeyakumar, R.B.; Bajhaiya, A.K. Microalgae as a Source of Mycosporine-like Amino Acids (MAAs); Advances and Future Prospects. *Int. J. Environ. Res. Public Health* **2021**, *18*, 12402. [CrossRef] [PubMed]
172. Sakamoto, T.; Hashimoto, A.; Yamaba, M.; Wada, N.; Yoshida, T.; Inoue-Sakamoto, K.; Nishiuchi, T.; Matsugo, S. Four chemotypes of the terrestrial cyanobacterium *Nostoc commune* characterized by differences in the mycosporine-like amino acids. *Phycol. Res.* **2019**, *67*, 3–11. [CrossRef]
173. Singh, S.P.; Klisch, M.; Sinha, R.P.; Häder, D.P. Effects of abiotic stressors on synthesis of the mycosporine-like amino acid shinorine in the cyanobacterium *Anabaena variabilis* PCC 7937. *Photochem. Photobiol.* **2008**, *84*, 1500–1505. [CrossRef]
174. Rosic, N. Molecular Mechanisms of Stress Tolerance in Cyanobacteria. In *Ecophysiology and Biochemistry of Cyanobacteria*; Rastogi, R.P., Ed.; Springer: Singapore, 2021; p. 131.
175. Shick, J.M.; Dunlap, W.C.; Pearse, J.S.; Pearse, V.B. Mycosporine-like Amino Acid Content in Four Species of Sea Anemones in the Genus *Anthopleura* Reflects Phylogenetic but Not Environmental or Symbiotic Relationships. *Biol. Bull.* **2002**, *203*, 315–330. [CrossRef]
176. Cruz, A.M.; Gonçalves, M.C.; Marques, M.S.; Veiga, F.; Paiva-Santos, A.C.; Pires, P.C. In Vitro Models for Anti-Aging Efficacy Assessment: A Critical Update in Dermocosmetic Research. *Cosmetics* **2023**, *10*, 66. [CrossRef]
177. Cohen, G.; Jakus, J.; Portillo, M.; Gvirtz, R.; Ogen-Shtern, N.; Silberstein, E.; Ayzenberg, T.; Rozenblat, S. In vitro, ex vivo, and clinical evaluation of anti-aging gel containing EPA and CBD. *J. Cosmet. Dermatol.* **2023**, *22*, 3047–3057. [CrossRef]
178. Ghimeray, A.K.; Jung, U.S.; Lee, H.Y.; Kim, Y.H.; Ryu, E.K.; Chang, M.S. In vitro antioxidant, collagenase inhibition, and in vivo anti-wrinkle effects of combined formulation containing *Punica granatum*, *Ginkgo biloba*, *Ficus carica*, and *Morus alba* fruits extract. *Clin. Cosmet. Investig. Dermatol.* **2015**, *8*, 389–396. [CrossRef]
179. de Almeida, A.; de Oliveira, J.; da Silva Pontes, L.V.; de Souza Júnior, J.F.; Gonçalves, T.A.F.; Dantas, S.H.; de Almeida Feitosa, M.S.; Silva, A.O.; de Medeiros, I.A. ROS: Basic Concepts, Sources, Cellular Signaling, and its Implications in Aging Pathways. *Oxidative Med. Cell. Longev.* **2022**, *2022*, 1225578. [CrossRef]
180. Sies, H.; Berndt, C.; Jones, D.P. Oxidative Stress. *Annu. Rev. Biochem.* **2017**, *86*, 715–748. [CrossRef] [PubMed]
181. Deng, J.; Cheng, W.; Yang, G. A novel antioxidant activity index (AAU) for natural products using the DPPH assay. *Food Chem.* **2011**, *125*, 1430–1435. [CrossRef]
182. Storey, A.; McArdle, F.; Friedmann, P.S.; Jackson, M.J.; Rhodes, L.E. Eicosapentaenoic Acid and Docosahexaenoic Acid Reduce UVB- and TNF-α-induced IL-8 Secretion in Keratinocytes and UVB-induced IL-8 in Fibroblasts. *J. Investig. Dermatol.* **2005**, *124*, 248–255. [CrossRef] [PubMed]
183. Kendall, A.C.; Pilkington, S.M.; Murphy, S.A.; Carratore, F.D.; Sunarwidhi, A.L.; Kiezel-Tsugunova, M.; Urquhart, P.; Watson, R.E.B.; Breitling, R.; Rhodes, L.E.; et al. Dynamics of the human skin mediator lipidome in response to dietary ω-3 fatty acid supplementation. *FASEB J.* **2019**, *33*, 13014–13027. [CrossRef]
184. Xie, C.-L.; Liu, Q.; Xia, J.-M.; Gao, Y.; Yang, Q.; Shao, Z.-Z.; Liu, G.; Yang, X.-W. Anti-Allergic Compounds from the Deep-Sea-Derived Actinomycete *Nesterenkonia flava* MCCC 1K00610. *Mar. Drugs* **2017**, *15*, 71. [CrossRef]
185. Zaky, A.A.; Simal-Gandara, J.; Eun, J.B.; Shim, J.H.; Abd El-Aty, A.M. Bioactivities, Applications, Safety, and Health Benefits of Bioactive Peptides From Food and By-Products: A Review. *Front. Nutr.* **2021**, *8*, 815640. [CrossRef]
186. Graves, P.R.; Haystead, T.A. Molecular biologist's guide to proteomics. *Microbiol. Mol. Biol. Rev.* **2002**, *66*, 39–63. [CrossRef]
187. Lauritano, C.; Ferrante, M.I.; Rogato, A. Marine natural products from microalgae: An-omics overview. *Mar. Drugs* **2019**, *17*, 269. [CrossRef]
188. Al-Amrani, S.; Al-Jabri, Z.; Al-Zaabi, A.; Alshekaili, J.; Al-Khabori, M. Proteomics: Concepts and applications in human medicine. *World J. Biol. Chem.* **2021**, *12*, 57–69. [CrossRef]
189. He, S.-J.; Li, J.; Zhou, J.-C.; Yang, Z.-Y.; Liu, X.; Ge, Y.-W. Chemical proteomics accelerates the target discovery of natural products. *Biochem. Pharmacol.* **2024**, *230*, 116609. [CrossRef]
190. Patwary, Z.P.; Zhao, M.; Wang, T.; Paul, N.A.; Cummins, S.F. A Proteomic Analysis for the Red Seaweed *Asparagopsis taxiformis*. *Biology* **2023**, *12*, 167. [CrossRef] [PubMed]
191. Brown, K.A.; Tucholski, T.; Eken, C.; Knott, S.; Zhu, Y.; Jin, S.; Ge, Y. High-Throughput Proteomics Enabled by a Photocleavable Surfactant. *Angew. Chem. Int. Ed. Engl.* **2020**, *59*, 8406–8410. [CrossRef]
192. Montone, C.M.; Capriotti, A.L.; Cavaliere, C.; La Barbera, G.; Piovesana, S.; Zenezini Chiozzi, R.; Laganà, A. Peptidomic strategy for purification and identification of potential ACE-inhibitory and antioxidant peptides in *Tetradesmus obliquus* microalgae. *Anal. Bioanal. Chem.* **2018**, *410*, 3573–3586. [CrossRef]
193. Schmollinger, S.; Mühlhaus, T.; Boyle, N.R.; Blaby, I.K.; Casero, D.; Mettler, T.; Moseley, J.L.; Kropat, J.; Sommer, F.; Strenkert, D.; et al. Nitrogen-Sparing Mechanisms in *Chlamydomonas* Affect the Transcriptome, the Proteome, and Photosynthetic Metabolism. *Plant Cell* **2014**, *26*, 1410–1435. [CrossRef] [PubMed]
194. Shang, C.; Zhu, S.; Wang, Z.; Qin, L.; Alam, M.A.; Xie, J.; Yuan, Z. Proteome response of *Dunaliella parva* induced by nitrogen limitation. *Algal Res.* **2017**, *23*, 196–202. [CrossRef]

195. Mühlroth, A.; Li, K.; Røkke, G.; Winge, P.; Olsen, Y.; Hohmann-Marriott, M.F.; Vadstein, O.; Bones, A.M. Pathways of Lipid Metabolism in Marine Algae, Co-Expression Network, Bottlenecks and Candidate Genes for Enhanced Production of EPA and DHA in Species of Chromista. *Mar. Drugs* **2013**, *11*, 4662. [CrossRef]
196. Andrade, L.M.; Tito, C.A.; Mascarenhas, C.; Lima, F.A.; Dias, M.; Andrade, C.J.; Mendes, M.A.; Nascimento, C.A.O. *Chlorella vulgaris* phycoremediation at low Cu(+2) contents: Proteomic profiling of microalgal metabolism related to fatty acids and CO(2) fixation. *Chemosphere* **2021**, *284*, 131272. [CrossRef]
197. Plouviez, M.; Dubreucq, E. Key Proteomics Tools for Fundamental and Applied Microalgal Research. *Proteomes* **2024**, *12*, 13. [CrossRef]

marine drugs

MDPI

Review

Biotechnological Potential of Macroalgae during Seasonal Blooms for Sustainable Production of UV-Absorbing Compounds

Nedeljka Rosic [1,2,*] and Carol Thornber [3]

1 Faculty of Health, Southern Cross University, Gold Coast, QLD 4225, Australia
2 Marine Ecology Research Centre, Southern Cross University, Lismore, NSW 2480, Australia
3 Department of Natural Resources Science, University of Rhode Island, 120 Flagg Road, Kingston, RI 02881, USA; thornber@uri.edu
* Correspondence: nedeljka.rosic@scu.edu.au

Abstract: Marine macroalgae (seaweeds) are important primary global producers, with a wide distribution in oceans around the world from polar to tropical regions. Most of these species are exposed to variable environmental conditions, such as abiotic (e.g., light irradiance, temperature variations, nutrient availability, salinity levels) and biotic factors (e.g., grazing and pathogen exposure). As a result, macroalgae developed numerous important strategies to increase their adaptability, including synthesizing secondary metabolites, which have promising biotechnological applications, such as UV-absorbing Mycosporine-Like Amino Acid (MAAs). MAAs are small, water-soluble, UV-absorbing compounds that are commonly found in many marine organisms and are characterized by promising antioxidative, anti-inflammatory and photoprotective properties. However, the widespread use of MAAs by humans is often restricted by their limited bioavailability, limited success in heterologous expression systems, and low quantities recovered from the natural environment. In contrast, bloom-forming macroalgal species from all three major macroalgal clades (Chlorophyta, Phaeophyceae, and Rhodophyta) occasionally form algal blooms, resulting in a rapid increase in algal abundance and high biomass production. This review focuses on the bloom-forming species capable of producing pharmacologically important compounds, including MAAs, and the application of proteomics in facilitating macroalgal use in overcoming current environmental and biotechnological challenges.

Keywords: macroalgae; algal bloom; *Sargassum*; *Ulva*; *Gracilaria*; photoprotection; ultraviolet radiation; mycosporine-like amino acids; sunscreens; proteomics

1. Introduction

Organisms are exposed to diverse levels of ultraviolet radiation (UVR: 280–400 nm) depending on the geographic location. In areas near the equator, the detected UVR levels are extremely high, while UVR levels measured at both poles are very low. Beyond latitude, the seasonal viability in UVR levels also needs to be considered regarding organisms' capacity to adapt to the periods of high UVR that can be much higher during summer times compared to winter [1]. Additional environmental factors, such as altitude and clouds, could also impact UV levels [2].

Prolonged exposure to UVR may lead to DNA damage, resulting in cellular mutations and long-lasting negative impacts. Humans are especially sensitive to prolonged UVR exposure, leading to more incidences of skin cancers [3–5]. Current sunscreens are missing the sufficient sustainable features that are needed for environmental protection. Therefore, natural products (NPs) that could be isolated from species exposed to high levels of UVR are currently very attractive options with clear industrial interest. Potentially promising sources of UV-absorbing NPs come from abundant marine species that are naturally exposed to high UVR, like algae (including macroalgae and microalgae), and can be used to generate sustainable and environmentally friendly sunscreens.

Marine macroalgae (seaweeds) form the base of many marine and estuarine food webs worldwide. Most members of this polyphyletic grouping belong to one of three main clades: Chlorophyta (green algae), Phaeophyceae (brown algae), and Rhodophyta (red algae). Macroalgae can be found from polar regions to the tropics and range from inhabiting the intertidal zone to nearly 300 m in depth [6,7] in tropical waters (primarily rhodoliths, a group of coralline red algae), although most live in shallower waters of 100 m or less [6]. Macroalgae are all dependent upon specific physical parameters (including light, temperature, salinity, and nutrients) for survival, growth, and reproduction. Macroalgae serve as habitats and/or food sources for a wide variety of marine organisms, modify wave action in coastal areas and serve in blue carbon sequestration pathways [8–11]. They are also part of a >USD 13 billion dollar global aquaculture industry for human uses, including direct consumption and biomedical and pharmaceutical industries, among others [12–15]. Macroalgae have evolved to occupy a diverse suite of ecological and environmental niches, with some capable of forming algal blooms. While some species are primarily adapted to cold temperate to polar regions, other groups thrive in tropical locations. Some species have adapted to live in highly stochastic intertidal environments, with diel swings in temperature, salinity and UV exposure, while others occupy much more constant environments in subtidal habitats. Species living in more stochastic environments have evolved with a wide array of defensive compounds and mechanisms. Species in intertidal environments subject to freezing temperatures have evolved to survive the freeze–thaw cycles [16], and most brown macroalgal species have phenolic compounds that protect against a variety of biotic and abiotic stressors [17,18]. Some intertidal species can reduce photosynthetic activity when emersed, which may reduce damage from excess light [19], even though they have increased access to CO_2 [20].

Many macroalgal species living in high-light environments contain mycosporine-like amino acids (MAAs), which are small, temperature- and light-stable, and water-soluble UV-absorbing compounds, with maximal absorption within the range of ~310–360 nm [12,21–23]. Strong photoprotective properties and the capacity to absorb light in the UV-A (315–400 nm) and UV-B (280–315 nm) ranges without the generation of harmful free radicals have been confirmed for various MAAs [24–26]. Dominant UV-A, which makes ~95% of UV energy reaching the Earth's surface at moderate levels, has a stimulating role in macroalgal growth and photosynthesis, while UV-B usually has a more harmful impact on marine macroalgae [27,28]. Although UV-A can enhance algal photosynthesis in moderate doses, high levels can reduce quantum yield [29] and inhibit photosynthesis [30], which can be particularly damaging to microscopic stages [31]. Although many species thrive in physiologically stressful habitats, there is still limited understanding and ability to predict the interactive impacts of multiple abiotic stressors on macroalgae [32,33].

In ancient times, human uses of macroalgae started independently across the globe; Romans used it as food for animals [34], macroalgal remnants have been found in hearths in southern Chile from 12,000 BCE, and medicinal uses of macroalgae were documented nearly 5000 years ago in traditional Chinese herbal medicine [35], as well as many other cultures. Since then, a wide range of applications have been developed, and in modern times, macroalgae have been used in human and animal food, pharmaceutical and other industries, and cosmetics for skin protection due to the presence of components with anti-aging properties, photoprotective, and specifically UV-absorbing capacities (Figures 1 and 2).

Macroalgae also have an ecologically important role as bioindicators of water quality and have been utilized for bioremediation strategies and the removal of waste products, including heavy metals [36,37]. Pesticides used in local agriculture negatively impact water quality and aquatic ecosystems [38,39], and various macroalgae, such as the red macroalga, *Gracilaria lemaneiformis*, were successfully used for the reduction in insecticide cypermethrin concentration [40], or the brown alga *Saccharina japonica*, for the removal of the herbicide glyphosate in saline waters [41].

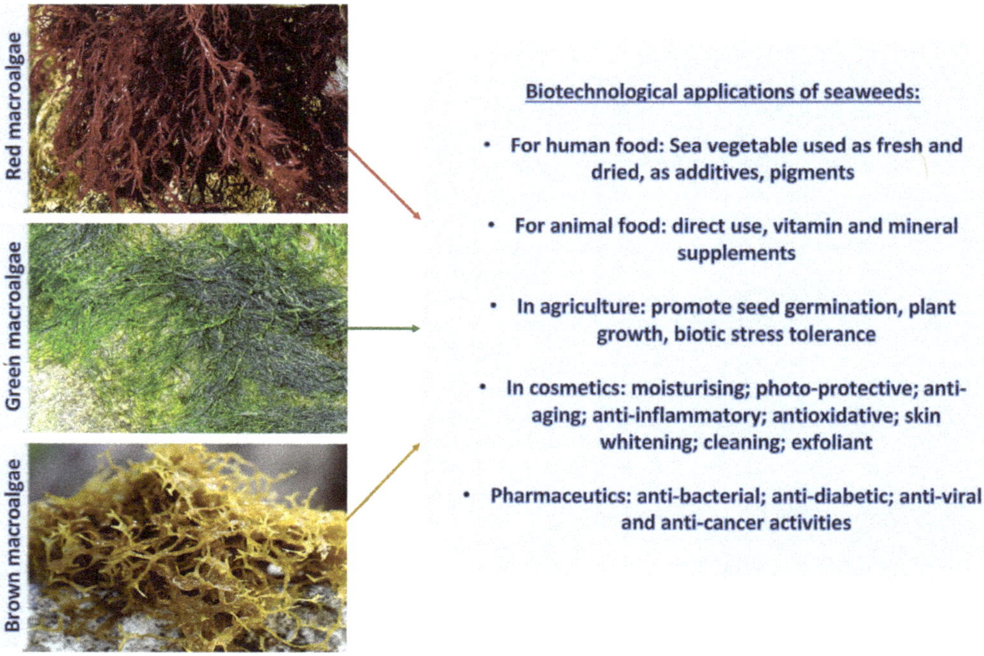

Figure 1. An overview of biotechnological applications of macroalgae in various industries.

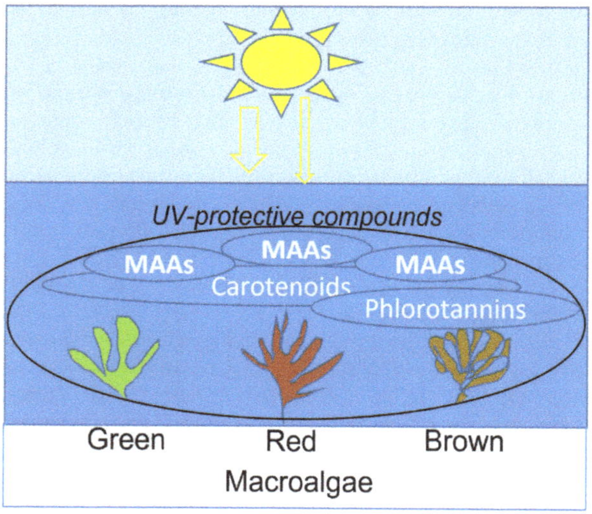

Figure 2. Distribution of main UV-absorbing compounds in macroalgae.

Human uses of macroalgae have typically focused on either cultivated macroalgae or those harvested from naturally growing populations. Relatively little research has focused on the vast pools of bloom-forming macroalgae, which are sometimes washed ashore (i.e., beach cast or wrack algae), as potential sources of compounds for human usage. Because some bloom-forming species are known to produce UV-absorbing compounds such as MAAs [12], blooms may represent an additional resource for the isolation and purification of these compounds. The aim of this review is to (1) provide an overview of macroalgae forming algal blooms, including the factors triggering these events; (2) determine the biotechnological capacity of selected macroalgal species; (3) evaluate the potential of using these species as a source of UV-absorbing compounds. Furthermore, this review assesses the current application of proteomics for evaluating and utilizing macroalgae capable of forming algal blooms as a sustainable resource for future sunscreens.

2. Macroalgal Blooms

2.1. Bloom Overview

Algal blooms are naturally occurring events described as a rapid increase in algal abundance in both microalgae and/or macroalgae, and usually lasting for weeks to months [42,43]. This review is focused specifically on blooms of macroalgae, as bloom-forming species can be found in all three major groups of marine macroalgae [44–46]. While many species can occur in both benthic (attached) and pelagic (drift) states, a subset of species typically reach sufficient quantities of drift biomass, characterized as harmful macroalgal blooms. Unlike benthic populations of macroalgae, blooms are characterized by large floating or drifting mats, leading to increased biomass via fragmentation and/or reproduction processes. While some environmental parameters, including light, temperature, and salinity, are well understood to trigger macroalgal booms [47], the exact initiations for a particular species (or group of species) to form a bloom are frequently dependent upon a complex interaction of these abiotic and other biotic factors [48].

Harmful macroalgal blooms can have significant adverse environmental and economic impacts on their surroundings. Blooms can impact coral reefs, reduce solar radiation for deeper-dwelling species, hamper gas exchange, and outcompete seagrass [49–52]. Blooms frequently occur in coastal areas, impeding marine aquaculture, fishing, recreation, and tourism activities. Bloom biomass can foul beaches, ruin fishing gear, and impede fishing, deter tourism, and impede coastal use by a variety of human stakeholders [53,54]. As the location of a particular bloom is impacted by tidal motion, wave energy, and wind direction, their distribution can shift rapidly, making sustained monitoring and/or removal efforts more challenging. In addition, blooms frequently deposit large amounts of biomass on and near coastlines (called beach-cast algae) and subsequently decompose over a period of days to weeks [10,55–57]. This decomposition, due to microbial activity, creates hypoxic conditions, harming fish and benthic marine invertebrate communities [58]. In addition, some species release hydrogen sulfide upon decomposition [59] and have a negative impact on carbon sediments [60]. Rather than only serving as a nuisance (or worse) to coastal communities, these beach-cast macroalgae can represent an important source of biomass for human usage. While current efforts typically focus on the removal of beach-cast algae and its subsequent deposition in landfills, the opportunity to use the algae for one (or more) technological applications remains understudied [57].

Most bloom-forming species are characterized as being able to survive increased levels of physical stress due to their presence at immediately above or below the waterline. In these environments, they are subjected to increased fluctuations in temperature and salinity (both high and low), as well as higher UV radiation exposure compared to their deeper-dwelling counterparts. Many species of red, green, and brown macroalgae with these characteristics have been documented in blooms, and these blooms are frequently deposited on shorelines, representing a potential for their harvest and utilization. Some of the most prevalent and well-studied blooms include those of the green sea lettuces (*Ulva* spp.), the brown algae *Sargassum* spp. (sometimes referred to as gulfweed), and the red algae *Gracilaria* spp. [45,48,61–63]. All three of these taxonomic groups have significant promise for the commercial utilization of bloom biomass [14,64–67]. Although not covered explicitly in this review, we recognize that many other macroalgal species can form blooms and/or be deposited on beaches in mass quantities, including the genera *Asparagopsis, Ecklonia, Dasysiphonia* [68–70], among many others (see review [57]), which indicates the widespread availability of bloom tissue for potential biotechnological uses.

2.2. *Ulva* Blooms

The genus *Ulva* contains approximately 100 taxonomically accepted species worldwide, which form thin blades and/or tubes and live predominately in shallow marine and estuarine environments (algaebase.org accessed on 7 June 2023). Blooms of several species of *Ulva* (also known as green tides) occur worldwide in predominantly temperate and subtropical coastal systems. *Ulva* blooms gained worldwide notoriety in 2008 during the Summer Olympics, when their presence threatened the Olympic sailing events [71]. In addition to the Yellow Sea and East China Sea, *Ulva* blooms have been reported near Brittany, France [72], Venice, Italy [73], California and Washington, USA [74], the New England region, USA; [46,75,76], South Africa [77], and the Gulf of California, Mexico [78].

Ulva blooms are frequently triggered by increases in nutrients on either localized or regional scales [79–81], and co-occurring *Ulva* species may react differently to the same environmental triggers [82]. *U. prolifera* blooms in the south Yellow Sea have been linked to the cultivation of nori (the red alga *Pyropia*—previously identified as *Porphyra*) [79,80]. In these blooms, the predominant mechanism of spread originates from attached *U. prolifera* on the floating *Pyropia* aquaculture rafts. These patches then dislodge and drift with prevailing surface currents into the northern Yellow Sea (see summary in Zhang et al. 2017 [80]), and their growth is enhanced by a combination of inorganic and organic nitrogen sources [83]. In Narragansett Bay, Rhode Island USA, *Ulva* blooms frequently contain both *U. compressa* and *U. lacinulata* (as *U. rigida*), and their growth rates are positively impacted by increases in dissolved inorganic nitrogen due, in part, to outputs from sewage treatment plants and changing rainfall patterns [82,84]. Although *U. lacinulata* and *U. compressa* co-occur in blooms, they vary in their growth rates, thermal tolerances, production of allelopathic chemicals, and susceptibility to herbivory [75,85].

Ulva can reproduce both sexually and vegetatively through fragmentation, with few top-down controls on its spread [75,86]. Most of the environments where *Ulva* blooms are found are in the intertidal/shallow subtidal, which are high-light (and coupled UV-A and UV-B) conditions [87]. In many ecosystems, *Ulva* bloom biomass is deposited on shorelines due to changes in wind, water currents, and/or tidal patterns [48,80], thereby exacerbating its impacts on coastal communities.

Ulva's ability to withstand high irradiance has been well documented [87,88], although DNA and photosystem II damage can result [89]. *Ulva* has physiological protective mechanisms that limit the accrual of DNA damage due to UV-B [90], and *Ulva* can undergo photoinhibition during periods of increased UV-B [91]. In addition, some *Ulva* generates higher concentrations of UV-B absorbing pigments than other genera [92]. However, they lack high levels of UV-absorbing compounds like mycosporine-like amino acids and phlorotannins that are found in red and brown macroalgae [93,94].

2.3. *Sargassum* Blooms

The brown algal genus *Sargassum* contains many species found in temperate and tropical systems across the globe. However, large floating mats of pelagic *Sargassum* are typically found in the North Atlantic Ocean. There are two species that are most abundant in this region: *Sargassum fluitans* and *S. natans* [66]. The aptly named Sargasso Sea has been reported from the 15th century onwards [45]; it occupies a wide swath of ocean in the North Atlantic sub-tropical gyre. Although the *Sargassum* in the Sargasso Sea is not typically considered to be a bloom due to its longevity, it is characterized by the same qualities of having drifting, large quantities of biomass similar to those found in macroalgal blooms. The *Sargassum* species in the Sargasso Sea are well recognized for their importance as a habitat for numerous fish and invertebrate species as well as food sources [95]. As a genus, *Sargassum* can reproduce sexually or asexually; pelagic *S. fluitans* and *S. natans* reproduce via the latter mechanism, with vegetative growth and division [66], which enhances their potential rate of biomass increase.

In contrast to the long-documented Sargasso Sea, it is only within the past fifteen years that the presence of a 'Great Atlantic *Sargassum* Belt' has been identified; this *Sargassum* belt originates in the equatorial Atlantic, not the Sargasso Sea [96]. The annual mega bloom ('golden tide') extends from West Africa to the Gulf of Mexico and contains over 20 million tons of *Sargassum* [45,97]; when it reaches coastlines in the Caribbean and Florida (USA) and is deposited on beaches, it can wreak havoc on the environment and economies of local communities [98,99]. Nutrient enrichment is most likely the cause of these blooms, including the introduction of nutrients from increased flooding in the Amazon basin as well as periodic upwelling along the western coast of Africa; as such, golden tides can be viewed as indicators of large-scale eutrophication [100,101]. Another species, *Sargassum horneri*, has been reported to form blooms in the Yellow Sea, indicating that the spread of blooms by this cosmopolitan genus is occurring [61].

By living at or close to the ocean's surface, *Sargassum* is subject to high levels of UV radiation. Many brown algae contain protective antioxidants, including phenolics, carotenoids, and/or isoprenoids [102] to combat the impacts of UV stressors. As a genus, *Sargassum* is no exception to this pattern [103,104]. *S. filipendula*, like other *Sargassum* species, has high antioxidant activity [102]. Species can also undergo structural changes due to UV exposure; *S. cymosum* increases the abundance of phenolic compounds and thickens its cell walls in response to UVR exposure [105]. However, assessments of the physiological properties of pelagic *Sargassum* can be challenging due to logistical constraints, as pelagic *Sargassum* does not grow well in traditional culturing conditions [106]. A recent study of *S. horneri* in the Yellow Sea [107] found that its photosynthetic activity is decreased by exposure to UVR, with higher tissue concentrations of malondialdehyde (MDA) in specimens exposed to UVR. At the same time, the production of carotenoids and UV-absorbing compounds was increased, indicating photoprotective mechanisms that allow *Sargassum* existence in high-light, high-UV conditions at the ocean's surface. Similarly, pelagic *S. natans* and *S. fluitans* can increase carotenoid production as a result of increased light exposure [108]. In addition, pelagic *S. natans* and *S. fluitans* can release large amounts of dissolved inorganic carbon (DOC), with a high concentration of phlorotannins, a class of polyphenolics [109].

2.4. Gracilaria Blooms

The red algal genus *Gracilaria* is typically found in intertidal and shallow subtidal estuarine and rocky habitats in tropical and temperate zones and is well known for its bloom-forming capabilities. This highly branched genus can fragment easily and persist while floating in nearshore habitats, increasing its potential for forming large-scale blooms in coastal systems. Two species, *Gracilaria tikvahiae* and *G. vermiculophylla*, have frequently been documented in blooms in the Atlantic Ocean, ranging from Florida to Maine USA [84,110,111], Portugal [112], northern Europe [113] and the Gulf of California, Mexico [63], among other locations. These *Gracilaria* blooms are frequently deposited on shorelines in large amounts [48,114]. Although most studies of *Gracilaria* blooms have focused on these two species, other species, such as *G. tenuistipitata* in Shenzhen Bay, China, have also been documented as forming blooms [115]. Like their bloom-forming counterparts in the green and brown algae, these species live at or near the ocean's surface and are thus subjected to high light and UV radiation levels. Some species of *Gracilaria* are also cultivated extensively, primarily for agar production or direct human consumption [116].

3. Macroalgae as a Source of UV-Absorbing Compounds

The biotechnological potential of macroalgae includes the range of molecules from polysaccharides, lipids, proteins, pigments and phenolic composites to various halogenated derivatives [117]. Macroalgae, like many other marine organisms, have been exposed to severe variations in environmental conditions that forced them to adjust, adapt and survive under various external pressures [118]. Macroalgae, like other sessile organisms, experience variable abiotic conditions, including temperature, light irradiance, salinity, and water turbidity, impacting their physiological performance [19,32,51,119–125]. These stressful conditions especially require adaptability to high levels of UVR, specifically ultraviolet

A (UVA; 320–400 nm) and ultraviolet B (UVB; 280–315 nm) [34]. Macroalgal adaptability to extremely high light conditions is facilitated via a range of secondary metabolites, such as different photoprotective pigments (i.e., chlorophyll and carotenoids), as well as UV-absorbing compounds, such as mycosporine-like amino acids (MAAs). Compared to terrestrial plants, macroalgal species have the advantageous capacity to produce biofuels and chemicals due to the renewable nature of these resources [126]. The presence of MAAs in macroalgae was confirmed in 486 species of red algae, 45 species of green algae, and 41 species of brown algae [127]. However, the variability in MAA content and profiles in macroalgae and other marine species are noted to be strongly influenced by the environmental conditions, symbiosis, nutrient bioavailability (e.g., ammonium availability), as well as seasonal changes, especially variation in irradiance levels [128–133]. MAAs biotechnological potential is well recognized due to their pharmacological properties, including antioxidant capacities, the ability to suppress singlet oxygen-induced damage [127,129,134,135], anti-inflammatory and anti-aging properties [136–138]. To be able to improve the biotechnological application and use of MAAs in cosmetics and for medical purposes, a number of challenges need to be overcome, including the lack of sufficient research regarding the steps necessary for obtaining purified MAA standards, as well as overcoming strong water solubility issues and the low yields of MAAs coming from natural resources. These challenges are in some parts overpowered by an application of various methods for MAA isolation and characterizations [139] and via a use of heterologous expression systems [127,140–142] and/or by stimulation of MAA synthesis via specific nutrient and UVR stimulative conditions [90,128,143–146]. Recovered MAA quantities isolated from different red algae including bloom-forming *Gracilaria* sp. were found to be significantly impacted by extraction solvent used, with the highest yield (increased up to 32.34%) obtained when 25% of ethanol was used compared to other extraction solvents [139]. In addition, there are still significant challenges in harvesting bloom biomass from beach cast areas as well as the water, rapidly separating species of interest from harvested blooms, purifying algal tissues for the extraction of compounds of interest, and conducting all steps with an economically viable approach [57,147,148].

Macroalgal species capable of forming algal blooms are especially attractive for biotechnological applications due to the potential economic benefits resulting from their high biomass (Figure 3). Various bioactive compounds can be utilized for different applications ranging from UV protection, food sources, and cosmetics to eco-friendly biopesticides (Table 1). MAA production is confirmed in all three major groups of macroalgae at different profiles and compositions [127]. In green algae, 45 species report detectable levels of MAAs, with the highest quantities identified in species from the class Prasiolales, with MAA content of more than 3.5 mg/g DW, such as reported in *Prasiola crispa* [149], while lower or no detectable MAA quantities are reported in some *Ulva* species [127]. Many macroalgal blooms are dominated by green macroalgae, with the majority of blooms resulting from *Ulva* species [150]. In the case of the bloom-forming genus *Ulva* [151] that leads to overwhelming green tides across the world, mycosporine-glycine and porphyra-334 are confirmed in detectable levels [127]. Biotechnological applications of ulvan beyond MAAs also includes other sulfated polysaccharide with promising anti-cancer, anti-viral, antioxidant and other pharmacological activities [152].

Increasing levels of golden algal blooms are reported, especially for *Sargassum* blooms [150]. Brown algae (the class Phaeophyceae) contain the pigments chlorophyll and carotenoids, including fucoxanthin, which is important for photo and antioxidant protection, as well as polyphenolic compounds phlorotannins and complex carbohydrates laminarin and mannitol [153,154]. MAAs are reported in over 40 species coming from brown algae, including some *Saragasumm* species, such as *Sargassum oligocystum* and *S. fluitans* [127]. The invasive species from the genus *Sargassum* (F. Sargassaceae) contain bioactive compounds demonstrating anti-bacterial, anti-inflammatory, antioxidant, anti-tumor, anti-viral, and other pharmacologically promising activities [155], while UVB absorption is confirmed in ethanol extract demonstrating [156] and UVA photoprotective properties [157]. *Sargassum* species are well distributed in tropical and subtropical climate regions and have been used as a source of food for a long time due to their nutrient values (a rich source of vitamins, proteins, and minerals). In addition, pharmacological bioactivities recognized in *Sargassum* species include antioxidant, anti-fouling, anti-microbial, and anti-tumor activities [14]. For example, UV protective, anti-microbial and anti-inflammatory activities are demonstrated in *Sargassum cristaefolium* with confirmed presence of MAA palythene [157,158]. Multiple other MAAs are confirmed in another *Saragassum* species (Table 2).

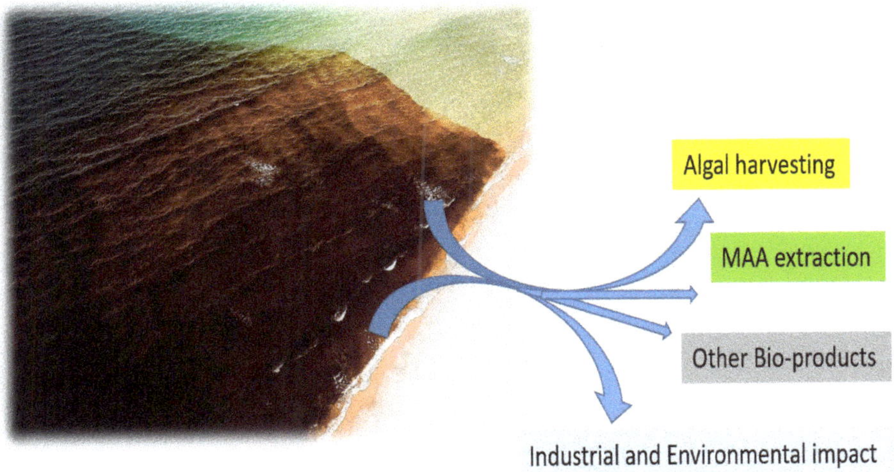

Figure 3. The most economically beneficial source of biomass for massive MAA and other bio-products manufacturing.

Red algae can produce promising bioactive compounds and pigments [14]. Therefore, red macroalgae capable of forming algal blooms are especially attractive due to the high growth and biomass available [117]. On the other hand, the species from the order Bonnemaisoniales, *Asparagopsis armata*, recognized as one of the most aggressive invasive macroalgal species and able to form blooms [159], has very promising biotechnology potential containing various secondary metabolites, including MAAs [117]. *Asparagopsis armata* is becoming very abundant in some countries, presenting a huge negative environmental and economic impact on local communities [160,161]. Although MAA levels are the highest in species coming from the genera *Porphyra* and *Bangia*, the species from the genus *Asparagopsis* are also prosperous with MAA [162], with MAA concentration and profile being directly influenced by nitrogen status [163]. Strong capacity for synthesis of UV-absorbing molecules is reported for *Asparagopsis armata* [163], but not for another invasive bloom-forming species *Dasysiphonia japonica* [44] (order Ceramiales), with species with low content of MAAs (1–2 mg g^{-1} DW). Rhodophyta orders characterized by the highest levels of MAAs of >2 mg g^{-1} DW include Bangiales, Gelidiales and Gracilariales [162]. In 23 red algal species analyzed, the most common MAAs are shinorine, palythine, asterina-330 and porphyra-334 [164]. However, even in red algae, the MAA content may vary from low levels (<1 mg g^{-1} of DW) to higher levels above 2 mg g^{-1} of DW in some species [165]. While the bloom-forming *G. vermiculophylla* contains various MAAs (Tables 1 and 2), the MAA content and profile can vary seasonally [166]. For cultivated *G. vermiculophylla*, the highest levels of porphyra-334 and shinorine are reported in November–January, while palythine and asterina-330 are highest from April to August. Similarly, when grown in seawater with elevated nutrient concentrations (150 uM of NH_4^+ and 15 uM of PO_4^{3-}) and increased UVA and UVB levels, MAA production can increase by 50% in *G. cornea* (old name *Hydropuntia cornea*) [167]. The targeted UV radiation also induces a substantial increase in the MAA levels in macroalga *Gracilaria gracilis* compared to control [143]. Similarly, elevated levels of NO_3^- can increase levels of MAAs in *G. tenuistipitata* [162,168]. Thus, as blooms of *Gracilaria* frequently occur during the summer months when UV exposure is higher and nutrient levels are frequently elevated as well [169], they may represent a high-quality source for obtaining the high biomass levels needed for MAA extraction in the future (Figure 3).

Table 1. Macroalgal species forming algal blooms were reported to accumulate high levels of UV-absorbing compounds.

Macroalgal Species	Biotechnology Use [Ref.]	MAAs
Green algae Order: Ulvales *Ulva* spp.	Human and animal nutrients; preservatives; pharmaceuticals; cosmeceuticals	MG, PR [127]
Brown algae Order: Fucales *Sargassum cristaefolium* *Sargassum oligocystum*	Photoprotective activity against UVR; Inhibited proinflammatory TNF-*α* and IL-6 expression while increasing IL-10 production in the BALB/*c* mice skin [157,158]	PE dominant MAAs PR, PI, SH 5 [127,170]
Red algae Order: Bonnemaisoniales *Asparagopsis armata*	High biofiltration capacity of nutrients; UV photoprotection [163] Exudate cocktail as a biopesticide for eco-friendly weed control [171] Preservatives, cosmeceuticals, biopharmaceuticals [117]	MAAs (accumulated only under a high ammonium-N availability) [163] AS, PR, PE, SH, UN [127]
Red algae Order: Gracilariales *Gracilaria vermiculophylla*	Increase in MAAs in freshly released spores increased under UVR 8 h [172]	AS, PE, PR, PI, SH, US, UN [127,166]

Abbreviations: Asterina-330 (AS), Mycosporine-glycine (MG), Palythine (PI), Palythene (PE), Porphyra-334 (PR), Shinorine (SH), Unidentified MAAs (UN), Usujirene (US). Red algae—Rhodophyta; Green algae—Chlorophyta; brown algae—Phaeophyceae.

Table 2. Chemical structure of common MAAs found in bloom forming macroalgal species.

UV-Protective Natural Products	Chemical Structure	Key Properties (Ref)	λ Max (nm) ε Coefficient (M^{-1} cm^{-1}) Molecular Mass (g/mol)
Mycosporine-glycine ($C_{10}H_{15}NO_6$)	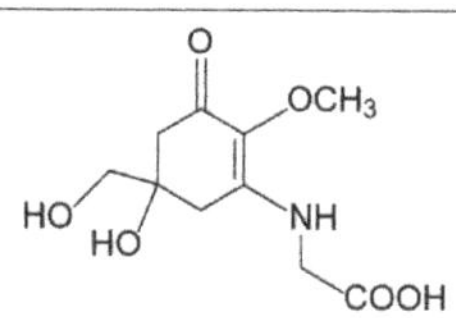	UV-absorbing, antioxidants [134]	310 nm 28,100 M^{-1} cm^{-1} 245 g/mol
Shinorine ($C_{13}H_{20}N_2O_8$)	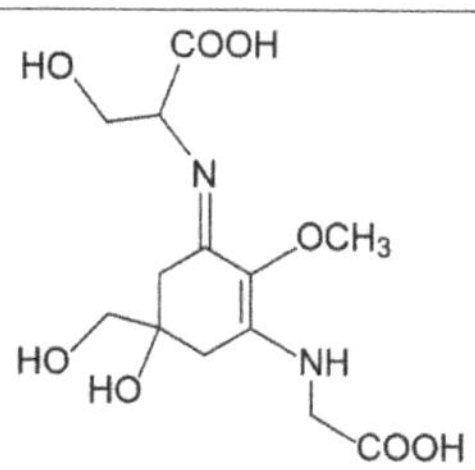	UV-absorbing, antioxidants [135,173]	334 nm 44,668 M^{-1} cm^{-1} 332 g/mol

Table 2. *Cont.*

UV-Protective Natural Products	Chemical Structure	Key Properties (Ref)	λ Max (nm) ε Coefficient (M^{-1} cm^{-1}) Molecular Mass (g/mol)
Usujirene ($C_{13}H_{20}N_2O_5$)	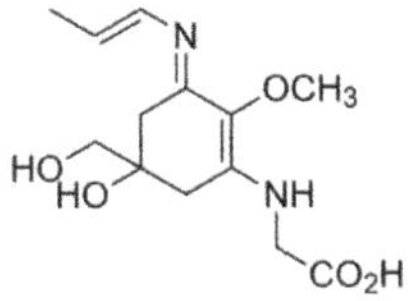	UV-absorbing, antioxidants [174]	357 nm 45,070 M^{-1} cm^{-1} 284 g/mol
Asterina-330 ($C_{12}H_{20}N_2O_6$)	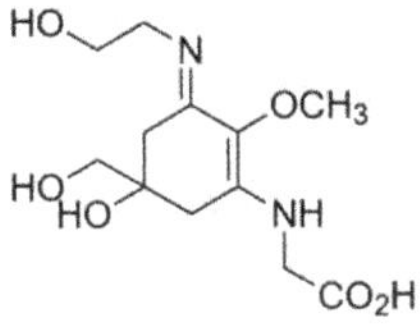	UV-absorbing, antioxidants [175]	330 nm 43,800 M^{-1} cm^{-1} 288 g/mol
Porphyra-334 ($C_{14}H_{22}N_2O_8$)	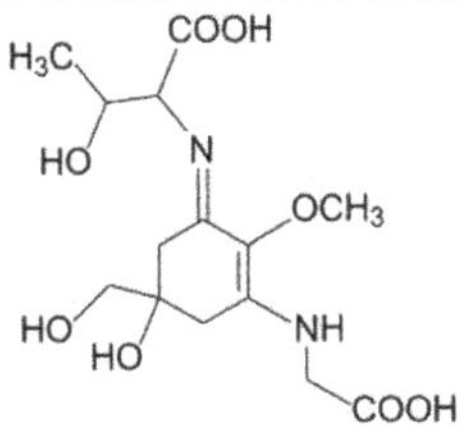	UV-absorbing, antioxidants [135,175]	334 nm 42,300 M^{-1} cm^{-1} 346 g/mol
Palythene ($C_{13}H_{20}N_2O_5$)	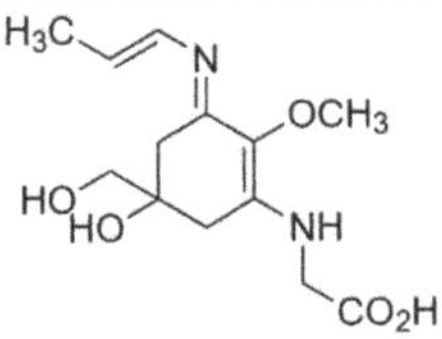	UV-absorbing, antioxidants [176]	360 nm 50,000 M^{-1} cm^{-1} 284 g/mol
Palythine ($C_{10}H_{16}N_2O_5$)	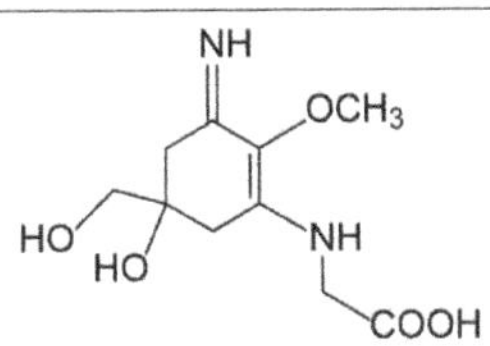	UV-absorbing, antioxidants [175,177]	320 nm 35,500–36,200 M^{-1} cm^{-1} 244 g/mol

4. Proteomics for Monitoring Macroalgal Blooms and Discovery of MAA Profiles

The use of transcriptomics and proteomics in bioinformatic pipelines enables faster discovery and functional characterization of novel marine natural products [178]. The exploration of the biosynthesis of MAAs employed genomic mining techniques [142,179], confirming that MAA production occurs via the shikimate pathway [180,181] and the pentose phosphate pathway [179]. Proteomic data analyses demonstrated that UV-induced MAA production mainly occurs via the shikimate pathway and is, therefore, more critical in photoprotection [182]. Four genes making a core of the MAA pathway (e.g., so-called mys cluster genes) were identified to be dehydroquinate synthase (DHQS), O-methyltransferase (O-MT), adenosine triphosphate (ATP) grasp, and nonribosomalpeptide synthetase (NRPS) in the cyanobacterium *Anabaena variabilis* [142]. Some of these genes are duplicated or have additional *mys*-cluster genes in different organisms [129,179,183,184]. Similar observations were reported in other algal groups with mechanisms such as horizontal gene transfer and acquisition of diverse MAA gene clusters playing a driving role in the development of species with high-temperature resilience [185], plus epigenetics mechanisms influencing gene expression patterns [186].

Various proteomic techniques can be applied to explore MAA synthesis and monitor macroalgal blooms that could be utilized for the biotechnological application of MAAs. These include high-throughput methods such as top-down proteomic methodologies that separate proteins and then complete individual characterization, such as mass spectrometry (MS)-based proteomics [187]. In bottom-up proteomics (also called 'shotgun' proteomics), the proteins first undergo the digestion process, producing a mixture of peptides that are analyzed using MS or LC/MS and compared to existing databases via automated analyses [188]. Most MAA analyses include the purification step, identification and quantification using high-performance liquid chromatography (HPLC) separation and identification based on retention times and UV spectra [189–191]. In addition to HPLC chromatography, the confirmatory analyses for improved MAA characterization and quantification also included the implementation of mass spectrometry, including various types of liquid chromatography (LC/MS) methodologies [139,192–196]. The different HPLC and MS techniques improved the discovery and characterization of MAAs, especially when purified MAAs were characterized by nuclear magnetic resonance (NMR) [173,197]. The use of ultrahigh-performance liquid chromatography (UHPLC) was also applied in MAA analyses [198], combined with hyphenated to orbitrap high-resolution tandem mass spectroscopy for feature-based molecular networking characterization and classification of MAAs, which is one of the most recent advancements in proteomics [199]. This approach incorporates the published MAA fragmentation patterns and uses in silico annotation tools that allow for more accurate identification, discovery, and classification of MAAs [199].

Furthermore, predicting algal bloom events could be critical so that the biotechnological industry can utilize these natural events and harvest a large amount of algal biomass as a source of bioproducts. Proteomic studies of differential expression of specific proteins involved in growth processes and stress response can be potentially useful as biomarkers for the prediction of future algal blooms [200]. In China's coastal waters, distinct genetics patterns were linked to harmful macroalgal blooms (HMBs) involving green and gold tides [201]. Due to the increased negative ecological impact of these HMBs, monitoring these changes and identifying driving regulatory mechanisms is becoming critically important to allow improved scientific forecasting of future algal blooms. Consequently, utilizing other omics datasets was recognized as a promising way to increase the modeling strength for predicting climate-driven algal blooms [202,203].

5. Conclusions

Macroalgal blooms are spontaneous and frequently brief occurrences usually triggered by anthropogenic factors; they are characterized by a sharp rise in drift macroalgal abundance. Consequently, large amounts of macroalgal biomass can be harvested from these blooms and utilized for biotechnological purposes, although there may be technological challenges associated with the proper harvesting of bloom-forming species. Several macroalgal species that have a strong capacity for both algal blooms and large MAA production were identified in this review. Environmental conditions, including seasonal periods of high UVR exposure and nutrient enrichment, which enhance algal growth, also enhance the accumulation of MAAs, which hold excessive biotechnological potential due to their enhanced photoprotective and other pharmacological properties. Advancing new trends, such as applying a novel molecular networking approach, presents a promising field that combines in silico tools with modern high-throughput chemistry methodologies for analyzing the clusters of MAAs chemistry based on their fragmentation patterns for improved characterization and classifications. Finally, advancements in proteomic techniques can empower our understanding of MAA structural diversity and functional significance for better utilization of natural phenomena of algal blooms and future biotechnological developments (i.e., sunscreens) in an environmentally sustainable manner.

Funding: This research received no external funding.

Institutional Review Board Statement: Not applicable, as this study did not involve humans or animals.

Data Availability Statement: The original data presented in the study are included in the article; further inquiries can be directed to the corresponding author.

Acknowledgments: The author would like to thank the two anonymous reviewers, as well as Isidora Skrlin, for their critical reviews of this paper.

Conflicts of Interest: The authors declare no conflict of interest.

References

1. Neale, R.E.; Hamilton, A.R.; Janda, M.; Gies, P.; Green, A.C. Seasonal variation in measured solar ultraviolet radiation exposure of adults in subtropical Australia. *Photochem. Photobiol.* **2010**, *86*, 445–448. [CrossRef] [PubMed]
2. Blumthaler, M.; Ambach, W.; Ellinger, R. Increase in solar UV radiation with altitude. *J. Photochem. Photobiol. B Biol.* **1997**, *39*, 130–134. [CrossRef]
3. Olsen, C.M.; Wilson, L.F.; Green, A.C.; Bain, C.J.; Fritschi, L.; Neale, R.E.; Whiteman, D.C. Cancers in Australia attributable to exposure to solar ultraviolet radiation and prevented by regular sunscreen use. *Aust. N. Z. J. Public Health* **2015**, *39*, 471. [CrossRef] [PubMed]
4. Climstein, M.; Doyle, B.; Stapelberg, M.; Rosic, N.; Hertess, I.; Furness, J.; Simas, V.; Walsh, J. Point prevalence of non-melanoma and melanoma skin cancers in Australian surfers and swimmers in Southeast Queensland and Northern New South Wales. *PeerJ* **2022**, *10*, e13243. [CrossRef] [PubMed]
5. Miller, I.J.; Stapelberg, M.; Rosic, N.; Hudson, J.; Coxon, P.; Furness, J.; Walsh, J.; Climstein, M. Implementation of artificial intelligence for the detection of cutaneous melanoma within a primary care setting: Prevalence and types of skin cancer in outdoor enthusiasts. *PeerJ* **2023**, *11*, e15737. [CrossRef] [PubMed]
6. Foster, M.S.; Filho, G.M.A.; Kamenos, N.A.; Riosmena-Rodríguez, R.; Steller, D.L. Rhodoliths and Rhodolith Beds. In *Research and Discoveries: The Revolution of Science through SCUBA*; Lang, M.A., Marinelli, R.L., Roberts, S.J., Taylor, P.R., Eds.; Smithsonian Contributions to the Marine Sciences; Smithsonian Institution Scholarly Press: Washington, DC, USA, 2013; pp. 143–155.
7. Littler, M.M.; Littler, D.S. The nature of crustose coralline algae and their interactions on reefs. In *Research and Discoveries: The Revolution of Science through SCUBA*; Lang, M.A., Marinelli, R.L., Roberts, S.J., Taylor, P.R., Eds.; Smithsonian Contributions to the Marine Sciences; Smithsonian Institution Scholarly Press: Washington, DC, USA, 2013; pp. 199–212.
8. Fulton, C.J.; Berkström, C.; Wilson, S.K.; Abesamis, R.A.; Bradley, M.; Åkerlund, C.; Barrett, L.T.; Bucol, A.A.; Chacin, D.H.; Chong-Seng, K.M.; et al. Macroalgal meadow habitats support fish and fisheries in diverse tropical seascapes. *Fish Fish.* **2020**, *21*, 700–717. [CrossRef]
9. Smale, D.A.; Pessarrodona, A.; King, N.; Burrows, M.T.; Yunnie, A.; Vance, T.; Moore, P. Environmental factors influencing primary productivity of the forest-forming kelp *Laminaria hyperborea* in the northeast Atlantic. *Sci. Rep.* **2020**, *10*, 12161. [CrossRef]
10. Zheng, X.; Como, S.; Huang, L.; Magni, P. Temporal changes of a food web structure driven by different primary producers in a subtropical eutrophic lagoon. *Mar. Environ. Res.* **2020**, *161*, 105128. [CrossRef]
11. Ouyang, X.; Kristensen, E.; Zimmer, M.; Thornber, C.; Yang, Z.; Lee, S.Y. Response of macrophyte litter decomposition in global blue carbon ecosystems to climate change. *Glob. Chang. Biol.* **2023**, *29*, 3806–3820. [CrossRef]
12. Ashkenazi, D.Y.; Figueroa, F.L.; Korbee, N.; García-Sánchez, M.; Vega, J.; Ben-Valid, S.; Paz, G.; Salomon, E.; Israel, Á.; Abelson, A. Enhancing Bioproducts in Seaweeds via Sustainable Aquaculture: Antioxidant and Sun-Protection Compounds. *Mar. Drugs* **2022**, *20*, 767. [CrossRef]
13. Moreira, A.; Cruz, S.; Marques, R.; Cartaxana, P. The underexplored potential of green macroalgae in aquaculture. *Rev. Aquac.* **2022**, *14*, 5–26. [CrossRef]
14. Pinteus, S.; Lemos, M.F.L.; Alves, C.; Neugebauer, A.; Silva, J.; Thomas, O.P.; Botana, L.M.; Gaspar, H.; Pedrosa, R. Marine invasive macroalgae: Turning a real threat into a major opportunity—The biotechnological potential of *Sargassum muticum* and *Asparagopsis armata*. *Algal Res.* **2018**, *34*, 217–234. [CrossRef]
15. Rebours, C.; Marinho-Soriano, E.; Zertuche-González, J.A.; Hayashi, L.; Vásquez, J.A.; Kradolfer, P.; Soriano, G.; Ugarte, R.; Abreu, M.H.; Bay-Larsen, I.; et al. Seaweeds: An opportunity for wealth and sustainable livelihood for coastal communities. *J. Appl. Phycol.* **2014**, *26*, 1939–1951. [CrossRef] [PubMed]
16. Davison, I.R.; Dudgeon, S.R.; Ruan, H.-M. Effect of freezing on seaweed photosynthesis. *Mar. Ecol. Prog. Ser.* **1989**, *58*, 123–131. [CrossRef]
17. van Hees, D.H.; Olsen, Y.S.; Wernberg, T.; Van Alstyne, K.L.; Kendrick, G.A. Phenolic concentrations of brown seaweeds and relationships to nearshore environmental gradients in Western Australia. *Mar. Biol.* **2017**, *164*, 74. [CrossRef]
18. Jormalainen, V.; Honkanen, T. Variation in natural selection for growth and phlorotannins in the brown alga *Fucus vesiculosus*. *J. Evol. Biol.* **2004**, *17*, 807–820. [CrossRef]
19. Williams, S.L.; Dethier, M.N. High and Dry: Variation in Net Photosynthesis of the Intertidal Seaweed *Fucus gardneri*. *Ecology* **2005**, *86*, 2373–2379. [CrossRef]

20. Beardall, J.; Beer, S.; Raven, J.A. Biodiversity of Marine Plants in an Era of Climate Change: Some Predictions Based on Physiological Performance. *J. Bot. Mar.* **1998**, *41*, 113–124. [CrossRef]
21. Dunlap, W.C.; Malcolm, S.J. Ultraviolet radiation-absorbing mycosporine-like amino acids in coral reef organisms: A biological and environmental perspective. *J. Phycol.* **1998**, *34*, 418–430. [CrossRef]
22. Rastogi, R.P.; Richa; Sinha, R.P.; Singh, S.P.; Häder, D.P. Photoprotective compounds from marine organisms. *J. Ind. Microbiol. Biotechnol.* **2010**, *37*, 537–558. [CrossRef]
23. Singh, S.P.; Kumari, S.; Rastogi, R.P.; Singh, K.L.; Sinha, R.P. Mycosporine-like amino acids (MAAs): Chemical structure, biosynthesis and significance as UV-absorbing/screening compounds. *Indian J. Exp. Biol.* **2008**, *46*, 7–17.
24. Kageyama, H.; Waditee-Sirisattha, R. Antioxidative, anti-inflammatory, and anti-aging properties of mycosporine-like amino acids: Molecular and cellular mechanisms in the protection of skin-aging. *Mar. Drugs* **2019**, *17*, 222. [CrossRef]
25. Rastogi, R.P.; Incharoensakdi, A. UV radiation-induced biosynthesis, stability and antioxidant activity of mycosporine-like amino acids (MAAs) in a unicellular cyanobacterium *Gloeocapsa* sp. CU2556. *J. Photochem. Photobiol. B Biol.* **2014**, *130*, 287–292. [CrossRef]
26. Wada, N.; Sakamoto, T.; Matsugo, S. Mycosporine-Like Amino Acids and Their Derivatives as Natural Antioxidants. *Antioxidants* **2015**, *4*, 603–646. [CrossRef]
27. Beardall, J.; Stojkovic, S.; Gao, K. Interactive effects of nutrient supply and other environmental factors on the sensitivity of marine primary producers to ultraviolet radiation: Implications for the impacts of global change. *Aquat. Biol.* **2014**, *22*, 5–23. [CrossRef]
28. Ji, Y.; Xu, Z.; Zou, D.; Gao, K. Ecophysiological responses of marine macroalgae to climate change factors. *J. Appl. Phycol.* **2016**, *28*, 2953–2967. [CrossRef]
29. Xu, J.; Gao, K. Use of UV-A Energy for Photosynthesis in the Red Macroalga *Gracilaria lemaneiformis*. *Photochem. Photobiol.* **2010**, *86*, 580–585. [CrossRef]
30. Zheng, Y.; Gao, K. Impacts of solar uv radiation on the photosynthesis, growth, and uv-absorbing compounds in *Gracilaria lemaneiformis* (rhodophyta) grown at different nitrate concentrations1. *J. Phycol.* **2009**, *45*, 314–323. [CrossRef] [PubMed]
31. Ji, Y.; Gao, K. Chapter Two—Effects of climate change factors on marine macroalgae: A review. In *Advances in Marine Biology*; Sheppard, C., Ed.; Academic Press: Cambridge, MA, USA, 2021; Volume 88, pp. 91–136.
32. Wahl, M.; Jormalainen, V.; Eriksson, B.K.; Coyer, J.A.; Molis, M.; Schubert, H.; Dethier, M.; Karez, R.; Kruse, I.; Lenz, M.; et al. Chapter Two—Stress Ecology in Fucus: Abiotic, Biotic and Genetic Interactions. In *Advances in Marine Biology*; Lesser, M., Ed.; Academic Press: Cambridge, MA, USA, 2011; Volume 59, pp. 37–105.
33. Harley, C.D.G.; Anderson, K.M.; Demes, K.W.; Jorve, J.P.; Kordas, R.L.; Coyle, T.A.; Graham, M.H. Effects of climate change on global seaweed communities. *J. Phycol.* **2012**, *48*, 1064–1078. [CrossRef] [PubMed]
34. Leandro, A.; Pereira, L.; Gonçalves, A.M.M. Diverse Applications of Marine Macroalgae. *Mar. Drugs* **2019**, *18*, 17. [CrossRef]
35. Mouritsen, O.G. *Seaweeds: Edible, Available, and Sustainable*; University of Chicago Press: Chicago, IL, USA, 2013.
36. Neveux, N.; Bolton, J.J.; Bruhn, A.; Roberts, D.A.; Ras, M. The Bioremediation Potential of Seaweeds: Recycling Nitrogen, Phosphorus, and Other Waste Products. In *Blue Biotechnology*; Wiley: Hoboken, NJ, USA, 2018; pp. 217–239. [CrossRef]
37. Henriques, B.; Lopes, C.B.; Figueira, P.; Rocha, L.S.; Duarte, A.C.; Vale, C.; Pardal, M.A.; Pereira, E. Bioaccumulation of Hg, Cd and Pb by *Fucus vesiculosus* in single and multi-metal contamination scenarios and its effect on growth rate. *Chemosphere* **2017**, *171*, 208–222. [CrossRef] [PubMed]
38. Rosic, N.; Bradbury, J.; Lee, M.; Baltrotsky, K.; Grace, S. The impact of pesticides on local waterways: A scoping review and method for identifying pesticides in local usage. *Environ. Sci. Policy* **2020**, *106*, 12–21. [CrossRef]
39. Mosalaei Rad, S.; Ray, A.; Barghi, S. Water Pollution and Agriculture Pesticide. *Clean Technol.* **2022**, *4*, 1088–1102. [CrossRef]
40. Wang, Z.; Yue, W. Removal of cypermethrin with seaweed *Gracilaria lemaneiformis*. *J. Ocean Univ. China* **2015**, *14*, 858–864. [CrossRef]
41. Tang, X.; Shen, L.; Liu, S.; Gao, J. Effective removal of the herbicide glyphosate by the kelp *Saccharina japonica* female gametophytes from saline waters and its mechanism elucidation. *Chemosphere* **2021**, *274*, 129826. [CrossRef]
42. Lyons, D.A.; Mant, R.C.; Bulleri, F.; Kotta, J.; Rilov, G.; Crowe, T.P. What are the effects of macroalgal blooms on the structure and functioning of marine ecosystems? A systematic review protocol. *Environ. Evid.* **2012**, *1*, 7. [CrossRef]
43. Hallegraeff, G.M.; Anderson, D.M.; Belin, C.; Bottein, M.-Y.D.; Bresnan, E.; Chinain, M.; Enevoldsen, H.; Iwataki, M.; Karlson, B.; McKenzie, C.H.; et al. Perceived global increase in algal blooms is attributable to intensified monitoring and emerging bloom impacts. *Commun. Earth Environ.* **2021**, *2*, 117. [CrossRef]
44. Young, C.S.; Lee, C.-S.; Sylvers, L.H.; Venkatesan, A.K.; Gobler, C.J. The invasive red seaweed, *Dasysiphonia japonica*, forms harmful algal blooms: Mortality in early life stage fish and bivalves and identification of putative toxins. *Harmful Algae* **2022**, *118*, 102294. [CrossRef]
45. Wang, M.; Hu, C.; Barnes, B.B.; Mitchum, G.; Lapointe, B.; Montoya, J.P. The great Atlantic *Sargassum* belt. *Science* **2019**, *365*, 83–87. [CrossRef]
46. Valiela, I.; McClelland, J.W.; Hauxwell, J.A.; Behr, P.J.; Hersh, D.; Foreman, K.H. Macroalgal blooms in shallow estuaries: Controls and ecophysiological and ecosystem consequences. *Limnol. Oceanogr.* **1997**, *42*, 1105–1118. [CrossRef]
47. Teichberg, M.; Fox, S.E.; Olsen, Y.S.; Valiela, I.; Martinetto, P.; Iribarne, O.; Muto, E.Y.; Petti, M.A.V.; Corbisier, T.N.; Soto- Jiménez, M.; et al. Eutrophication and macroalgal blooms in temperate and tropical coastal waters: Nutrient enrichment experiments with *Ulva* spp. *Glob. Chang. Biol.* **2010**, *16*, 2624–2637. [CrossRef]

48. Thornber, C.S.; Guidone, M.; Deacutis, C.; Green, L.; Ramsay, C.N.; Palmisciano, M. Spatial and temporal variability in macroalgal blooms in a eutrophied coastal estuary. *Harmful Algae* **2017**, *68*, 82–96. [CrossRef]
49. Qi, L.; Hu, C.; Xing, Q.; Shang, S. Long-term trend of *Ulva prolifera* blooms in the western Yellow Sea. *Harmful Algae* **2016**, *58*, 35–44. [CrossRef]
50. Schein, A.; Courtenay, S.C.; Crane, C.S.; Teather, K.L.; van den Heuvel, M.R. The Role of Submerged Aquatic Vegetation in Structuring the Nearshore Fish Community Within an Estuary of the Southern Gulf of St. Lawrence. *Estuaries Coasts* **2012**, *35*, 799–810. [CrossRef]
51. Thomsen, M.S.; McGlathery, K. Effects of accumulations of sediments and drift algae on recruitment of sessile organisms associated with oyster reefs. *J. Exp. Mar. Biol. Ecol.* **2006**, *328*, 22–34. [CrossRef]
52. Young, C.S.; Peterson, B.J.; Gobler, C.J. The Bloom-Forming Macroalgae, *Ulva*, Outcompetes the Seagrass, *Zostera marina*, Under High CO_2 Conditions. *Estuaries Coasts* **2018**, *41*, 2340–2355. [CrossRef]
53. Liu, D.; Keesing, J.K.; Xing, Q.; Shi, P. World's largest macroalgal bloom caused by expansion of seaweed aquaculture in China. *Mar. Pollut. Bull.* **2009**, *58*, 888–895. [CrossRef]
54. Teichberg, M.; Fricke, A.; Bischof, K. Increased physiological performance of the calcifying green macroalga *Halimeda opuntia* in response to experimental nutrient enrichment on a Caribbean coral reef. *Aquat. Bot.* **2013**, *104*, 25–33. [CrossRef]
55. Castaldelli, G.; Welsh, D.T.; Flachi, G.; Zucchini, G.; Colombo, G.; Rossi, R.; Fano, E.A. Decomposition dynamics of the bloom forming macroalga *Ulva rigida* C. Agardh determined using a 14C-carbon radio-tracer technique. *Aquat. Bot.* **2003**, *75*, 111–122. [CrossRef]
56. Conover, J.; Green, L.A.; Thornber, C.S. Biomass decay rates and tissue nutrient loss in bloom and non-bloom-forming macroalgal species. *Estuar. Coast. Shelf Sci.* **2016**, *178*, 58–64. [CrossRef]
57. Harb, T.B.; Chow, F. An overview of beach-cast seaweeds: Potential and opportunities for the valorization of underused waste biomass. *Algal Res.* **2022**, *62*, 102643. [CrossRef]
58. Lyons, D.A.; Arvanitidis, C.; Blight, A.J.; Chatzinikolaou, E.; Guy-Haim, T.; Kotta, J.; Orav-Kotta, H.; Queirós, A.M.; Rilov, G.; Somerfield, P.J.; et al. Macroalgal blooms alter community structure and primary productivity in marine ecosystems. *Glob. Chang. Biol.* **2014**, *20*, 2712–2724. [CrossRef]
59. Lomstein, B.A.; Guldberg, L.B.; Neubauer, A.-T.A.; Hansen, J.; Donnelly, A.; Herbert, R.A.; Viaroli, P.; Giordani, G.; Azzoni, R.; de Wit, R.; et al. Benthic decomposition of *Ulva lactuca*: A controlled laboratory experiment. *Aquat. Bot.* **2006**, *85*, 271–281. [CrossRef]
60. Liu, S.; Jiang, Z.; Wu, Y.; Deng, Y.; Chen, Q.; Zhao, C.; Cui, L.; Huang, X. Macroalgae bloom decay decreases the sediment organic carbon sequestration potential in tropical seagrass meadows of the South China Sea. *Mar. Pollut. Bull.* **2019**, *138*, 598–603. [CrossRef]
61. Xiao, J.; Wang, Z.; Song, H.; Fan, S.; Yuan, C.; Fu, M.; Miao, X.; Zhang, X.; Su, R.; Hu, C. An anomalous bi-macroalgal bloom caused by *Ulva* and *Sargassum* seaweeds during spring to summer of 2017 in the western Yellow Sea, China. *Harmful Algae* **2020**, *93*, 101760. [CrossRef]
62. Fidai, Y.A.; Dash, J.; Tompkins, E.L.; Tonon, T. A systematic review of floating and beach landing records of *Sargassum* beyond the Sargasso Sea. *Environ. Res. Commun.* **2020**, *2*, 122001. [CrossRef]
63. Piñón-Gimate, A.; Soto-Jiménez, M.F.; Ochoa-Izaguirre, M.J.; García-Pagés, E.; Páez-Osuna, F. Macroalgae blooms and δ15N in subtropical coastal lagoons from the Southeastern Gulf of California: Discrimination among agricultural, shrimp farm and sewage effluents. *Mar. Pollut. Bull.* **2009**, *58*, 1144–1151. [CrossRef]
64. Joniver, C.F.H.; Photiades, A.; Moore, P.J.; Winters, A.L.; Woolmer, A.; Adams, J.M.M. The global problem of nuisance macroalgal blooms and pathways to its use in the circular economy. *Algal Res.* **2021**, *58*, 102407. [CrossRef]
65. López-Contreras, A.; Núñez, P.; García, B.; Driegen, J.; Lwanga, E.; Domin, P.; Gurrola, M.P.; Rosas-Luis, R.; Verde-Gómez, Y.; Vrije, T. *Sargassum in Mexico: From Environmental Problem to Valuable Resource*; Wageningen Food & Biobased Research: Wageningen, The Netherlands, 2022. [CrossRef]
66. Marx, U.C.; Roles, J.; Hankamer, B. *Sargassum* blooms in the Atlantic Ocean—From a burden to an asset. *Algal Res.* **2021**, *54*, 102188. [CrossRef]
67. Oxenford, H.A.; Cox, S.-A.; van Tussenbroek, B.I.; Desrochers, A. Challenges of Turning the *Sargassum* Crisis into Gold: Current Constraints and Implications for the Caribbean. *Phycology* **2021**, *1*, 27–48. [CrossRef]
68. Lorbeer, A.J.; Tham, R.; Zhang, W. Potential products from the highly diverse and endemic macroalgae of Southern Australia and pathways for their sustainable production. *J. Appl. Phycol.* **2013**, *25*, 717–732. [CrossRef]
69. Newton, C.; Bracken, M.E.; McConville, M.; Rodrigue, K.; Thornber, C.S. Invasion of the red seaweed *Heterosiphonia japonica* spans biogeographic provinces in the Western North Atlantic Ocean. *PLoS ONE* **2013**, *8*, e62261. [CrossRef] [PubMed]
70. Manilal, A.; Sujith, S.; Sabarathnam, B.; Kiran, G.S.; Selvin, J.; Shakir, C.; Lipton, A.P. Bioactivity of the red algae *Asparagopsis taxiformis* collected from the Southwestern coast of India. *Braz. J. Oceanogr.* **2010**, *58*, 93–100. [CrossRef]
71. Shimada, S.; Nagano, M.; Hiraoka, M.; Ichihara, K.; Mineur, F.; Zhu, W. Phylogeographic analysis of the genus *Ulva* (Ulvales, Chlorophyta), including bloom sample in Qingdao, China. *Coast. Mar. Sci.* **2010**, *34*, 117–122.
72. Quillien, N.; Nordström, M.C.; Guyonnet, B.; Maguer, M.; Le Garrec, V.; Bonsdorff, E.; Grall, J. Large-scale effects of green tides on macrotidal sandy beaches: Habitat-specific responses of zoobenthos. *Estuar. Coast. Shelf Sci.* **2015**, *164*, 379–391. [CrossRef]
73. Viaroli, P.; Bartoli, M.; Azzoni, R.; Giordani, G.; Mucchino, C.; Naldi, M.; Nizzoli, D.; Tajé, L. Nutrient and iron limitation to *Ulva* blooms in a eutrophic coastal lagoon (Sacca di Goro, Italy). *Hydrobiologia* **2005**, *550*, 57–71. [CrossRef]

74. Nelson, T.A.; Nelson, A.V.; Tjoelker, M. Seasonal and Spatial Patterns of "Green Tides" (Ulvoid Algal Blooms) and Related Water Quality Parameters in the Coastal Waters of Washington State, USA. *Bot. Mar.* **2003**, *46*, 263–275. [CrossRef]
75. Guidone, M.; Thornber, C.S.; Van Alstyne, K.L. Herbivore impacts on two morphologically similar bloom-forming *Ulva* species in a eutrophic bay. *Hydrobiologia* **2015**, *753*, 175–188. [CrossRef]
76. Fox, S.E.; Teichberg, M.; Valiela, I.; Heffner, L. The Relative Role of Nutrients, Grazing, and Predation as Controls on Macroalgal Growth in the Waquoit Bay Estuarine System. *Estuaries Coasts* **2012**, *35*, 1193–1204. [CrossRef]
77. Human, L.R.D.; Adams, J.B.; Allanson, B.R. Insights into the cause of an *Ulva lactuca* Linnaeus bloom in the Knysna Estuary. *S. Afr. J. Bot.* **2016**, *107*, 55–62. [CrossRef]
78. Chávez-Sánchez, T.; Piñón-Gimate, A.; Serviere-Zaragoza, E.; López-Bautista, J.M.; Casas-Valdez, M. *Ulva* blooms in the southwestern Gulf of California: Reproduction and biomass. *Estuar. Coast. Shelf Sci.* **2018**, *200*, 202–211. [CrossRef]
79. Huo, Y.; Zhang, J.; Chen, L.; Hu, M.; Yu, K.; Chen, Q.; He, Q.; He, P. Green algae blooms caused by *Ulva prolifera* in the southern Yellow Sea: Identification of the original bloom location and evaluation of biological processes occurring during the early northward floating period. *Limnol. Oceanogr.* **2013**, *58*, 2206–2218. [CrossRef]
80. Zhang, J.; Zhao, P.; Huo, Y.; Yu, K.; He, P. The fast expansion of *Pyropia* aquaculture in "Sansha" regions should be mainly responsible for the *Ulva* blooms in Yellow Sea. *Estuar. Coast. Shelf Sci.* **2017**, *189*, 58–65. [CrossRef]
81. Ober, G.T.; Thornber, C.S. Divergent responses in growth and nutritional quality of coastal macroalgae to the combination of increased pCO_2 and nutrients. *Mar. Environ. Res.* **2017**, *131*, 69–79. [CrossRef] [PubMed]
82. Green-Gavrielidis, L.A.; Thornber, C.S. Will Climate Change Enhance Algal Blooms? The Individual and Interactive Effects of Temperature and Rain on the Macroalgae Ulva. *Estuaries Coasts* **2022**, *45*, 1688–1700. [CrossRef]
83. Li, H.; Zhang, Y.; Han, X.; Shi, X.; Rivkin, R.B.; Legendre, L. Growth responses of *Ulva prolifera* to inorganic and organic nutrients: Implications for macroalgal blooms in the southern Yellow Sea, China. *Sci. Rep.* **2016**, *6*, 26498. [CrossRef] [PubMed]
84. Thornber, C.S.; DiMilla, P.; Nixon, S.W.; McKinney, R.A. Natural and anthropogenic nitrogen uptake by bloom-forming macroalgae. *Mar. Pollut. Bull.* **2008**, *56*, 261–269. [CrossRef] [PubMed]
85. Green-Gavrielidis, L.A.; MacKechnie, F.; Thornber, C.S.; Gomez-Chiarri, M. Bloom-forming macroalgae (*Ulva* spp.) inhibit the growth of co-occurring macroalgae and decrease eastern oyster larval survival. *Mar. Ecol. Prog. Ser.* **2018**, *595*, 27–37. [CrossRef]
86. Young, C.S.; Lowell, A.; Peterson, B.; Gobler, C.J. Ocean acidification and food limitation combine to suppress herbivory by the gastropod *Lacuna vincta*. *Mar. Ecol. Prog. Ser.* **2019**, *627*, 83–94. [CrossRef]
87. Beer, S. Photosynthetic traits of the ubiquitous and prolific macroalga *Ulva* (Chlorophyta): A review. *Eur. J. Phycol.* **2022**, *58*, 390–398. [CrossRef]
88. Cruces, E.; Rautenberger, R.; Cubillos, V.M.; Ramírez-Kushel, E.; Rojas-Lillo, Y.; Lara, C.; Montory, J.A.; Gómez, I. Interaction of Photoprotective and Acclimation Mechanisms in *Ulva rigida* (Chlorophyta) in Response to Diurnal Changes in Solar Radiation in Southern Chile. *J. Phycol.* **2019**, *55*, 1011–1027. [CrossRef]
89. Pescheck, F.; Campen, H.; Nichelmann, L.; Bilger, W. Relative sensitivity of DNA and photosystem II in *Ulva intestinalis* (Chlorophyta) under natural solar irradiation. *Mar. Ecol. Prog. Ser.* **2016**, *555*, 95–107. [CrossRef]
90. Bischof, K.; Peralta, G.; Kräbs, G.; van de Poll, W.H.; Pérez-Lloréns, J.L.; Breeman, A.M. Effects of solar UV-B radiation on canopy structure of *Ulva* communities from southern Spain. *J. Exp. Bot.* **2002**, *53*, 2411–2421. [CrossRef] [PubMed]
91. Figueroa, F.L.; Domínguez-González, B.; Korbee, N. Vulnerability and acclimation to increased UVB radiation in three intertidal macroalgae of different morpho-functional groups. *Mar. Environ. Res.* **2014**, *97*, 30–38. [CrossRef] [PubMed]
92. Han, Y.-S.; Han, T. UV-B induction of uv-b protection in *Ulva pertusa* (chlorophyta). *J. Phycol.* **2005**, *41*, 523–530. [CrossRef]
93. Pescheck, F.; Bilger, W. Compensation of lack of UV screening by cellular tolerance in green macroalgae (Ulvophyceae) from the upper eulittoral. *Mar. Biol.* **2018**, *165*, 132. [CrossRef]
94. Pescheck, F.; Bischof, K.; Bilger, W. Screening of ultraviolet-a and ultraviolet-b radiation in marine green macroalgae (chlorophyta). *J. Phycol.* **2010**, *46*, 444–455. [CrossRef]
95. Laffoley, D.; Roe, H.; Angel, M.V.; Ardron, J.; Bates, N.; Boyd, L.L.; Brooke, S.; Buck, K.; Carlson, C.; Causey, B.; et al. *The Protection and Management of the Sargasso Sea: The Golden Floating Rainforest of the Atlantic Ocean: Summary Science and Supporting Evidence Case*; Sargasso Sea Alliance, Government of Bermuda: Bermuda, UK, 2011.
96. Putman, N.F.; Goni, G.J.; Gramer, L.J.; Hu, C.; Johns, E.M.; Trinanes, J.; Wang, M. Simulating transport pathways of pelagic *Sargassum* from the Equatorial Atlantic into the Caribbean Sea. *Prog. Oceanogr.* **2018**, *165*, 205–214. [CrossRef]
97. Oviatt, C.A.; Huizenga, K.; Rogers, C.S.; Miller, W.J. What nutrient sources support anomalous growth and the recent *Sargassum* mass stranding on Caribbean beaches? A review. *Mar. Pollut. Bull.* **2019**, *145*, 517–525. [CrossRef]
98. Chávez, V.; Uribe-Martínez, A.; Cuevas, E.; Rodríguez-Martínez, R.E.; van Tussenbroek, B.I.; Francisco, V.; Estévez, M.; Celis, L.B.; Monroy-Velázquez, L.V.; Leal-Bautista, R.; et al. Massive Influx of Pelagic *Sargassum* spp. on the Coasts of the Mexican Caribbean 2014–2020: Challenges and Opportunities. *Water* **2020**, *12*, 2908. [CrossRef]
99. Rodríguez-Muñoz, R.; Muñiz-Castillo, A.I.; Euán-Avila, J.I.; Hernández-Núñez, H.; Valdés-Lozano, D.S.; Collí-Dulá, R.C.; Arias-González, J.E. Assessing temporal dynamics on pelagic *Sargassum* influx and its relationship with water quality parameters in the Mexican Caribbean. *Reg. Stud. Mar. Sci.* **2021**, *48*, 102005. [CrossRef]
100. Aquino, R.; Noriega, C.; Mascarenhas, A.; Costa, M.; Monteiro, S.; Santana, L.; Silva, I.; Prestes, Y.; Araujo, M.; Rollnic, M. Possible Amazonian contribution to *Sargassum* enhancement on the Amazon Continental Shelf. *Sci. Total Environ.* **2022**, *853*, 158432. [CrossRef]

101. Lapointe, B.E.; Brewton, R.A.; Herren, L.W.; Wang, M.; Hu, C.; McGillicuddy, D.J.; Lindell, S.; Hernandez, F.J.; Morton, P.L. Nutrient content and stoichiometry of pelagic *Sargassum* reflects increasing nitrogen availability in the Atlantic Basin. *Nat. Commun.* **2021**, *12*, 3060. [CrossRef] [PubMed]
102. Polo, L.K.; Chow, F. Variation of antioxidant capacity and antiviral activity of the brown seaweed *Sargassum filipendula* (Fucales, Ochrophyta) under UV radiation treatments. *Appl. Phycol.* **2022**, *3*, 260–273. [CrossRef]
103. Liu, L.; Heinrich, M.; Myers, S.; Dworjanyn, S.A. Towards a better understanding of medicinal uses of the brown seaweed *Sargassum* in Traditional Chinese Medicine: A phytochemical and pharmacological review. *J. Ethnopharmacol.* **2012**, *142*, 591–619. [CrossRef]
104. Lann, K.L.; Ferret, C.; VanMee, E.; Spagnol, C.; Lhuillery, M.; Payri, C.; Stiger-Pouvreau, V. Total phenolic, size-fractionated phenolics and fucoxanthin content of tropical Sargassaceae (Fucales, Phaeophyceae) from the South Pacific Ocean: Spatial and specific variability. *Phycol. Res.* **2012**, *60*, 37–50. [CrossRef]
105. Polo, L.K.; de L. Felix, M.R.; Kreusch, M.; Pereira, D.T.; Costa, G.B.; Simioni, C.; Ouriques, L.C.; Chow, F.; Ramlov, F.; Maraschin, M.; et al. Photoacclimation Responses of the Brown Macroalga *Sargassum cymosum* to the Combined Influence of UV Radiation and Salinity: Cytochemical and Ultrastructural Organization and Photosynthetic Performance. *Photochem. Photobiol.* **2014**, *90*, 560–573. [CrossRef]
106. Magaña-Gallegos, E.; García-Sánchez, M.; Graham, C.; Olivos-Ortiz, A.; Siuda, A.N.S.; van Tussenbroek, B.I. Growth rates of pelagic *Sargassum* species in the Mexican Caribbean. *Aquat. Bot.* **2023**, *185*, 103614. [CrossRef]
107. Xu, Z.; Li, L.; Jiang, H.; Yan, F.; Liu, L.; Zang, S.; Ma, Y.; Wu, H. Photosynthetic responses of a golden tide alga (*Sargassum horneri*) to ultraviolet radiation. *Front. Mar. Sci.* **2022**, *9*, 978376. [CrossRef]
108. Vázquez-Delfín, E.; Freile-Pelegrín, Y.; Salazar-Garibay, A.; Serviere-Zaragoza, E.; Méndez-Rodríguez, L.C.; Robledo, D. Species composition and chemical characterization of *Sargassum* influx at six different locations along the Mexican Caribbean coast. *Sci. Total Environ.* **2021**, *795*, 148852. [CrossRef]
109. Powers, L.C.; Hertkorn, N.; McDonald, N.; Schmitt-Kopplin, P.; Del Vecchio, R.; Blough, N.V.; Gonsior, M. *Sargassum* sp. Act as a Large Regional Source of Marine Dissolved Organic Carbon and Polyphenols. *Glob. Biogeochem. Cycles* **2019**, *33*, 1423–1439. [CrossRef]
110. Whitehouse, L.N.A.; Lapointe, B.E. Comparative ecophysiology of bloom-forming macroalgae in the Indian River Lagoon, Florida: *Ulva lactuca, Hypnea musciformis*, and *Gracilaria tikvahiae*. *J. Exp. Mar. Biol. Ecol.* **2015**, *471*, 208–216. [CrossRef]
111. Tyler, A.C.; McGlathery, K.J. Uptake and release of nitrogen by the macroalgae *Gracilaria vermiculophylla* (Rhodophyta). *J. Phycol.* **2006**, *42*, 515–525. [CrossRef]
112. Cacabelos, E.; Engelen, A.H.; Mejia, A.; Arenas, F. Comparison of the assemblage functioning of estuary systems dominated by the seagrass *Nanozostera noltii* versus the invasive drift seaweed *Gracilaria vermiculophylla*. *J. Sea Res.* **2012**, *72*, 99–105. [CrossRef]
113. Thomsen, M.; Staehr, P.; Nyberg, C.; Schwaerter, S.; Krause-Jensen, D.; Silliman, B. *Gracilaria vermiculophylla* (Ohmi) Papenfuss, 1967 (Rhodophyta, Gracilariaceae) in northern Europe, with emphasis on Danish conditions, and what to expect in the future. *ECU Publ.* **2007**, *2*, 83–94. [CrossRef]
114. Gómez, M.; Barreiro, F.; López, J.; Lastra, M.; de la Huz, R. Deposition patterns of algal wrack species on estuarine beaches. *Aquat. Bot.* **2013**, *105*, 25–33. [CrossRef]
115. Wang, C.; Lei, A.; Zhou, K.; Hu, Z.; Hao, W.; Yang, J. Growth and nitrogen uptake characteristics reveal outbreak mechanism of the opportunistic macroalga *Gracilaria tenuistipitata*. *PLoS ONE* **2014**, *9*, e108980. [CrossRef]
116. Torres, P.; Santos, J.P.; Chow, F.; dos Santos, D.Y.A.C. A comprehensive review of traditional uses, bioactivity potential, and chemical diversity of the genus *Gracilaria* (Gracilariales, Rhodophyta). *Algal Res.* **2019**, *37*, 288–306. [CrossRef]
117. Félix, R.; Dias, P.; Félix, C.; Cerqueira, T.; Andrade, P.B.; Valentão, P.; Lemos, M.F.L. The biotechnological potential of *Asparagopsis armata*: What is known of its chemical composition, bioactivities and current market? *Algal Res.* **2021**, *60*, 102534. [CrossRef]
118. Rosic, N. Molecular Mechanisms of Stress Tolerance in Cyanobacteria. In *Ecophysiology and Biochemistry of Cyanobacteria*; Rastogi, R.P., Ed.; Springer Nature: Singapore, 2021; pp. 131–153. [CrossRef]
119. Tirtawijaya, G.; Negara, B.F.; Lee, J.-H.; Cho, M.-G.; Kim, H.K.; Choi, Y.-S.; Lee, S.-H.; Choi, J.-S. The Influence of Abiotic Factors on the Induction of Seaweed Callus. *J. Mar. Sci. Eng.* **2022**, *10*, 513. [CrossRef]
120. Rosic, N.; Remond, C.; Mello-Athayde, M.A. Differential impact of heat stress on reef-building corals under different light conditions. *Mar. Environ. Res.* **2020**, *158*, 104947. [CrossRef]
121. Rosic, N.N.; Pernice, M.; Dove, S.; Dunn, S.; Hoegh-Guldberg, O. Gene expression profiles of cytosolic heat shock proteins Hsp70 and Hsp90 from symbiotic dinoflagellates in response to thermal stress: Possible implications for coral bleaching. *Cell Stress Chaperones* **2011**, *16*, 69–80. [CrossRef]
122. Rosic, N.N.; Pernice, M.; Dunn, S.; Dove, S.; Hoegh-Guldberg, O. Differential regulation by heat stress of novel cytochrome P450 genes from the dinoflagellate symbionts of reef-building corals. *Appl. Environ. Microbiol.* **2010**, *76*, 2823–2829. [CrossRef] [PubMed]
123. Singh, S.P.; Klisch, M.; Sinha, R.P.; Häder, D.P. Effects of abiotic stressors on synthesis of the mycosporine-like amino acid shinorine in the cyanobacterium *Anabaena variabilis* PCC 7937. *Photochem. Photobiol.* **2008**, *84*, 1500–1505. [CrossRef] [PubMed]
124. Voolstra, C.; Miller, D.; Ragan, M.; Hoffmann, A.; Hoegh-Guldberg, O.; Bourne, D.; Ball, E.; Ying, H.; Foret, S.; Takahashi, S.; et al. The ReFuGe 2020 Consortium—Using "omics" approaches to explore the adaptability and resilience of coral holobionts to environmental change. *Front. Mar. Sci.* **2015**, *2*, 68. [CrossRef]

125. Ewere, E.E.; Rosic, N.; Bayer, P.E.; Ngangbam, A.; Edwards, D.; Kelaher, B.P.; Mamo, L.T.; Benkendorff, K. Marine heatwaves have minimal influence on the quality of adult Sydney rock oyster flesh. *Sci. Total Environ.* **2021**, *795*, 148846. [CrossRef]
126. Wei, N.; Quarterman, J.; Jin, Y.-S. Marine macroalgae: An untapped resource for producing fuels and chemicals. *Trends Biotechnol.* **2013**, *31*, 70–77. [CrossRef] [PubMed]
127. Sun, Y.; Zhang, N.; Zhou, J.; Dong, S.; Zhang, X.; Guo, L.; Guo, G. Distribution, Contents, and Types of Mycosporine-Like Amino Acids (MAAs) in Marine Macroalgae and a Database for MAAs Based on These Characteristics. *Mar. Drugs* **2020**, *18*, 43. [CrossRef] [PubMed]
128. Oren, A.; Gunde-Cimerman, N. Mycosporines and mycosporine-like amino acids: UV protectants or multipurpose secondary metabolites? *FEMS Microbiol. Lett.* **2007**, *269*, 1–10. [CrossRef]
129. Rosic, N.N. Mycosporine-Like Amino Acids: Making the Foundation for Organic Personalised Sunscreens. *Mar. Drugs* **2019**, *17*, 638. [CrossRef]
130. Rosic, N.N.; Dove, S. Mycosporine-like amino acids from coral dinoflagellates. *Appl. Environ. Microbiol.* **2011**, *77*, 8478–8486. [CrossRef] [PubMed]
131. Navarro, N.P.; Figueroa, F.L.; Korbee, N. Mycosporine-like amino acids vs carrageenan yield in *Mazzaella laminarioides* (Gigartinales; Rhodophyta) under high and low UV solar irradiance. *Phycologia* **2017**, *56*, 570–578. [CrossRef]
132. Korbee, N.; Huovinen, P.; Figueroa, F.L.; Aguilera, J.; Karsten, U. Availability of ammonium influences photosynthesis and the accumulation of mycosporine-like amino acids in two *Porphyra* species (Bangiales, Rhodophyta). *Mar. Biol.* **2005**, *146*, 645–654. [CrossRef]
133. Schneider, G.; Figueroa, F.L.; Vega, J.; Avilés, A.; Chaves, P.; Horta, P.A.; Korbee, N.; Bonomi-Barufi, J. Physiological and biochemical responses driven by different UV-visible radiation in *Osmundea pinnatifida* (Hudson) Stackhouse (Rhodophyta). *Photochem. Photobiol. Sci.* **2020**, *19*, 1650–1664. [CrossRef]
134. Dunlap, W.C.; Yamamoto, Y. Small-molecule antioxidants in marine organisms: Antioxidant activity of mycosporine-glycine. *Comp. Biochem. Physiol. Part B Biochem. Mol. Biol.* **1995**, *112*, 105–114. [CrossRef]
135. Gacesa, R.; Lawrence, K.P.; Georgakopoulos, N.D.; Yabe, K.; Dunlap, W.C.; Barlow, D.J.; Wells, G.; Young, A.R.; Long, P.F. The mycosporine-like amino acids porphyra-334 and shinorine are antioxidants and direct antagonists of Keap1-Nrf2 binding. *Biochimie* **2018**, *154*, 35–44. [CrossRef] [PubMed]
136. Suh, S.S.; Hwang, J.; Park, M.; Seo, H.H.; Kim, H.S.; Lee, J.H.; Moh, S.H.; Lee, T.K. Anti-inflammation activities of mycosporine-like amino acids (MAAs) in response to UV radiation suggest potential anti-skin aging activity. *Mar. Drugs* **2014**, *12*, 5174–5187. [CrossRef]
137. Rosic, N. Genome Mining as an Alternative Way for Screening the Marine Organisms for Their Potential to Produce UV-Absorbing Mycosporine-like Amino Acid. *Mar. Drugs* **2022**, *20*, 478. [CrossRef]
138. Becker, K.; Hartmann, A.; Ganzera, M.; Fuchs, D.; Gostner, J.M. Immunomodulatory Effects of the Mycosporine-Like Amino Acids Shinorine and Porphyra-334. *Mar. Drugs* **2016**, *14*, 119. [CrossRef]
139. Sun, Y.; Han, X.; Hu, Z.; Cheng, T.; Tang, Q.; Wang, H.; Deng, X.; Han, X. Extraction, Isolation and Characterization of Mycosporine-like Amino Acids from Four Species of Red Macroalgae. *Mar. Drugs* **2021**, *19*, 615. [CrossRef]
140. Rosic, N.; Climstein, M.; Boyle, G.M.; Thanh Nguyen, D.; Feng, Y. Exploring Mycosporine-like Amino Acid UV-Absorbing Natural Products for a New Generation of Environmentally Friendly Sunscreens. *Mar. Drugs* **2023**, *21*, 253. [CrossRef]
141. Jesus, A.; Sousa, E.; Cruz, M.T.; Cidade, H.; Lobo, J.M.S.; Almeida, I.F. UV Filters: Challenges and Prospects. *Pharmaceuticals* **2022**, *15*, 263. [CrossRef] [PubMed]
142. Balskus, E.P.; Walsh, C.T. The genetic and molecular basis for sunscreen biosynthesis in cyanobacteria. *Science* **2010**, *329*, 1653–1656. [CrossRef] [PubMed]
143. Ghedifa, A.B.; Vega, J.; Korbee, N.; Mensi, F.; Figueroa, F.L.; Sadok, S. Effects of light quality on the photosynthetic activity and biochemical composition of *Gracilaria gracilis* (Rhodophyta). *J. Appl. Phycol.* **2021**, *33*, 3413–3425. [CrossRef]
144. Schneider, G.; Figueroa, F.L.; Vega, J.; Avilés, A.; Horta, P.A.; Korbee, N.; Bonomi-Barufi, J. Effects of UV–visible radiation on growth, photosynthesis, pigment accumulation and UV-absorbing compounds in the red macroalga *Gracilaria cornea* (Gracilariales, Rhodophyta). *Algal Res.* **2022**, *64*, 102702. [CrossRef]
145. Franklin, L.A.; Kräbs, G.; Kuhlenkamp, R. Blue light and UV-A radiation control the synthesis of mycosporine-like amino acids in *Chondrus crispus* (Florideophyceae). *J. Phycol.* **2001**, *37*, 257–270. [CrossRef]
146. Kräbs, G.; Bischof, K.; Hanelt, D.; Karsten, U.; Wiencke, C. Wavelength-dependent induction of UV-absorbing mycosporine-like amino acids in the red alga *Chondrus crispus* under natural solar radiation. *J. Exp. Mar. Biol. Ecol.* **2002**, *268*, 69–82. [CrossRef]
147. Pardilhó, S.; Cotas, J.; Pacheco, D.; Gonçalves, A.M.M.; Bahcevandziev, K.; Pereira, L.; Figueirinha, A.; Dias, J.M. Valorisation of marine macroalgae waste using a cascade biorefinery approach: Exploratory study. *J. Clean. Prod.* **2023**, *385*, 135672. [CrossRef]
148. Mandalka, A.; Cavalcanti, M.I.; Harb, T.B.; Toyota Fujii, M.; Eisner, P.; Schweiggert-Weisz, U.; Chow, F. Nutritional Composition of Beach-Cast Marine Algae from the Brazilian Coast: Added Value for Algal Biomass Considered as Waste. *Foods* **2022**, *11*, 1201. [CrossRef]
149. Karsten, U.; Escoubeyrou, K.; Charles, F. The effect of re-dissolution solvents and HPLC columns on the analysis of mycosporine-like amino acids in the eulittoral macroalgae *Prasiola crispa* and *Porphyra umbilicalis*. *Helgol. Mar. Res.* **2009**, *63*, 231–238. [CrossRef]

150. Bermejo, R.; Green-Gavrielidis, L.; Gao, G. Editorial: Macroalgal blooms in a global change context. *Front. Mar. Sci.* **2023**, *10*, 1204117. [CrossRef]
151. Guidone, M.; Thornber, C.S. Examination of *Ulva* bloom species richness and relative abundance reveals two cryptically co-occurring bloom species in Narragansett Bay, Rhode Island. *Harmful Algae* **2013**, *24*, 1–9. [CrossRef]
152. Kidgell, J.T.; Magnusson, M.; de Nys, R.; Glasson, C.R.K. Ulvan: A systematic review of extraction, composition and function. *Algal Res.* **2019**, *39*, 101422. [CrossRef]
153. Schmitz, C.; Ramlov, F.; de Lucena, L.A.F.; Uarrota, V.; Batista, M.B.; Sissini, M.N.; Oliveira, I.; Briani, B.; Martins, C.D.L.; Nunes, J.M.d.C.; et al. UVR and PAR absorbing compounds of marine brown macroalgae along a latitudinal gradient of the Brazilian coast. *J. Photochem. Photobiol. B Biol.* **2018**, *178*, 165–174. [CrossRef] [PubMed]
154. Haugan, J.A.; Liaaen-Jensen, S.v. Algal carotenoids 54. Carotenoids of brown algae (Phaeophyceae). *Biochem. Syst. Ecol.* **1994**, *22*, 31–41. [CrossRef]
155. Rushdi, M.I.; Abdel-Rahman, I.A.M.; Saber, H.; Attia, E.Z.; Abdelraheem, W.M.; Madkour, H.A.; Hassan, H.M.; Elmaidomy, A.H.; Abdelmohsen, U.R. Pharmacological and natural products diversity of the brown algae genus *Sargassum*. *RSC Adv.* **2020**, *10*, 24951–24972. [CrossRef]
156. Chen, B.; Chen, H.; Qu, H.; Qiao, K.; Xu, M.; Wu, J.; Su, Y.; Shi, Y.; Liu, Z.; Wang, Q. Photoprotective effects of *Sargassum thunbergii* on ultraviolet B-induced mouse L929 fibroblasts and zebrafish. *BMC Complement. Med. Ther.* **2022**, *22*, 144. [CrossRef]
157. Prasedya, E.S.; Syafitri, S.M.; Geraldine, B.; Hamdin, C.D.; Frediansyah, A.; Miyake, M.; Kobayashi, D.; Hazama, A.; Sunarpi, H. UVA Photoprotective Activity of Brown Macroalgae *Sargassum cristafolium*. *Biomedicines* **2019**, *7*, 77. [CrossRef]
158. Prasedya, E.S.; Martyasari, N.W.R.; Abidin, A.S.; Pebriani, S.A.; Ilhami, B.T.K.; Frediansyah, A.; Sunarwidhi, A.L.; Widyastuti, S.; Sunarpi, H. Macroalgae *Sargassum cristaefolium* Extract Inhibits Proinflammatory Cytokine Expression in BALB/C Mice. *Science* **2020**, *2020*, 9769454. [CrossRef]
159. Silva, C.O.; Simões, T.; Félix, R.; Soares, A.; Barata, C.; Novais, S.C.; Lemos, M.F.L. *Asparagopsis armata* Exudate Cocktail: The Quest for the Mechanisms of Toxic Action of an Invasive Seaweed on Marine Invertebrates. *Biology* **2021**, *10*, 223. [CrossRef]
160. Silva, C.O.; Lemos, M.F.L.; Gaspar, R.; Gonçalves, C.; Neto, J.M. The effects of the invasive seaweed *Asparagopsis armata* on native rock pool communities: Evidences from experimental exclusion. *Ecol. Indic.* **2021**, *125*, 107463. [CrossRef]
161. Streftaris, N.; Zenetos, A. Alien Marine Species in the Mediterranean—The 100 'Worst Invasives' and their Impact. *Mediterr. Mar. Sci.* **2006**, *7*, 87–118. [CrossRef]
162. Vega, J.; Schneider, G.; Moreira, B.R.; Herrera, C.; Bonomi-Barufi, J.; Figueroa, F.L. Mycosporine-Like Amino Acids from Red Macroalgae: UV-Photoprotectors with Potential Cosmeceutical Applications. *Appl. Sci.* **2021**, *11*, 5112. [CrossRef]
163. Lopez Figueroa, F.; Bueno, A.; Korbee, N.; Santos, R.; Mata, L.; Schuenhoff, A. Accumulation of Mycosporine-like Amino Acids in *Asparagopsis armata* Grown in Tanks with Fishpond Effluents of Gilthead Sea Bream, Sparus aurata. *J. World Aquac. Soc.* **2008**, *39*, 5. [CrossRef]
164. Orfanoudaki, M.; Hartmann, A.; Miladinovic, H.; Nguyen Ngoc, H.; Karsten, U.; Ganzera, M. Bostrychines A-F, Six Novel Mycosporine-Like Amino-Acids and a Novel Betaine from the Red Alga *Bostrychia scorpioides*. *Mar. Drugs* **2019**, *17*, 356. [CrossRef] [PubMed]
165. Suh, S.; Pham, C.; Smith, J.; Mesinkovska, N.A. The banned sunscreen ingredients and their impact on human health: A systematic review. *Int. J. Dermatol.* **2020**, *59*, 1033–1042. [CrossRef]
166. Barceló-Villalobos, M.; Figueroa, F.L.; Korbee, N.; Álvarez-Gómez, F.; Abreu, M.H. Production of Mycosporine-Like Amino Acids from *Gracilaria vermiculophylla* (Rhodophyta) Cultured Through One Year in an Integrated Multi-trophic Aquaculture (IMTA) System. *Mar. Biotechnol.* **2017**, *19*, 246–254. [CrossRef]
167. Álvarez-Gómez, F.; Korbee, N.; Figueroa, F.L. Effects of UV Radiation on Photosynthesis, Antioxidant Capacity and the Accumulation of Bioactive Compounds in *Gracilariopsis longissima, Hydropuntia cornea* and *Halopithys incurva* (Rhodophyta). *J. Phycol.* **2019**, *55*, 1258–1273. [CrossRef]
168. Barufi, J.B.; Mata, M.T.; Oliveira, M.C.; Figueroa, F.L. Nitrate reduces the negative effect of UV radiation on photosynthesis and pigmentation in *Gracilaria tenuistipitata* (Rhodophyta): The photoprotection role of mycosporine-like amino acids. *Phycologia* **2012**, *51*, 636–648. [CrossRef]
169. Guo, C.; Zhang, G.; Sun, J.; Leng, X.; Xu, W.; Wu, C.; Li, X.; Pujari, L. Seasonal responses of nutrient to hydrology and biology in the southern Yellow Sea. *Cont. Shelf Res.* **2020**, *206*, 104207. [CrossRef]
170. Karsten, U.; Sawall, T.; Wiencke, C. A survey of the distribution of UV-absorbing substances in tropical macroalgae. *Phycol. Res.* **1998**, *46*, 271–279. [CrossRef]
171. Duarte, B.; Carreiras, J.; Feijão, E.; de Carvalho, R.C.; Matos, A.R.; Fonseca, V.F.; Novais, S.C.; Lemos, M.F.L. Potential of *Asparagopsis armata* as a Biopesticide for Weed Control under an Invasive Seaweed Circular-Economy Framework. *Biology* **2021**, *10*, 1321. [CrossRef] [PubMed]
172. Roleda, M.Y.; Nyberg, C.D.; Wulff, A. UVR defense mechanisms in eurytopic and invasive *Gracilaria vermiculophylla* (Gracilariales, Rhodophyta). *Physiol. Plant* **2012**, *146*, 205–216. [CrossRef] [PubMed]
173. Ngoennet, S.; Nishikawa, Y.; Hibino, T.; Waditee-Sirisattha, R.; Kageyama, H. A Method for the Isolation and Characterization of Mycosporine-Like Amino Acids from Cyanobacteria. *Methods Protoc.* **2018**, *1*, 46. [CrossRef]

174. Pliego-Cortés, H.; Bedoux, G.; Boulho, R.; Taupin, L.; Freile-Pelegrín, Y.; Bourgougnon, N.; Robledo, D. Stress tolerance and photoadaptation to solar radiation in *Rhodymenia pseudopalmata* (Rhodophyta) through mycosporine-like amino acids, phenolic compounds, and pigments in an Integrated Multi-Trophic Aquaculture system. *Algal Res.* **2019**, *41*, 101542. [CrossRef]
175. Torres, P.; Santos, J.P.; Chow, F.; Pena Ferreira, M.J.; dos Santos, D.Y.A.C. Comparative analysis of in vitro antioxidant capacities of mycosporine-like amino acids (MAAs). *Algal Res.* **2018**, *34*, 57–67. [CrossRef]
176. Nishida, Y.; Kumagai, Y.; Michiba, S.; Yasui, H.; Kishimura, H. Efficient Extraction and Antioxidant Capacity of Mycosporine-Like Amino Acids from Red Alga Dulse *Palmaria palmata* in Japan. *Mar. Drugs* **2020**, *18*, 502. [CrossRef]
177. Lawrence, K.P.; Gacesa, R.; Long, P.F.; Young, A.R. Molecular photoprotection of human keratinocytes in vitro by the naturally occurring mycosporine-like amino acid palythine. *Br. J. Dermatol.* **2018**, *178*, 1353–1363. [CrossRef]
178. Xie, C.-L.; Liu, Q.; Xia, J.-M.; Gao, Y.; Yang, Q.; Shao, Z.-Z.; Liu, G.; Yang, X.-W. Anti-Allergic Compounds from the Deep-Sea-Derived Actinomycete *Nesterenkonia flava* MCCC 1K00610. *Mar. Drugs* **2017**, *15*, 71. [CrossRef]
179. Miyamoto, K.T.; Komatsu, M.; Ikeda, H. Discovery of gene cluster for mycosporine-like amino acid biosynthesis from Actinomycetales microorganisms and production of a novel mycosporine-like amino acid by heterologous expression. *Appl. Environ. Microbiol.* **2014**, *80*, 5028–5036. [CrossRef]
180. Shick, J.M.; Dunlap, W.C. Mycosporine-like amino acids and related gadusols: Biosynthesis, accumulation, and UV-protective functions in aquatic organisms. *Annu. Rev. Physiol.* **2002**, *64*, 223–262. [CrossRef]
181. Shick, J.M.; Romaine-Lioud, S.; Ferrier-Pagès, C.; Gattuso, J.P. Ultraviolet-B radiation stimulates shikimate pathway-dependent accumulation of mycosporine-like amino acids in the coral *Stylophora pistillata* despite decreases in its population of symbiotic dinoflagellates. *Limnol. Oceanogr.* **1999**, *44*, 1667–1682. [CrossRef]
182. Pope, M.A.; Spence, E.; Seralvo, V.; Gacesa, R.; Heidelberger, S.; Weston, A.J.; Dunlap, W.C.; Shick, J.M.; Long, P.F. O-methyltransferase is shared between the pentose phosphate and shikimate pathways and is essential for mycosporine-like amino acid biosynthesis in *Anabaena variabilis* ATCC 29413. *ChemBioChem* **2015**, *16*, 320–327. [CrossRef] [PubMed]
183. Rosic, N.N. Phylogenetic analysis of genes involved in mycosporine-like amino acid biosynthesis in symbiotic dinoflagellates. *Appl. Microbiol. Biotechnol.* **2012**, *94*, 29–37. [CrossRef] [PubMed]
184. D'Agostino, P.M.; Woodhouse, J.N.; Liew, H.T.; Sehnal, L.; Pickford, R.; Wong, H.L.; Burns, B.P.; Neilan, B.A. Bioinformatic, phylogenetic and chemical analysis of the UV-absorbing compounds scytonemin and mycosporine-like amino acids from the microbial mat communities of Shark Bay, Australia. *Environ. Microbiol.* **2019**, *21*, 702–715. [CrossRef] [PubMed]
185. Shoguchi, E. Gene clusters for biosynthesis of mycosporine-like amino acids in dinoflagellate nuclear genomes: Possible recent horizontal gene transfer between species of Symbiodiniaceae (Dinophyceae). *J. Phycol.* **2022**, *58*, 1–11. [CrossRef]
186. de Mendoza, A.; Bonnet, A.; Vargas-Landin, D.B.; Ji, N.; Li, H.; Yang, F.; Li, L.; Hori, K.; Pflueger, J.; Buckberry, S.; et al. Recurrent acquisition of cytosine methyltransferases into eukaryotic retrotransposons. *Nat. Commun.* **2018**, *9*, 1341. [CrossRef]
187. Angel, T.E.; Aryal, U.K.; Hengel, S.M.; Baker, E.S.; Kelly, R.T.; Robinson, E.W.; Smith, R.D. Mass spectrometry-based proteomics: Existing capabilities and future directions. *Chem. Soc. Rev.* **2012**, *41*, 3912–3928. [CrossRef]
188. Zhang, Y.; Fonslow, B.R.; Shan, B.; Baek, M.-C.; Yates, J.R., III. Protein Analysis by Shotgun/Bottom-up Proteomics. *Chem. Rev.* **2013**, *113*, 2343–2394. [CrossRef]
189. Callone, A.I.; Carignan, M.; Montoya, N.G.; Carreto, J.I. Biotransformation of mycosporine like amino acids (MAAs) in the toxic dinoflagellate *Alexandrium tamarense*. *J. Photochem. Photobiol. B Biol.* **2006**, *84*, 204–212. [CrossRef]
190. Jeffrey, S.; MacTavish, H.; Dunlap, W.; Vesk, M.; Groenewoud, K. Occurrence of UVA-and UVB-absorbing compounds in 152 species (206 strains) of marine microalgae. *Mar. Ecol. Prog. Ser.* **1999**, *189*, 35–51. [CrossRef]
191. Yamamoto, R.; Takizawa, K.; Miyabe, Y.; Mune Mune, M.A.; Kishimura, H.; Kumagai, Y. Mycosporine-like Amino Acids from Red Alga Dulse (*Devaleraea inkyuleei*): Monthly Variation and Improvement in Extraction. *Phycology* **2023**, *3*, 394–404. [CrossRef]
192. Rosic, N.N.; Braun, C.; Kvaskoff, D. Extraction and Analysis of Mycosporine-Like Amino Acids in Marine Algae. In *Natural Products from Marine Algae: Methods and Protocols*; Stengel, D.B., Connan, S., Eds.; Springer: New York, NY, USA, 2015; pp. 119–129. [CrossRef]
193. Whitehead, K.; Hedges, J.I. Analysis of mycosporine-like amino acids in plankton by liquid chromatography electrospray ionization mass spectrometry. *Mar. Chem.* **2002**, *80*, 27–39. [CrossRef]
194. Hartmann, A.; Becker, K.; Karsten, U.; Remias, D.; Ganzera, M. Analysis of mycosporine-like amino acids in selected algae and cyanobacteria by hydrophilic interaction liquid chromatography and a novel MAA from the red alga *Catenella repens*. *Mar. Drugs* **2015**, *13*, 6291–6305. [CrossRef] [PubMed]
195. Geraldes, V.; de Medeiros, L.S.; Jacinavicius, F.R.; Long, P.F.; Pinto, E. Development and validation of a rapid LC-MS/MS method for the quantification of mycosporines and mycosporine-like amino acids (MAAs) from cyanobacteria. *Algal Res.* **2020**, *46*, 101796. [CrossRef]
196. Parailloux, M.; Godin, S.; Fernandes, S.C.M.; Lobinski, R. Untargeted Analysis for Mycosporines and Mycosporine-Like Amino Acids by Hydrophilic Interaction Liquid Chromatography (HILIC)—Electrospray Orbitrap MS2/MS3. *Antioxidants* **2020**, *9*, 1185. [CrossRef]
197. Orfanoudaki, M.; Hartmann, A.; Karsten, U.; Ganzera, M. Chemical profiling of mycosporine-like amino acids in twenty-three red algal species. *J. Phycol.* **2019**, *55*, 393–403. [CrossRef] [PubMed]
198. Zwerger, M.; Ganzera, M. Fast and Efficient Separation of Eleven Mycosporine-like Amino Acids by UHPLC-DAD and Their Quantification in Diverse Red Algae. *Mar. Drugs* **2022**, *20*, 395. [CrossRef]

199. Zwerger, M.J.; Hammerle, F.; Siewert, B.; Ganzera, M. Application of feature-based molecular networking in the field of algal research with special focus on mycosporine-like amino acids. *J. Appl. Phycol.* **2023**, *35*, 1377–1392. [CrossRef]
200. Chan, L.L.; Hodgkiss, I.J.; Wan, J.M.; Lum, J.H.; Mak, A.S.; Sit, W.H.; Lo, S.C. Proteomic study of a model causative agent of harmful algal blooms, *Prorocentrum triestinum* II: The use of differentially expressed protein profiles under different growth phases and growth conditions for bloom prediction. *Proteomics* **2004**, *4*, 3214–3226. [CrossRef]
201. Xiao, J.; Wang, Z.; Liu, D.; Fu, M.; Yuan, C.; Yan, T. Harmful macroalgal blooms (HMBs) in China's coastal water: Green and golden tides. *Harmful Algae* **2021**, *107*, 102061. [CrossRef] [PubMed]
202. Hennon, G.M.M.; Dyhrman, S.T. Progress and promise of omics for predicting the impacts of climate change on harmful algal blooms. *Harmful Algae* **2020**, *91*, 101587. [CrossRef] [PubMed]
203. Yu, Z.; Tang, Y.; Gobler, C.J. Harmful algal blooms in China: History, recent expansion, current status, and future prospects. *Harmful Algae* **2023**, *129*, 102499. [CrossRef] [PubMed]

MDPI AG
Grosspeteranlage 5
4052 Basel
Switzerland
Tel.: +41 61 683 77 34

Marine Drugs Editorial Office
E-mail: marinedrugs@mdpi.com
www.mdpi.com/journal/marinedrugs

www.ingramcontent.com/pod-product-compliance
Lightning Source LLC
Chambersburg PA
CBHW040317270926
51850CB00007BC/1505